Springer Mineralogy

The Springer Mineralogy series publishes a broad portfolio of scientific books on Mineralogy, Petrology, Crystallography and Gemology. Springer Mineralogy is aiming at researchers, students, and everyone interested in mineralogy and petrology. The series includes peer-reviewed monographs, edited volumes, textbooks, and conference proceedings

Taras L. Panikorovskii · Nikita V. Chukanov

Vesuvianite-Supergroup Minerals

Springer

Taras L. Panikorovskii
Laboratory of Nature-Inspired Technologies and Environmental Safety of the Arctic
Nanomaterial Research Center Kola Science Centre, Russian Academy of Sciences
Apatity, Russia

Department of Crystallography
Institute of Earth Sciences
St. Petersburg State University
St. Petersburg, Russia

Nikita V. Chukanov
Federal Research Center of Problems of Chemical Physics and Medicinal Chemistry
Russian Academy of Sciences
Chernogolovka, Russia

ISSN 2366-1585 ISSN 2366-1593 (electronic)
Springer Mineralogy
ISBN 978-3-032-06969-6 ISBN 978-3-032-06970-2 (eBook)
https://doi.org/10.1007/978-3-032-06970-2

This Springer imprint is published by the registered company Springer Nature Switzerland AG
The registered company address is: Gewerbestrasse 11, 6330 Cham, Switzerland

Competing Interests The authors have no competing interests to declare that are relevant to the content of this manuscript.

Introduction

Vesuvianite-supergroup minerals (VSMs) attract attention of petrologists, mineralogists, and crystallographers during the last 200 years owing to their diversity in habit, chemical composition, mineral associations, formation conditions, and high-level complexity of the modular crystal structure (Pallas 1793; Werner 1795; Klaproth 1797; Jannasch 1884; Prendel 1887; Wherry and Chapin 1908; Warren and Modell 1931; Kurbatov 1946; Kononova V. A. 1960; Arem and Burnham 1969; Rucklidge et al. 1975; Fitzgerald et al. 1986a; Groat et al. 1992a; Armbruster et al. 2002; Panikorovskii et al. 2016a). VSMs are important components of different contact metasomatic and regional-metamorphic rocks including skarns, rodingites, and metagabbroids (Chatterjee 1962; Batieva et al. 1987; Groat et al. 1992b). These minerals are widespread in a wide range of geological settings due to the high-level stability of their crystal structures, which are described within the rod polytypism concept (Armbruster and Gnos 2000a). The rod polytypism is the most important feature of the crystal structures of VSMs and leads to symmetry lowering by the schemes: *P*4/*nnc* → *P*4/*n* and *P*4/*nnc* → *P*4*nc*, which makes these minerals sensitive indicators of crystallization conditions (Gnos and Armbruster 2006).

The VSMs are of interest for U-Pb dating of rocks (Romer 1992a; Zhang et al. 2022a; Stifeeva et al. 2024a) and as prototypes of materials with technologically important properties. In particular, vesuvianite-type materials are potential components for immobilization of dangerous waste products including radioactive materials due to their high isomorphic capacity and structure stability (Li et al. 2009; Elmi et al. 2011; Malczewski and Dziurowicz 2015; Xiaoyan et al. 2015). Synthetic analogues of vesuvianite (Ito and Arem 1970; Olesch 1979; Hochella et al. 1982; Smart et al. 2023) are considered as adsorbents for industrial rust removal devices (Wang et al. 2015; Liu 2016; Yang 2016). Acicular crystals and whiskers of vesuvianite show pronounced properties of light conductors (Galuskin et al. 2007b; Panikorovskii et al. 2017e) and the non-centrosymmetric *P*4*nc* vesuvianite variety has nonlinear optical properties (Ohkawa 1994; Armbruster and Gnos 2000b; Panikorovskii et al. 2016c).

In addition to obvious practical aspects: the using of single-crystal X-ray diffraction (SCXRD) data for 2D or 3D "thermodynamic mapping" of deposits and using as XRD geothermometer (Panikorovskii et al. 2024a), the study of crystal chemical features of minerals of this supergroup expands the theory of the formation of modular crystal structures, as well as the regularities of stress relaxation in complex structures.

Investigation of chemical and structural diversity of VSMs via a multimethodical approach based on modern physical and chemical methods, including X-ray diffraction analysis, solid-state NMR, differential scanning calorimetry, Mössbauer, photoelectron and IR spectroscopy serves the purpose to develop a modern classification of VSMs and expands knowledge of mechanisms of their formation, prospects of synthesis and practical use as compounds with complex crystal structures and specific properties.

This book contains results of the most comprehensive investigations of the vesuvianite supergroup minerals ever studied. Over 4000 chemical analyses were obtained for 170 samples from numerous localities over the world. All of these samples were analyzed using X-ray diffraction analysis. Solid-state NMR, Mössbauer and infrared spectroscopy were also used, which made it possible to find and determine dependencies previously hidden from researchers, as well as expand the statistics on VSMs crystal structures, significantly changing some of the ideas about the crystal chemistry of minerals in this supergroup. This made it possible to find fast and reliable ways to use minerals of the vesuvianite supergroup in geology, as an indicator of mineral formation conditions (Justiniano et al. 2025a) and a diffraction geothermometer (Panikorovskii et al. 2024b), and also to formulate ideas and provisions for the nomenclature of this supergroup.

As a result of the systematic study of these samples, including chemical analysis, investigation of their crystal structures and physical properties, five new mineral species belonging to the vesuvianite supergroup (namely, alumovesuvianite, cyprine, magnesiovesuvianite, milanriederite, and manaevite-(Ce)) were discovered recently and approved by the Commission on New Minerals, Nomenclature and Classification of the International Mineralogical Association (IMA CNMNC). In these minerals, the cations Al^{3+}, Cu^{2+} (at the *Y*1 site with the five-fold coordination), Mg^{2+} (at two different sites of the crystal structures), and rare-earth elements (REE^{3+}) substituting Ca play a species-defining role. In addition, a Fe^{2+}-dominant member of the vesuvianite group was studied in detail for the first time, and reinvestigation of vesuvianite from its type locality has shown that it is Fe^{3+}-dominant. Also, the wiluite formula was revised and it was shown that Mg^{2+} ions prevail in the *Y*1 position instead of Fe^{3+}, unlike the initial treatment.

The substitution scheme $Ca^{2+} + Si^{4+} \leftrightarrow REE^{3+} + Al^{3+}$ was described for VSMs for the first time. The limits of the substitution scheme $(H_4O_4)^{4-} \leftrightarrow (SiO_4)^{4-}$ at tetrahedrally coordinated sites were established. Mechanisms of sodium incorporation into the VSMs crystal structures were described. The mechanisms of compensation of structural tensions caused by the incorporation of rare-earth elements into the vesuvianite structure through the ordering of tetragonal-pyramidal and octahedral-coordinated cations were considered in detail.

The results of crystal structures refinement for VSMs (unit cell parameters, atomic coordinates, and interatomic distances), as well as their powder X-ray diffraction patterns, are included in the databases of structural and powder data ICSD, AMCSD, and PDF. The resulting complex of information significantly expands data on the crystal chemistry of the VSMs and could be used to develop the nomenclature of this supergroup of minerals. The obtained data on the structural stability and limits of isomorphic substitutions can be useful in modelling the processes of mineral formation and the design of new functional materials with vesuvianite-based structures. Materials of this book can be used for accurate identification of mineral species belonging to the vesuvianite supergroup.

Contents

Chapter 1
Mineralogical, Petrological and Structural Diversity of Vesuvianite-Supergroup Minerals

Despite the efforts of many researchers, vesuvianite remained a mineral that was insufficiently studied in chemical, structural and genetic terms for two centuries (Arem 1973). The mineral was named after the Vesuvius volcano, where it was found for the first time as brownish-green crystals in the solidified lava (Werner 1795). The first description of the vesuvianite structure demonstrated its high complexity and modular character (Warren and Modell 1931). Petrological significance of the VSMs is determined by their wide distribution (Fig. 1.1) and rock-forming role in rocks formed under various geological conditions (Deer et al. 1962). The studies of different structural varieties demonstrate the possibility of use VSMs as a sensitive indicators of *PT* rock forming conditions (Gnos and Armbruster 2006; Justiniano et al. 2025b).

High-temperature crystallization conditions lead to the formation of long-range disordered structure (space group *P*4/*nnc*) whereas low-symmetry VSMs (space groups *P*4/*n* or *P*4*nc*) occur in low-temperature mineral assemblages (Allen and Burnham 1992). VSMs studied in this work demonstrate significant variability of geological settings and chemical compositions which made it possible to reveal chemical and morphological indicators of crystallization conditions as well as extend this group with five new mineral species approved by the IMA Commission on New Minerals, Nomenclature and Classification: cyprine (Panikorovskii et al. 2017f), magnesiovesuvianite (Panikorovskii et al. 2017e), alumovesuvianite (Panikorovskii et al. 2017a), milanriederite (Chukanov et al. 2019) and manaevite-(Ce) (Moiseev et al. 2020).

T. L. Panikorovskii and N. V. Chukanov, *Vesuvianite-Supergroup Minerals*, Springer Mineralogy, https://doi.org/10.1007/978-3-032-06970-2_1

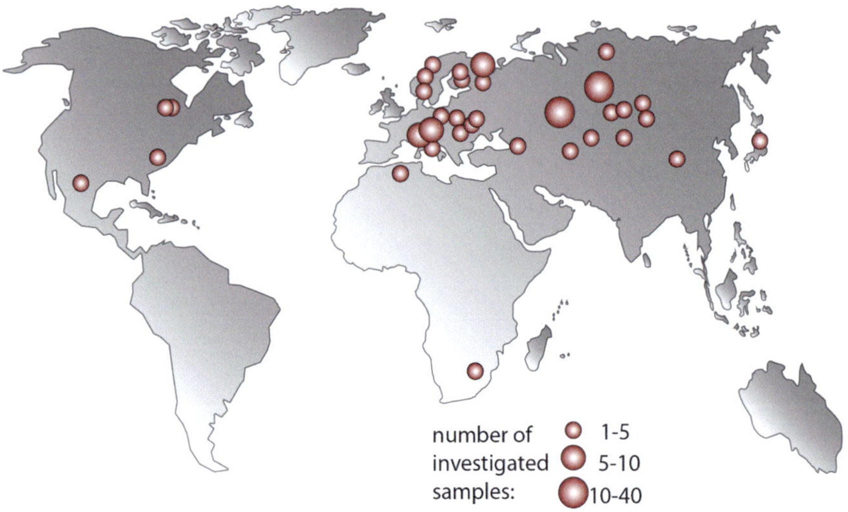

Fig. 1.1 Geography and number of VSMs samples studied in this work

1.1 Petrological Diversity

VSMs occur in mineral assemblages of different genetic types (volcanic, hydrothermal, metasomatic and regional metamorphic) related to various igneous rocks, including granitic, intermediate, basic and ultrabasic ones. Minerals of this supergroup never occur in primary (igneous or sedimentary) rocks but are products of their metamorphic, metasomatic or hydrothermal alteration. Most commonly, vesuvianite occurs in calc-schists formed by regional metamorphism of greenschist facies, which are widespread in the West Alphs (Chatterjee 1962; Trommsdorff 1968a). The estimated temperature of formation of these rocks mainly composed of VSMs, epidote, grossular, prehnite, chlorite, calcite, and dolomite is in the range 300–500 °C (Gnos and Armbruster 2006). The higher-temperature mineral associations (>500 °C) in the trap formation include, in addition to vesuvianite, also wollastonite, monticellite, garnets of the grossular-andradite series, melilites, quartz, calcite, diopside, spinel, plagioclase, and zoisite (Keusen 1972; Stalder and Aufdenblatten 1995a, a; Gnos and Armbruster 2006). Below, specific features of genesis of the VSMs depending on the average formation temperature are described (in the order from higher to lower temperatures).

1.1.1 Volcanic Rocks

Vesuvianite from the Somma-Vesuvius volcanic complex in the Campania region of southern Italy typically occurs within various types of xenoliths embedded in alkaline igneous rocks. It is commonly associated with Plinian eruptive events and is found as ejecta within pyroclastic deposits (Balassone et al. 2011, 2012, 2014, 2019, 2022a). The type locality of vesuvianite (see Fig. 1.2) pertains to metasomatically altered xenoliths, specifically skarns and related metasomatites, which are characterized as zoned, coarse-grained silicate rocks rich in Ca, Mg, Fe, and Mn. At Somma-Vesuvius, these rocks exhibit considerable complexity in their mineralogical and textural features. Mono- or bimineral zoning is prevalent, along with composite ejecta displaying sharp contacts between skarns and marbles or hornfelses (Gilg et al. 2001).

The mineral associations within the xenoliths typically include highly variable amounts of vesuvianite, wollastonite, clinopyroxene (such as diopside, hedenbergite, and "fassaite"), anorthite, phlogopite, clinoamphiboles, garnets (predominantly from the grossular-andradite series), forsterite, and humite-group minerals. Additionally, other minerals observed in minor to accessory roles include melilite (gehlenite), feldspathoids (such as leucite, nepheline, sodalite, haüyne, nosean, scapolites,

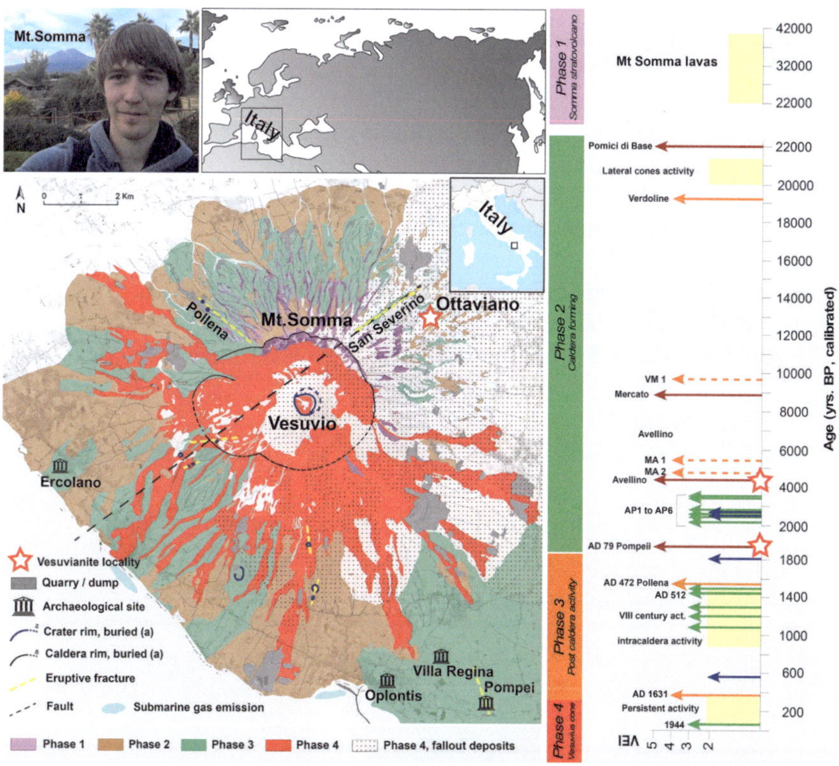

Fig. 1.2 View of the Vesuvius volcano and its geological map. Reprinted from (Sbrana et al. 2020) under the creative commons attribution (CC BY) license 4.0

davyne, and balliranoite), fluorapatite, fluorite, cuspidine, zircon, perovskite, calcite, baddeleyite, spinel, magnetite, *REE*-minerals, and sulfides (Gilg et al. 2001).

It is likely that the vesuvianite-bearing calc-silicate ejecta are associated with the Plinian eruptions of Avellino (3800 BC) and Pompeii (79 AD), which produced a significant volume of skarn ejecta. Investigations into fluid inclusions and the isotopic compositions of Pb, Nd, and Sr have been conducted on vesuvianite crystals, as detailed in Gilg et al. (2001). According to this work, the formation of skarns may be attributed to the assimilation of carbonate wall rocks by alkaline magma at moderate depths (below 5 km), leading to the exsolution of CO_2-rich vapor and complex saline melts from the contaminated magma that subsequently interacted with the carbonate rocks to form skarns.

Together with the skarn occurrences, vesuvianite has been also observed in thermally metamorphosed carbonate nodules. Its origin was interpreted as the result of recrystallization and in part devolatilization without significant metasomatic interaction with aqueous fluids or silicate melts (Gilg et al. 2001). In these calcitic rocks, vesuvianite can be variably associated with wollastonite, clinopyroxene, sodalite and humite; clinopyroxene, sodalite, cuspidine, forsterite, nepheline and davyne. Vesuvianite was also found in sanidine ejecta in association with fluorite, fluorapatite and sulfides (Werner 1795) (see Fig. 1.3).

The distinctive geochemical feature of VSMs of volcanic origin is a very high crystallization temperature and low pressure. As a result, there are a large number of impurities in the chemical composition, depending on the composition of the contact rocks. Also, the *P*4/*nnc* symmetry is characteristic of samples of volcanic genesis. Vesuvianite from the Vesuvius volcano was formed at a temperature of 800–890 °C and a pressure below 1 kbar (Balassone et al. 2011). Our studies confirmed minor impurities of boron which occurs in the additional $T1$ and $T2$ positions of the crystal structure, as well as the prevalence of Fe^{3+} in the position of the tetragonal pyramid $Y1$ (Panikorovskii et al. 2017b).

Fig. 1.3 Photo of well-shaped brown vesuvianite crystals in sanidinite ejecta. Specimens No. 19150 E7681 (**a**) and No. 9791 D242 (**b**) from the Real Museo Mineralogico of the Centro Musei delle Scienze Naturali e Fisiche Università di Napoli Federico II Italy. 1—vesuvianite, 2—sanidine, 3—calcite, 4—hedenbergite. Photos by Roberto Appiani

1.1.2 Skarns

One of the most common geological formations bearing VSMs is the classic skarn. In skarn rocks, VSMs are usually associated with grossular or/and andradite, epidote–clinozoisite, calcite, diopside, scapolite and clinohumite. These minerals were formed as a result of alteration processes that occur when hydrothermal fluids interact with a protolith of either igneous or sedimentary origin. In many instances, skarns are associated with the intrusion of granitic plutons, typically located in and around faults or shear zones that intrude into carbonate layers composed of either dolomite or limestone. VSMs related to skarns at contact aureoles typically crystallized at high temperatures above 600 °C and low pressures below 1–3 kbar (Gnos and Armbruster 2006). The estimated temperature of vesuvianite crystallization in skarns is in the range 400–800 °C (Allen and Burnham 1992).

The Pitkäranta ore field, Karelia, Russia (Fig. 1.4a) includes a wide variety of ore-bearing skarns, including the Lupikko district, where fluorovesuvianite was discovered (Britvin et al. 2003). In the skarn dumps of the Lupikko district, one of the main minerals is vesuvianite, which often forms large monomineral aggregates. It occurs as perfect crystals in the form of quadrangular prisms up to 3.5 cm long and has various colors: brown–red, brown, dark green, light green, or black. In the same area, there is the Be-Sn polymetallic deposit Hopunvaara, which also belongs to the skarn-greisen type and is confined to the contact zone of the rapakivi granites of the Salmi batholith with the carbonate horizons of the Pitkäranta suite. This ore occurrence is characterized by the formation of the following types of metasomatites: magnesian and calcareous skarns, fluorite-vesuvianite-magnetite metasomatites, aposkarn metasomatites (greisens, propylites and feldspar metasomatites) and late quartz-carbonate metasomatites (Ivashchenko 2021). The formation of skarns and fluorite-vesuvianite-magnetite metasomatites is associated with the magmatic stage, while all aposkarn metasomatites and quartz-carbonate metasomatites were formed in the postmagmatic stage.

Fluorite-vesuvianite-magnetite metasomatites of the Hopunvaara ore occurrence form sheet and pipe-shaped complex branching bodies in marbles and calciphyres of the Pitkäranta suite (Fig. 1.4b). A characteristic feature of these rocks is a rhythmic fine-banded structure. Beryllium mineralization is associated with them, with almost all Be being found as an isomorphic impurity in vesuvianite up to 0.85% BeO (Stifeeva et al. 2024a).

Rock-forming vesuvianite (Fig. 1.5) of diopside-garnet skarns with diverse ore mineralization (molybdenite, chalcopyrite, sphalerite, pyrite, cassiterite, scheelite) forms short-prismatic crystals up to 1 cm in length of brown color with well-developed tetragonal prism faces {110} and {111}. Among the features of the chemical composition, it is worth to note significant contents of F (up to 2.0 wt%), B_2O_3 (up to 0.9 wt%), BeO (up to 0.85% wt%) and Cl (up to 0.50 wt%). It should be noted that the formation of fluorovesuvianite appears at the late stage of hydrothermal alteration of primary rocks, and its origin is in no way related to the formation of rock-forming vesuvianite (Britvin et al. 2003).

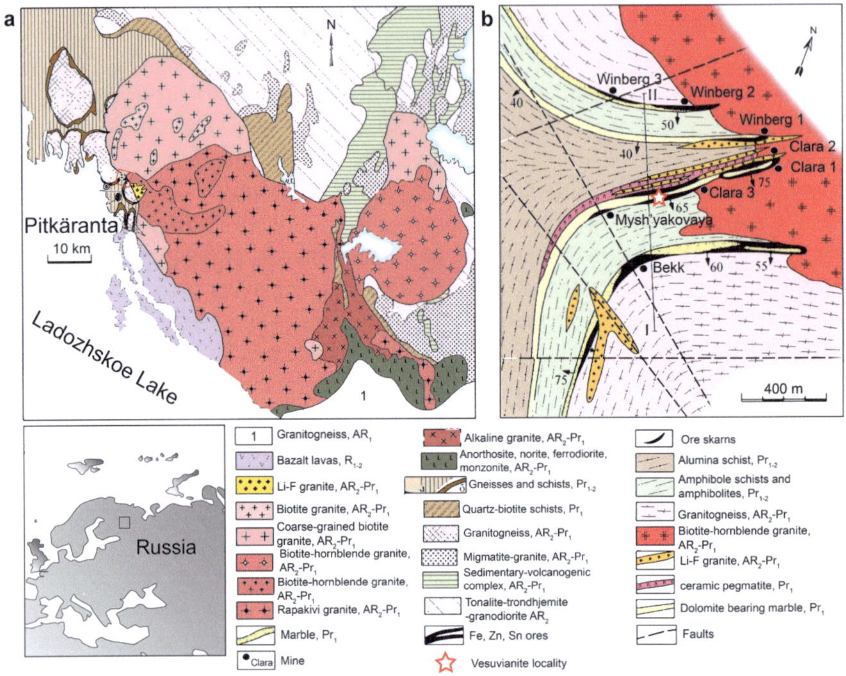

Fig. 1.4 Geological structure of the Salma batholith and location of deposits of the Pitkäranta ore district (**a**), Geological scheme of the Hopunvaara vesuvianite-bearing skarn, Pitkäranta ore district (**b**) (after Stifeeva et al. (2024a) with modifications)

The high-temperature genesis of the Hapunvaara vesuvianite is consistent with the Cl– ↔ OH– substitution at the O10 position in the structure channels (Galuskin et al. 2005), as well as the introduction of Al^{3+} into the tetrahedral $T1$ position due to the substitution of OH groups. The introduction of Be^{2+} into the VSMs is most expected at the $T1$ position. In a studied sample, the occupancy of this position is ($\square_{0.56}Be_{0.44}$), which corresponds to 1.76 Be atoms per formula unit or 1.60 wt%. For low-temperature vesuvianite with a crystallization temperature below 400 °C, this position is vacant (Stifeeva et al. 2024a). Thus, the temperature range of 550–800 °C corresponds to the conditions of the formation of vesuvianite-containing metasomatites of the Hopunvaara ore occurrence (Britvin et al. 2003).

1.1.3 Metamorphosed Volcanogenic-Sedimentary Rocks of the Trap Formation

A lot of different chemical varieties of VSMs found in the contact-metamorphic Achtaragdite rocks of the Wiluy trap formation (Fig. 1.6) formed at the contact

Fig. 1.5 Vesuvianite-bearing skarn marked by star on the Fig. 1.4: general view (**a**), contact of dolomite with vesuvianite-calcite rock (**b**), columnar aggregates of short-prismatic vesuvianite crystals (1) in association with calcite (2) from the Pitkäranta ore district, Karelia, Russia (**c**)

Early Palaeozoic marlstones (mergel) with different-aged (260–100 Ma) gabbroid intrusions (Oleynikov 1979; Kravchinsky et al. 2002). In this locality, Cl-bearing vesuvianite, Si-deficient vesuvianite, wiluite, Si-deficient wiluite and F-bearing vesuvianite were found in addition to vesuvianite sensu stricto (Galuskin et al. 2003b;

Panikorovskii et al. 2016a). According to Galuskin (2005), variability of VSMs chemical composition is connected with multistage achtaragdite rock formation.

Large prismatic wiluite crystals (Fig. 1.7a) were formed at the stage of postmagmatic Ca-skarnoids formation in a high-temperature mellite-bearing association (in the temperature range of 550–900 °C). The formation of hydrogrossular pseudomorphs after a mineral of the mayenite supergroup, known as achtaragdite, occurred at the stage of rodingitization of early skarns (Galuskina et al. 1998a; Galuskin and Galuskina 2000, 2003; Galuskin 2005). Rodingitized skarns are located within the differentiated Erbeyek intrusion, the geochronological age of which corresponds to the Lower Triassic (~260 Ma) (Oleynikov 1979; Oleynikov and Pavlushin 2000).

The Erbeyek differentiated massif forms a skarnification zone with a thickness of tens to hundreds of meters in the exocontact parts with terrigenous-carbonate deposits of the Kholomolokh suite of the Upper Cambrian. Formations of the exogenous zone are represented by wiluite-akhtaragdite, grossular, andradite and pyroxene

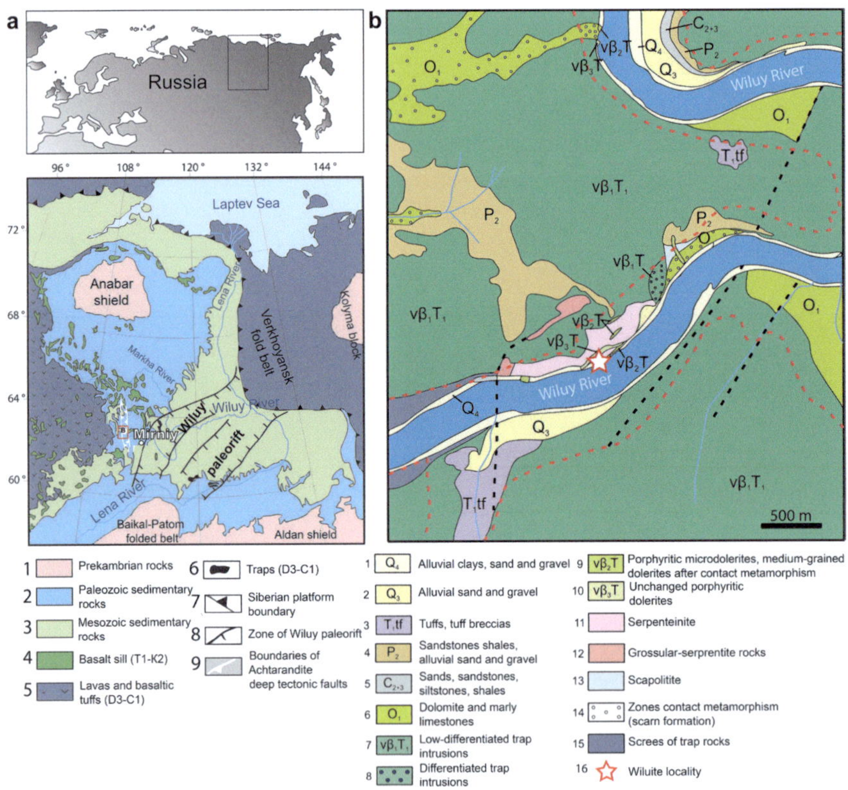

Fig. 1.6 Simplified geological map of the north-eastern part of the Siberian platform after Geological map of the USSR 1980 (**a**); schematic geological map of the wiluite locality before flooding, the outline of the modern river is shown by the red dashed line after (Galuskin 2005) with modifications (**b**)

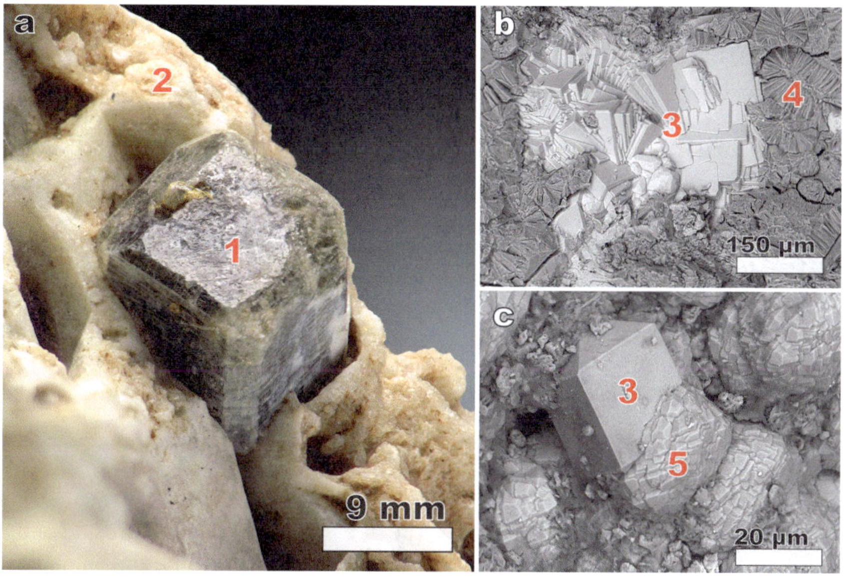

Fig. 1.7 Photograph of wiluite (1) in "achtaragdite" (2) rock together with crusts of "hydrovesuvianite" (**a**), SEM image of "achtaragdite" rock consisting of "hydrovesuvianite" (3), chlorite (4) and "hydrogrossular" (5)

contact rocks. The host akhtaragdite rocks are polymineral pseudomorphs, in which the main part is made up of "hydrogrossular" and "hydrovesuvianite" (Fig. 1.7b, c). The temperature of wiluite formation is in the range of 550–850 °C, while "hydrovesuvianite" crystallized at temperatures of 300–400 °C (Panikorovskii et al. 2016a).

1.1.4 Altered Alkaline Rocks

VSMs are not widespread in alkali rocks, but all such finds are interesting from the viewpoint of their crystal chemical diversity. Vesuvianite with even higher REE_2O_3 contents (up to 21 wt%) was described in Seward Peninsula, Alaska, USA (Himmelberg and Miller 1980) and the San-Benito locality in California, USA (Crook and Oswald 1979; Fitzgerald et al. 1987).

Vesuvianite enriched in Na was found in alkaline rocks of Western Keivy, Kola Peninsula (Chumakov et al. 1948; Batieva and Bel'kov 1984; Batieva et al. 1987). In these rocks, amphibolized gabbroids are altered by extremely Na-enriched alkali fluids derived from alkaline granites and/or nepheline syenite rocks of the Sakharyok massif (Fig. 1.8). The final product of their alteration is nepheline-microcline-phlogopite-plagioclase-clinopyroxene metasomatic rock with Na-bearing vesuvianite. Vesuvianite forms short-prismatic yellowish brown poikiloblasts (Fig. 1.9), with central parts filled by inclusions of Na-Al-bearing diopside, phlogopite, apatite, and albite (Panikorovskii et al. 2016b).

Among the features of the chemical composition of vesuvianite from the alkaline rocks of the Sakharyok massif, it is worth noting the high content of Na_2O (up to 0.98 wt%), TiO_2 (up to 2.76 wt%), and F (up to 2.59 wt%). More details on the mechanisms of Na incorporation into the crystal structure of vesuvianite are given in Chap. 4.

Vesuvianite from gabbroids of the West Keivy mounts, metasomatically altered at the contact with alkaline intrusions, formed from clinopyroxene and plagioclase at 500–600 °C and 2–4 kbar (Batieva et al. 1987).

1.1.5 Skarnoids

The Kovdor massif of ultramafic, alkaline rocks and carbonatites is located in the eastern part of Baltic Shield (Fig. 1.10a). The massif is a multiphase intrusion of the central type with a total area of 45 km^2, which broke through biotite and amphibole gneisses and granite gneisses of the Belomorian block (AR_2-PR_1). The formation of the massif occurred 360–420 million years ago (Rimskaya-Korsakova and Krasnova 2002). In plan, the massif has a zonal structure, the center of which is composed of olivinites, later phlogopite-diopside (Phlogopite complex), melilite and other metasomatic rocks, and the periphery of the massif is composed of turyaites and melteigite-urtites (Fig. 1.10b). The rocks of the massif and the host gneisses are penetrated by various veins of syenites, ijolites, shonkinites and calcite carbonatites. In the western part of the massif, on the border of olivinites and foidolites, there is a zonal stock of phoscorites (external zone) and magnetite-carbonate rocks (axial zone).

Various VSMs occur in the rocks of the Kovdor alkaline-ultrabasic massif associated with the skarnoids of the Phlogopite complex (Fig. 1.10c). Below, the term "skarnoid" is applied to a metasomatic rock similar in mineral composition to skarns, but intermediate in origin between skarns and hornfelses. Skarnoids are formed at the contacts of carbonate rocks containing initial silicate or aluminosilicate material with an intrusion. Skarnoids differ from typical skarns by the absence of metasomatic zoning (Ivanyuk et al. 2002).

The host apomelilite rocks include garnet-petolite-vesuvianite, phlogopite-monticellite, calcite-diopside-hastingsite and wollastonite-monticellite-melilite rocks. Garnet-pectolite-vesuvianite rocks form veins and pockets which are usually no more than 30 cm across. Andradite, monticellite, pectolite, thomsonite, apatite, phlogopite and calcite, as well as relict melilite are found in association with vesuvianite. Vein formations of this composition usually have a zonal structure: the outer zone 1–2 mm thick is composed of cryptocrystalline "hydrovesuvianite". Closer to the center are zones of diopside-vesuvianite, diopside-hastingsite, and (at the center) pectolite composition. Towards the phlogopite body, various apomelilite rocks are replaced by calcite-diopside-phlogopite and phlogopite-forsterite-diopside rocks of

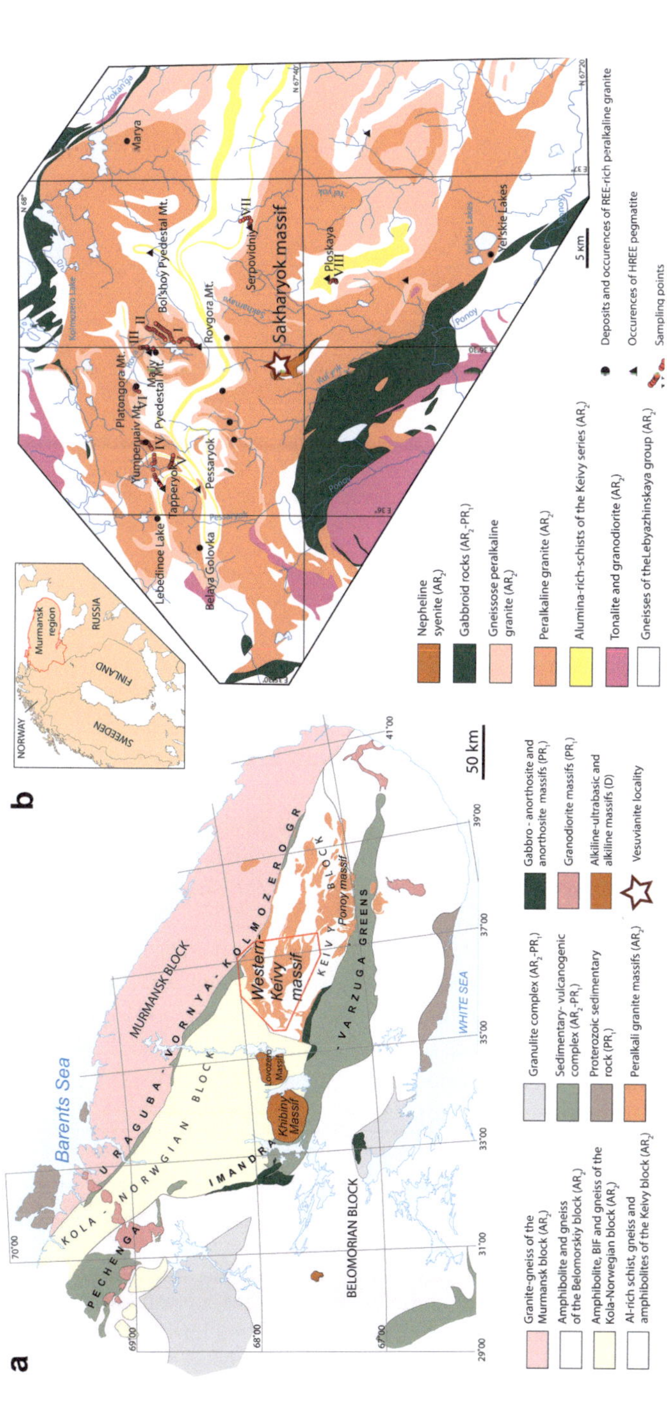

Fig. 1.8 Simplified geological schemes of Baltic Shield drawn using data by Gaskelberg et al. (1978) and Balagansky et al. (1996) (**a**), and alkaline granite massifs of the West Keivy mounts, Kola Peninsula drawn using data by Gaskelberg et al. (1978)

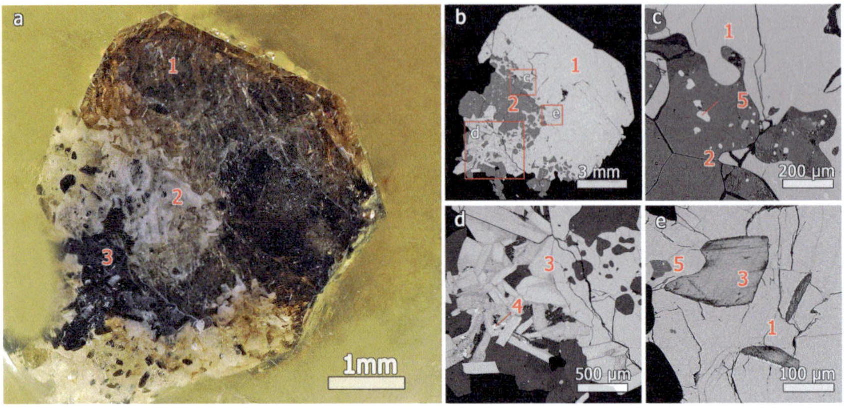

Fig. 1.9 Na-bearing vesuvianite and associated minerals from alkaline gabbroid of Western Keivy, Kola Peninsula, Russia (sample 50,578 FM): sample photo (**a**); BSE-image of the (001) section (**b**); prehnite inclusions in plagioclase associated with vesuvianite (**c**); aggregates of biotite with fluorapatite inclusions (**d**); biotite inclusions in vesuvianite (**e**). 1—vesuvianite, 2—plagioclase (No. 20–22), 3—biotite, 4—fluorapatite, 5—prehnite

the phlogopite complex, which are later formations, since relics of all the rocks considered above are often observed in them (Kaverin et al. 1988).

At least three different types of VSMs are known there: fine-grained aggregates of apple-green vesuvianite (Fig. 1.11a) with inclusions of pargasite and phlogopite and relics of natrolitized urtites (Seredkin 2001; Ivanyuk et al. 2002) and short-prismatic dark brown crystals of a *REE*-bearing vesuvianite up to 5 mm long (Fig. 1.11c) with dark orange fringes manaevite-(Ce) (1–2 mm in thickness), interspersed in coarse-grained calcite aggregates in association with andradite (Kaverin et al. 1988).

Three generations of vesuvianite were found in the skarnoids of the Kovdor phlogopite complex, the first of which is associated with the skarnoids themselves (Fig. 1.11a), and the second and third ones are confined to areas of their reworking by late calcite carbonatites (Fig. 1.11b, c). The estimated temperatures of formation of all varieties of vesuvianite lie within 280–380 °C and are in good agreement with the data obtained using a magnetite-ilmenite geothermometer (Panikorovskii et al. 2024a). The obtained data allow us to propose the use of the VSMs as a diffraction geothermometer (for more details see Chap. 4).

1.1.6 Rodingites

Gemstone-quality green VSMs are common in rodingites, i.e. calc-silicate metasomatic rocks, formed at the contact zone with serpentinized intrusions of basic rocks (typically, diabase), which usually are represented by veins hosted by ultrabasic rocks. These rocks are abundant in Urals, Russia (Koksharov 1853; Fedorov 1890;

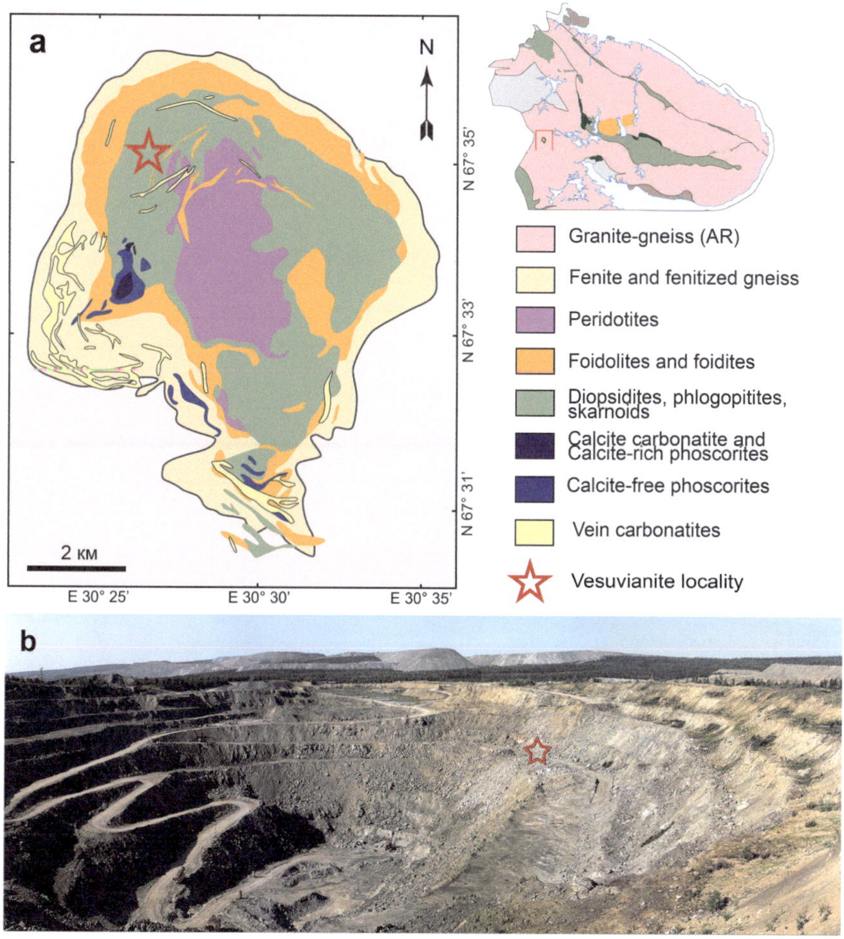

Fig. 1.10 Position of vesuvianite-bearing rocks in the Kovdor massif (after Mikhailova et al. (2016) with modifications) (**a**) Kovdor massif with the vesuvianite-containing rock marked with an asterisk (Photo by G.Yu. Ivanyuk) (**b**) and general view of the Mica open-pit mine (**c**)

Sjögren 1895; Kashinskii A.A. 1909; Antonov 2003a; Erokhin 2009). It is worth to mention also several other similar deposits: Asbestos in Quebec, Canada (Allen 1985; Fitzgerald et al. 1986a). Georgetown in California, USA (Allen 1985; Valley et al. 1985; Phillips et al. 1987) and Hindubagh in Pakistan (Allen and Burnham 1992). Rocks almost entirely composed of vesuvianite often occur in rodingites and were described with the names vesuvianite, californite, or vesuvianite jad (Kunz 1903; Clarke and Steiger 1905).

This paragenetic assemblage bearing low-symmetry vesuvianite and alumovesuvianite varieties with *P*4/*n* and *P*4*nc* symmetry has the temperature of crystallization in the range of 250–450 °C (Allen and Burnham 1992; Krivovichev et al. 2013).

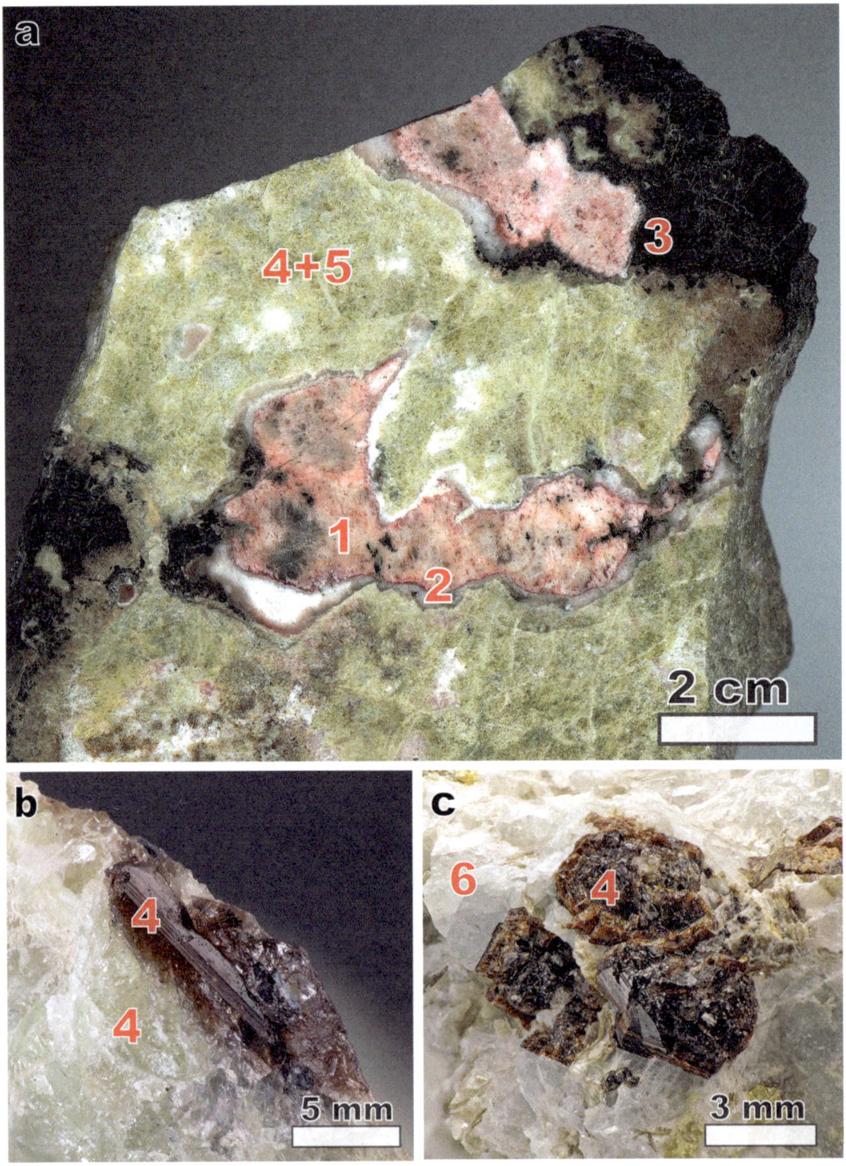

Fig. 1.11 Apple-green vesuvianite from apourtite pargasite-vesuvianite rock (**a**), green and brown *REE*-bearing vesuvianite from the vesuvianite-glagolevite rock (**b**); manaevite-(Ce) rims around zonal crystals of *REE*-enriched VSMs (**c**). Legend: 1—nepheline, 2—gonnardite, 3—biotite, 4—vesuvianite, 5—pargasite, 6—calcite

The absence of fluorine, boron, rare earth elements in the chemical composition is the typical feature of VSMs from rodingites According to Groat (1988a), chemical formulae of such samples are usually well calculated on 147 charges.

Jeffrey mine in Asbestos, Québec, Canada (Normand and Williams-Jones 2007; Horváth et al. 2013) is a classic locality of rodingites (Fig. 1.12). The mine is located in serpentinized ultramafic rocks that form a part of the Ordovician Asbestos ophiolite belt of the Appalachians and comprises serpentinized dunite, chromitite, pyroxenite and peridotite, all intruded by diorite, syenite, granite, gabbro and ultramafic to intermediate volcanic rocks, which was obducted onto the margin of the Laurentian craton during the ~ 470–460 Ma Taconic orogeny. Dunite and peridotite of chrysotyle asbestos deposits were heavily serpentinized in successive stages soon after crystallization, especially where sheared and intruded by granitic to dioritic magmas. In a later phase of serpentinization, shear zones and fractures channelled the asbestos-forming solutions. Later diorite dike intrusions into the serpentimized dunite and peridotite rocks resulted in replacement mineralization to form rodingites (Wares and Martin 1980). The estimate pressure and temperature of rodingite formation are 2.5–4 kbar and 400 °C (Normand and Williams-Jones 2007).

Cavernous rodingites with perfect crystals of VSMs occurring in cavities, in association with chlorite, calcite, prehnite, diopside and grossular (Fig. 1.13), are the result of recrystallization of primary rodingites (Kurbatov 1914; Kornetova et al. 1986). In Jeffrey Mine, green and crimson and almost colorless gem-quality VSMs occur in

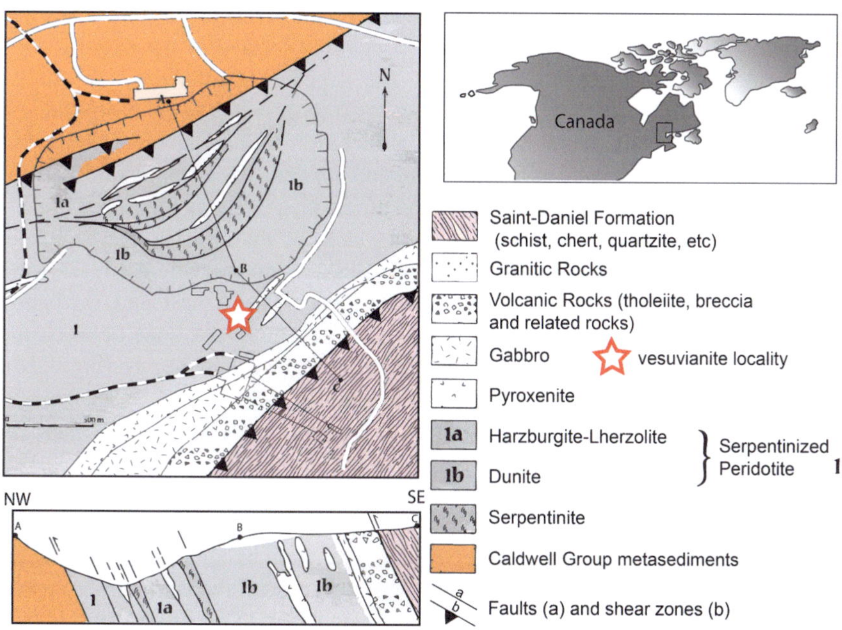

Fig. 1.12 Geologic scheme of Jeffrey Mine, Val-des-Sources, Les Sources RCM, Estrie, Québec, Canada (after Horváth et al. 2013) with modifications

Fig. 1.13 Crystals of Mn^{3+}-bearing alumovesuvianite (1) with colorless vesuvianite (2) and diopside (3) light-brown plates (**a**), druse of pink alumovesuvianite (**b**), druse of green alumovesuvianite (**c**), polychromatic alumovesuvianite crystal (**d**). All samples from Jeffrey Mine, Asbestos, Canada

the rodingite zone at the contact of a gabbroid rock with host serpentinite. The main minerals of the rodingite are diopside, alumovesuvianite, vesuvianite and prehnite.

VSMs from the Urals rodingites are generally similar to those from Canadian rodingites. The Bazhenov asbestos-bearing region is located in the eastern slope of Middle Urals, near the city of Asbest. The deposit is confined to the dunite-harzburgite massif, which is a part of the Asbest-Alapaevsk belt of gabbro-peridotite intrusions, which extents in a submeridional direction for a distance of about 180 km. The Bazhenov dunite-harzburgite massif and the spacially and structurally associated Asbest gabbro massif are located in an area with a complex block and folded-nappe structure (Erokhin et al. 2023). In the area of the deposit, pre-Paleozoic gneiss-migmatite complexes, ophiolites ($O–S_1$), island-arc formations ($S_2–D_2$) and granitoids of different ages ($C_1–P_2$) are developed. The Asbest and Bazhenov massifs are situated in a narrow (about 10 km wide) interblock zone between the Murzinsky-Aduisky block in the west and the Reftinsky block in the east (Fig. 1.14).

Within the Bazhenov ophiolite complex, rodingites are found exclusively in serpentinites and serpentinized ultrabasites of the dunite-peridotite massif and are confined to weakened tectonic zones. According to morphological features, two types of rodingites are distinguished there: vein (dyke) bodies and boudins (lenticular bodies). The main rock-forming minerals in rodingites of the Bazhenov ophiolite complex are grossular, diopside, vesuvianite, and clinochlore, as well as (in some rocks) zoisite and prehnite. Metamorphism of rocks of the Bazhenov complex (including rodingites) evolved from pumpellyite-actinolite ($T = 390$ °C, $P = 4$ kbar) to zeolite ($T = 220–50$ °C, $P = 2$ kbar) facies (Antonov 2003a).

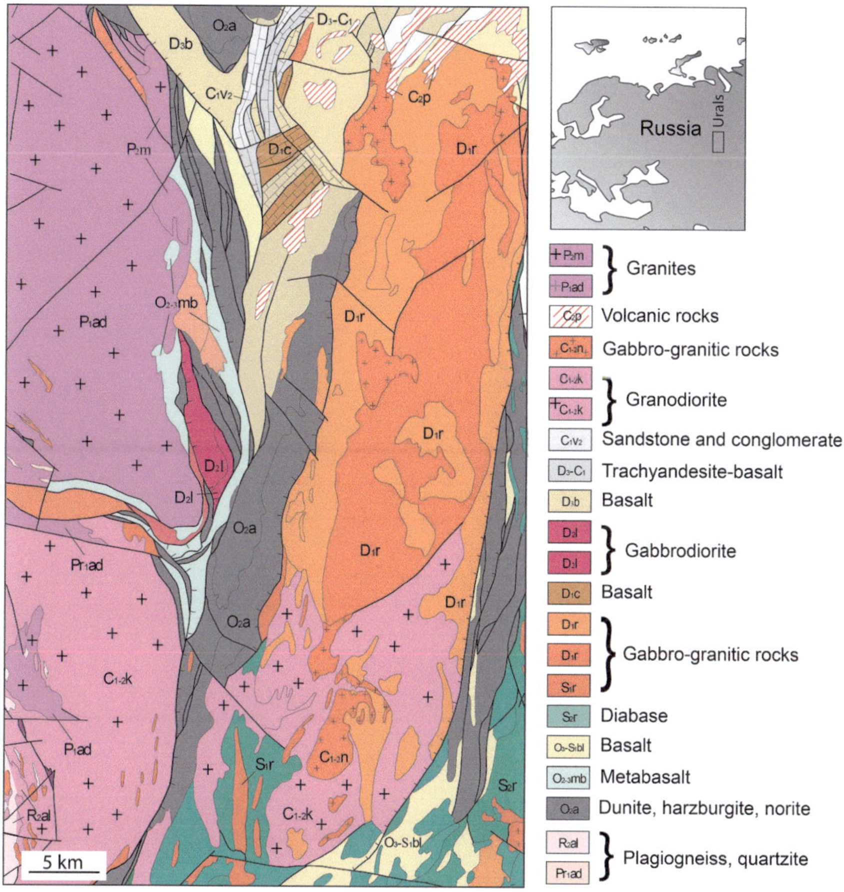

Fig. 1.14 Geological scheme of the Bazhenov (Bazhenovskiy) ophiolite complex according to Erokhin (2017a) with modifications

In the calcium-silicate vein rocks of the Bazhenov deposit, vesuvianite is one of the main minerals. In massive rodingites, vesuvianite is most often represented by small (up to 1 mm in diameter) isometric or slightly elongated grains. Dark marsh-green or light apple-green varieties of vesuvianite are most common (Fig. 1.15a, c). They associatr with clinochlore and/or calcite. A characteristic feature of the Bazhenov vesuvianite is a specific appearance of their crystals which are long-prismatic with a wide variety of crystal forms.

A Cr-bearing variety of vesuvianite was found among late chlorite veinlets in massive chromitites, which are rare at the deposit. Samples of Cr-bearing vesuvianite from the neighboring the Monetnaya Dacha deposit, occurring in a similar mineral association, were studied by us (Fig. 1.15b) (Krivovichev et al. 2013). Cr-bearing vesuvianite occurs there as small long-prismatic crystals that form small brushes from pale to emerald green. The chromium content in this variety is about 1 wt%.

Fig. 1.15 Aggregates of gem-quality low-temperature vesuvianite crystals from rodingites of the Bazhenovskoe deposit, Middle Urals (**a** sample 79,478), Cr-bearing vesuvianite from Monertnaya Dacha, Middle Urals (**b**, sample 17,342 FM), and vesuvianite from the Bazhenovskoe deposit, (**c**, A. A. Agakhanov's sample). 1—vesuvianite, 2—clinochlore, 3—chromite, 4—calcite

All samples of vesuvianite from rodingites have *P*4/*n* or *P*4*nc* symmetry and the estimated temperature and pressure of their formation are in the ranges of 250–450 °C and 2–4 kbar.

1.1.7 Contact-Metasomatic Rocks Close to Rodingites

Perfect vesuvianite crystals, rich in crystalline forms and often reaching significant sizes, are found in numerous mines in the Southern and Middle Urals. Gustav Rose, Barbot de Marny, Nikolai Koksharov, Pavel Eremeev, Mikhail Kurbatov and other famous researchers happily measured and described these crystals using goniometers (Koksharov 1853; Kurbatov 1924, 1925a, b). In the atlas of V. Goldschmidt, many vesuvianite crystals from deposits in the Southern Urals are described (Goldschmidt 1918). These deposits were developed by several mines, among which the most famous are the Shishimskaya (also known as the Polyakovsky mine or Barbotova pit), Nikolae-Maximilianovskaya (Leuchtenbergite), Zelentsovskaya, Eremeevskaya and Akhmatovskaya mines (in Russian mine—Kop'').

A number of low-temperature metasomatic vesuvianite deposits in the Ural Mountains do not fit into the concept of classical skarns, but their genesis also does not fully fit into the concept of classical rodingites, in the formation of which ultramafic rocks should participate. In particular, vesuvianite-bearing metasomatites of Southern Urals were formed at the contact of gabbro RF_2 and carbonate rocks RF_1; granites and granite gneisses are also located in the contact zone (Fig. 1.16). These rocks are confined to xenoliths of host rocks represented by dolomites RF_1, less often quartzites and hornfelses. Massifs of gabbro and granitoids, stretching for 70 km, were formed at about the same time, their age is 1.385–1.395 Ma (Kholodnov et al. 2010).

The content of rare and rare earth elements in the VSMs demonstrates a relationship with the content of these elements in paragenetic minerals and host gabbroids (Stativko et al. 2024). This suggests that the VSMs of the South Urals mines were formed as a result of contact metasomatism, and their compositional features depend on the mineral paragenesis and the influence of the frame rocks: gabbroids of the Kusinsky-Kopansky intrusive complex (Stativko et al. 2024). This entire group of mines was described by Mushketov (1877) Fig. 1.16 shows an out-of-scale image of the Shishim mine.

The Zelentsovskaya mine is located in one of the failures in the collapse zone of the underground workings of the Kusinsky deposit of ilmenite and magnetite (titanomagnetite) ores. In the sides of the failures, one can observe outcrops of gabbroids of the Kusinsky-Kopansky intrusive complex with titanomagnetite ore mineralization, as well as rounded bodies of silicate-carbonate rocks. The following zones of the body of silicate-carbonate rocks are distinguished (in the direction from its center of to the periphery): carbonate, brucite-carbonate, brucite-clinohumite-vesuvianite-carbonate, carbonate-vesuvianite-garnet-olivine, hornfels, amphibole-epidotized gabbro. Vesuvianite is usually found in the form of small, up to 5 mm, well-faceted green crystals together with calcite and garnet of the andradite-grossular series.

The Nikolae-Maximilianovskaya mine is confined to chlorite, chlorite-amphibole, talc-amphibole and amphibole rocks developed along carbonates in the contact zone of the gabbroids of the Kusinskaya intrusion and granite-gneisses of the Gubensky

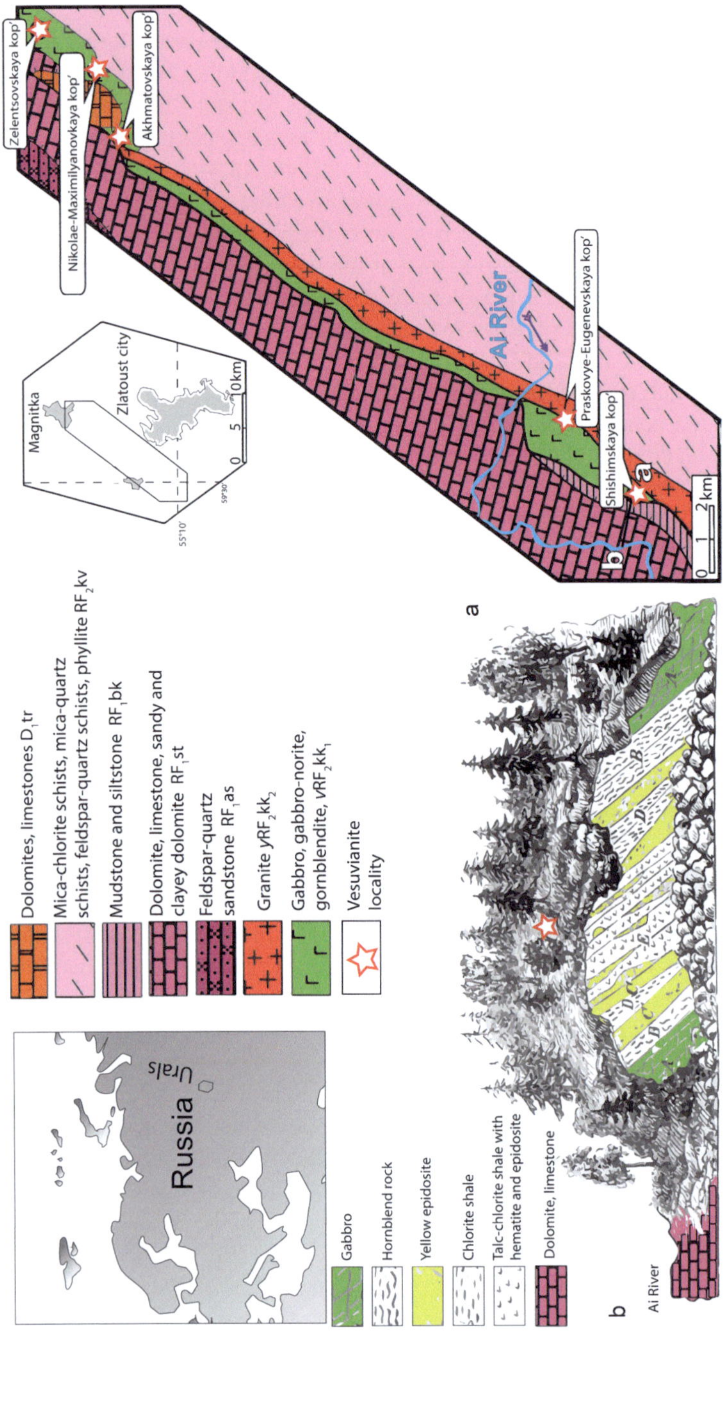

Fig. 1.16 Geological scheme of Shishimskie and Chuvashskie Mts (after Stativko et al. 2024) and schematic profile of Shishiskaya mine (after Mushketov (1877) with modifications)

massif. Epidote, diopside, clinochlore, brucite, magnesioludwigite, magnetite, ilmenite, spinel, titanite, perovskite, vesuvianite, forsterite, grossular, andradite, scapolite, apatite, xanthophyllite, calcite and other minerals are present here. Zircon is very rare. Unlike other mines, the Nikolae-Maksimilianovskaya mine contains an abundance of ferruginous spinel (Bocharnikova et al. 2011). The vesuvianite group minerals of the Nikolae-Maksimilianovskaya mine form large, well-formed crystals of a greenish-brown color (Fig. 1.17b) in association with calcite, clinochlore, and grossular.

In the Akhmatovskaya mine, dolomites at the contact with the Kusa gabbroids have been transformed into pyroxene-garnet-chlorite-amphibole rocks. The mineral paragenesis is represented by diopside, spinel, garnet of the grossular-andradite series, titanite, zircon, chlorite, calcite, apatite, olivine, magnetite, epidote, scapolite, vesuvianite, perovskite, etc. Vesuvianite with a high TiO_2 content was first identified here (Myasnikov 1940). Crystals from the Nyazemsky Mountains, presumably from the Akhmatovskaya mine, are shown in Fig. 1.17c. A distinctive feature of these samples is the high Ti content of up to 2 atoms per formula unit. Zoned short-prismatic crystals of the VSMs with a wiluite-type faceting are often encountered, where the more titanium zones correlate with brown shades, and the more ferruginous ones with green color.

The Shishimskaya mine was founded in 1833 by mining engineer N. P. Barbot-de-Marni. In its current form, it is a small, partially collapsed quarry, about 100 m long. The quarry exposed medium- to coarse-grained gabbroids, in which intensively amphibolized and epidotized areas, silicate-carbonate rocks and various contact-metasomatic formations, composed of pyroxene-chlorite, chlorite-serpentinite, chlorite, amphibole-chlorite, garnet-chlorite, garnet-vesuvianite, garnet rocks, are developed. Individual mine workings reveal either a complete section of the contact zone—from gabbro to marbled limestones through amphibole, amphibole-chlorite, chlorite-serpentine rocks with garnet-vesuvianite and garnet lenses and interlayers (Fig. 1.17), or certain types of contact formations—garnet-leuchtenbergite schists and amphibole-chlorite rocks.

Among the minerals of the Shishimskaya mine there are: garnet (andradite, grossular, almandine), vesuvianite, pyroxene (diopside), amphiboles (actinolite, hornblende with crystals up to 10–15 cm in size), feldspars (basic plagioclase, albite), chlorites (leuchtenbergite), mica (xanthophyllite), serpentine, epidote, tourmaline, clinohumite, hematite, magnetite, (including "titanomagnetite"), perovskite, Fe-bearing spinel, pyrite, calcite, the so-called "talc-apatite" (apatite decomposition products). The VSMs are represented by different types of crystals. Pistachio-green brushes in association with calcite on the surface of garnet-diopside rock with individual crystals up to 1.5 cm in size are typical. Such crystals are faceted by tetragonal dipyramid {111} and tetragonal prisms, {110} and {100}. Small polyhedral crystals are also found in the lower-temperature association with prehnite and clinochlore (Fig. 1.17a, d).

Despite the literary dispute about the origin of VSMs from various works, our studies using X-ray structural analysis showed that even for the "skarn" association with calcite, diopside and andradite, the symmetry of the crystals is always *P*4/*n*

Fig. 1.17 Gem-quality low-temperature vesuvianite crystals from metosomatic rocks of South Urals: multifaced green vesuvianite crystal from Polyakovskiy mine (Shishimskaya pit) (**a**, sample 32,801 FM), prismatic brown Ti-bearing vesuvianite from the Nikolae-Maximilyanovkaya mine (**b**, sample 67,560 FM), brown Ti-bearing vesuvianite from Nyazemskie Mts. (possible from the Akhmatovskaya mine) (**c**, 67,964 FM), crystals of green vesuvianite crystals from Barborova Yama (Shishimskaya mine) (**d**, 23,700 FM). 1—vesuvianite, 2—calcite, 3—clinochlore, 4—Fe bearing grossular, 5—prehnite

and their supposed formation temperature is in the range of 300–450 °C. It should be noted that samples in association with clinochlore and prehnite, as well as all polyhedral VSMs have the same symmetry but are characterized by a higher degree of ordering of cations in the channels of the crystal structure, which most likely corresponds to lower temperatures of their formation, presumably in the range of 250–400 °C (Popov and Rassomakhin 2023).

1.2 Physical Properties and Morphology

Morphology and physical properties of vesuvianite regularly change depending on chemical composition, *PT*-conditions and growth kinetics.

1.2.1 Crystal Habit

Vesuvianite has a tetragonal symmetry, so it is usually easily diagnosed by its crystal shape. The general appearance of vesuvianite crystals is quite diverse. Often the approximate formation temperature of VSMs can be determined by their habit. Their high-temperature varieties of these minerals crystallized in skarns, hornfelses and contact zones of alteration of alkaline rocks as well as VSMs of volcanic origin, which crystallized at temperatures above 500 °C, are characterized by short-prismatic or equant habit, a small variety of simple forms, the presence of striation on the faces of the main prisms. In some cases, thin columnar crystals are formed with significant development along the *c* axis, which often form parallel-columnar or radial aggregates (Fig. 1.18a–d).

Among the high-temperature varieties of VSMs, the short-prismatic habit is most common (Fig. 1.18a). Short-prismatic crystals of vesuvianite, hongheite, and wiluite are characteristic of the trap association in the Wiluy River basin and the skarns of Julia (Khakassia, Russia), Luppiko and Hopunvaara (Karelia, Russia), Gukunvaara, Outokumpu, (both in Finland), Ariccia in Italy, Sierra de Cruces (Mexico), Frag el Ma, Marrakesh-Safi Region (Morocco), Kristiansand in Norway, Hodruša-Hámre in Slovakia, Hulta Mines in Sweden, Visp in Switzerland, numerous localities in USA, including Warren, Cumberland County and Maine, York County, and many other objects.

Very rare are dipyramidal vesuvianite crystals (Fig. 1.18b, c) were found in rocks of different genesis (Kurbatov 1946; Eldjarn et al. 2005). Dipyramidal crystals are characteristic of skarns of Fushan Mine in China, Lanfall in Norway, Plumas County, California in USA and also metasomatic rocks of Banat Mountains, Caraş-Severin in Romania. Big columnar or prismatic crystals of VSMs were found in cavities of limestone skarns (Fig. 1.18d). Such aggregates are characteristic of skarns of Lupikko, Karelia (Russia), Ais (Namibia), Alchuri (Pakistan), Corral de Los Lobos 2 and Carlés Mine, (both—Spain).

In skarns where vesuvianite forms a significant volume of rock it can be found in the form of radial aggregates (Fig. 1.18e). Massive radial aggregates of vesuvianite are characteristic of the skarns of Ak-Chagyl (Kazakhstan), Hazlov (Czech Republic), Monforte, Portalegre (Portugal), Monsant Mines, Barcelona, Catalonia (Spain), and Ward Mine, Nevada (USA).

Most typical associations of high-temperature VSMs usually include calcite, grossular or/and andradite, epidote–clinozoisite, calcite, pyroxenes of the diopside-hedenbergite series, scapolite and clinohumite. Later chlorite may also be present in

Fig. 1.18 Morphological varieties of VSMs of hight-temperature (>500 °C) origin: short-prismatic vesuvianite crystals from Kristiansand, Norway (**a**, sample 14,001 FM); Short-prismatic dipyramydal vesuvianite crystals from Banat Mts., Caraş-Severin, Romania (**b**, sample 51,109 FM); dipyramidal vesuvianite crystals from Lanfall, Norway (**c**, E. V. Galuskin's sample); columnar crystals from Lupikko, Karelia, Russia (**d**, sample 189/12409 SM); radial aggregates of vesuvianite from Ak-Chagyl, Kazakhstan (**e**, sample 38,476 FM). 1—vesuvianite, 2—calcite, 3—grossular, 4—gehlenite, 5—malachite

the association. Minerals of this group can often form a "shirt" (Fig. 1.19a). Often chlorite zones alternate with vesuvianite ones (Fig. 1.19b–d), which means that the stability fields of these two minerals overlap. Chlorite usually forms at temperatures of 100–350 °C (Cheng et al. 2024), however, higher-temperature chlorite associations of 500–550 °C have also been noted in nature (de Caritat et al. 1993; Walker 1993). Thus, the presence of alternating zones of chlorite and vesuvianite in skarns is explainable.

In secondary steps of mineralization vesuvianite sometimes occurs in microcrystalline (usually, with chlorite) aggregates, which replace primary vesuvianite and other associated minerals, in particular grossular.

For low-temperature varieties of vesuvianite with a crystallization temperature of 200–500 °C, several habit types can also be distinguished. Short prismatic "wiluite-like" crystals are most common, but they are characterized by a much greater variety of simple forms (Fig. 1.20a) compared to crystals of the same habit, but of the high-temperature origin. Similar crystals of VSMs can be found in the mines of

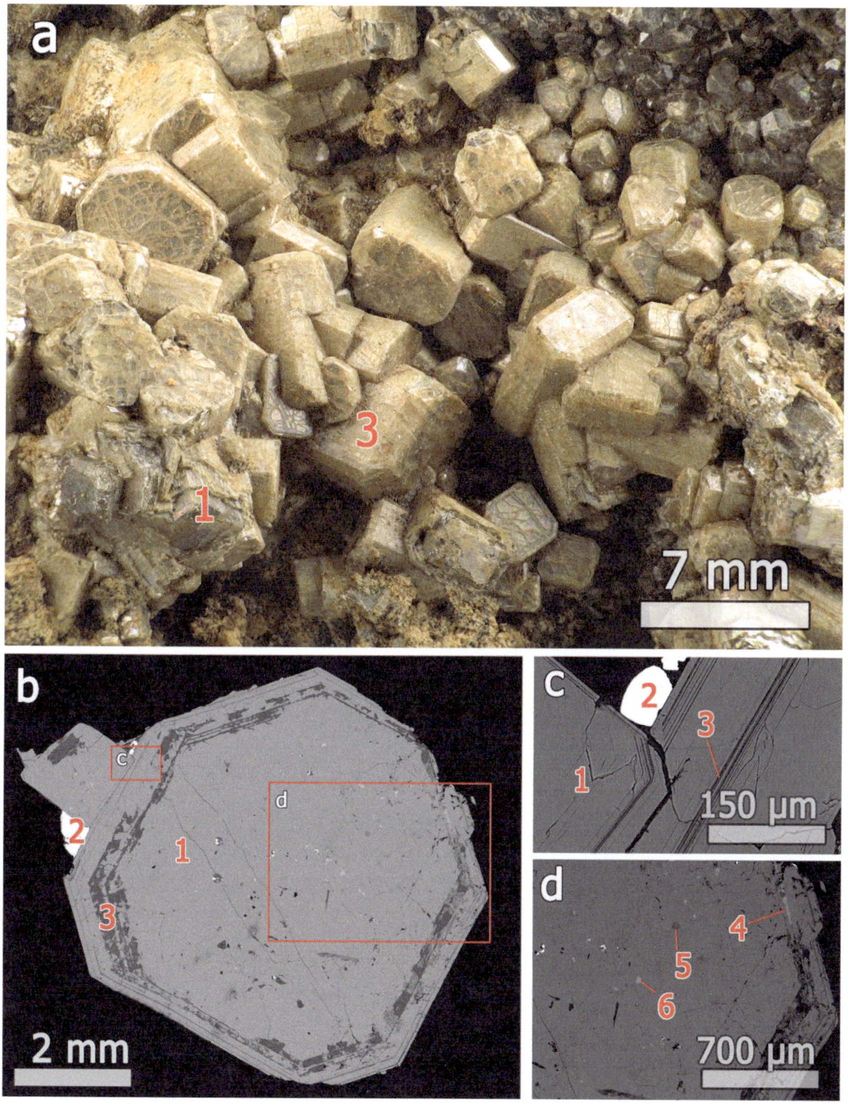

Fig. 1.19 Vesuvianite from "chlorite shirt" of VSM crystals from Gukunvara, Finland (sample No. 16908 FM): general view (**a**); BSE-image of the (001) section (**b**); chlorite inclusions with rhythmic zoning (**c**); inclusions of fluorite, diopside and a nigerite-group mineral (**d**). 1—vesuvianite, 2—sphalerite, 3—clinochlore, 4—fluorite, 5—Na- and Al-bearing diopside, 6—a nigerite-group mineral

Southern Urals (the Shishimskaya and Akhmatovskaya mines), as well as Diopsidrinne, Mayrhofen, Schwaz District, Tyrol, Austria; Rocher de Farinole, Olmeta-di-Capocorso, Bastia; Haute-Corse, Corsica, France; Aosta Valley, Italy; Flor de Peru I claim, Mt. Ullpac, Pampa Blanca District, Castrovirreyna Province, Huancavelica, Peru; Nová Anton vein, Hodruša-Hámre mines, Hodruša-Hámre, Žarnovica District, Banská Bystrica Region, Slovakia; Zermatt—Saas Fee area, Valais; Felskinn, Fee Glacier, Saas-Fee, Visp, Valais and Maderanertal valley, Uri (both—Switzerland).

Fig. 1.20 Morphological varieties of VSMs of low-temperature (below 500 °C) origin: Brown Ti-bearing vesuvianite with clinochlore from Nyazemskie Mts. (possible from the Akhmatovskaya mine) (**a**, 67,964 FM); acicular crystals of holotype magnesiovesuvianite from Tuydo, Macedonia (**b**, sample 59,783 FM); long-prismatic crystals of Cr-bearing vesuvianite from Kluchevskiy mine, Middle Urals, Russia (**c**, sample 55,291 FM); long-prismatic and splitted vesuvianite from Bellecombe, Châtillon, Italy (**d**, sample 79,621 FM); needle-like vesuvianite from Varaita Valley, Piedmont, Italy (**e**, sample 13,963 FM); druse of holotype alumovesuvianite on diopside from Jeffrey mine, Canada (**f**, N. V. Chukanov's sample);. 1—vesuvianite, 2—magnesivesuvianite, 3—grossular, 4—clinochlore, 5—alumovesuvianite

Acicular crystals are common for low-temperature VSMs including fluorvesuvianite, (Britvin et al. 2003), magnesuiovesuvianite (Panikorovskii et al. 2017e) and vesuvianite (Horváth and Gault 1990) (Fig. 1.20b). Due to the high refractive index, the thinnest vesuvianite needles of VSMs with a thickness of less than 100 μm have the properties of light guides (crystals with fibre-optic effect) (Galuskin et al. 2007b). Despite the fact that F-rich vesuvianite crystals with fibre-optic effect have the *P4/nnc* symmetry, the inferred crystallization temperature for them is 250–350 °C (Galuskin et al. 2003a; Britvin et al. 2003), while F-free magnesiovesuvianite is characterized by the *P4/n* symmetry (Panikorovskii et al. 2017e). Acicular fibres of vesuvianite, fluorovesuvianite and magnesiovesuvianite were noted in hydrothermally altered datolite-bearing skarns of Tas–Khayakhtakh Mountains (Polar Yakutia, Russia), in voids of hydrothermally altered skarns of Lupikko (Karelia, Russia), and rodingites of Tuydo (Macedonia).

Long prismatic crystals of vesuvianite (Fig. 1.20c–e) are most common in rodingites and contact-metosomatic rocks close to rodingites, as well as in regionally metamorphic rocks of the greenschist facies of metamorphism (Gnos and Armbruster 2006). Crystals of this habit almost always have *P4/n* or *P4nc* symmetry and are distributed in the mines of the Southern Urals; Bazhenovskoye deposit, Asbest, Sverdlovsk Oblast, Russia; Loja, Persenbeug-Gottsdorf, Melk District, Lower Austria, Austria; Jeffrey Mine, Canada; Rocher de Farinole, Olmeta-di-Capocorso, Bastia, Haute-Corse, Corsica, France; Tsangli mine, Eretria, Farsala, Larissa, Thessaly, Greece; Varaita Valley, Piedmont, Italy; Bellecombe, Châtillon, Aosta Valley, Italy; Zermatt—Saas Fee area, Valais, Switzerland. In rodingites, druses of VSMs, including alumovesuvianite (Panikorovskii et al. 2017a) are quite common (Fig. 1.20f).

1.2.2 Crystal Forms

Crystal morphology of VSMs is a good indicator of crystallization conditions to be used without application of analytical methods. The first studies of crystalline forms of vesuvianite were published in the works of Romée de Lisle (1783) and Haüy (1801). Numerous works on vesuvianite goniometry and crystal morphology were published in the period from the middle of XIX century to the beginning of XX century (Koksharov 1853; Zepharowich 1864; Kurbatov 1914, 1922, 1924, 1925a; Kurbatov and Solodovnikova 1923). These authors determined 103 simple forms of vesuvianite crystals, but according to Kurbatov (1946), only 53 simple forms are reliable. According to our data, only the most abundant six simple forms determine the morphological diversity of vesuvianite crystals: {100}, {001}, {110}, {111}, {131}, and {011}. They are often joined by forms {132}, {331} and {210}.

High-temperature VSMs are characterized by a short-prismatic or dipyramidal habit (Fig. 1.21a, b) sometimes close to isometric (Lyachovich 1954). The distinctive feature of such type of VSMs, usually characterized by significant contents of F or B (Borovikova and Kurazhkovskaya 2006) and high symmetry *P4/nnc* (Galuskin et al. 2003a), is the presence of the prisms {100} and {110}, dipyramids {111} and {101}, and base pinacoid {001} as the main (often, the only) forms (Goldschmidt 1918; Barth 1963; Annenkova and et al. 1972). Isometric multiface high-temperature vesuvianite crystals from the vicinity of Mount Vesuvius are a rare exception (Fig. 1.21a).

The most characteristic VSMs crystals from skarns are those of short prismatic habit, as for example for the Gukunvara skarns (Fig. 1.21c). For akhtarandite rocks of the Wiluy trap formation, two types of shapes of VSMs crystals are most common (Fig. 1.21b, c). During the study, the following simple forms were found for crystals from the Wilyui River basin: {001}, {111}, {331}, {131}, {110}, {010}, and {120}. The samples from Vilyui are characterized by the presence of vicinal faces (Prendel 1887); for more details see Sect. 1.3.2. Dipyramidal crystals of vesuvianite (Fig. 1.21d) are rare and were found in the skarns of Lanfall, Norway; Plumas County, California, USA and also in skarns of volcanic origin in Case Collina, Italy.

The increased elongation along the *c* axis compared to high-temperature VSMs varieties (Fig. 1.20a), the presence of numerous simple forms and growth striations on the main prisms {110} and {100} (Fig. 1.20c, e) as well as a wide diversity of crystal habit types are common for low-temperature VSMs (Gnos and Armbruster 2006; Panikorovskii et al. 2017d).

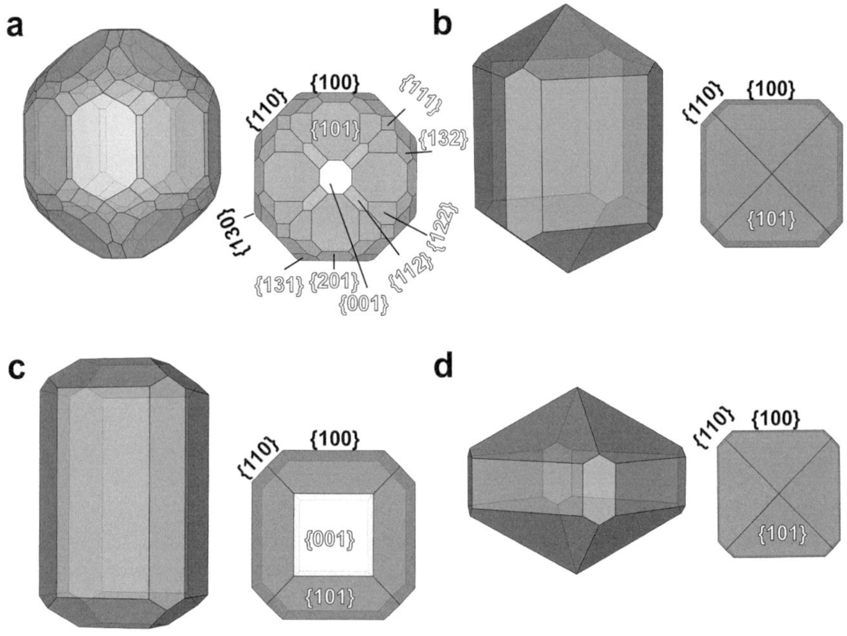

Fig. 1.21 The habit of VSMs crystals of hight-temperature (above 500 °C) origin from different localities: Vesuvius volcano (**a**, after Zepharowich (1864), achtaragdite rock of Mt. Shishimskaya, Wiluy River (**b**, after Koksharov (1853), vesuvianite from from skarn Gukunvara, Finland (**c**, our data); dipyramydal vesuvianite from skarn Case Collina, Italy (**d**, our data)

For low-temperature VSMs, prismatic habit is common, but the number of forms is much more diverse (Fig. 1.22a). Samples from Urals, described in goniometric works by Kurbatov, demonstrate especially wide diversity in habit (Kurbatov 1946). According to his data, these crystals have up to 31 various simple forms. The most common simple forms of low temperature vesuvianite from the Polyakovsky mine, Bazhenovskiy deposit, Urals and Wessels mine, RSA shown in Fig. 1.22b–d are prisms {110}, {100}, {120}, dipyramids {121}, {131}, {122}, {221} and base pinacoid {001}.

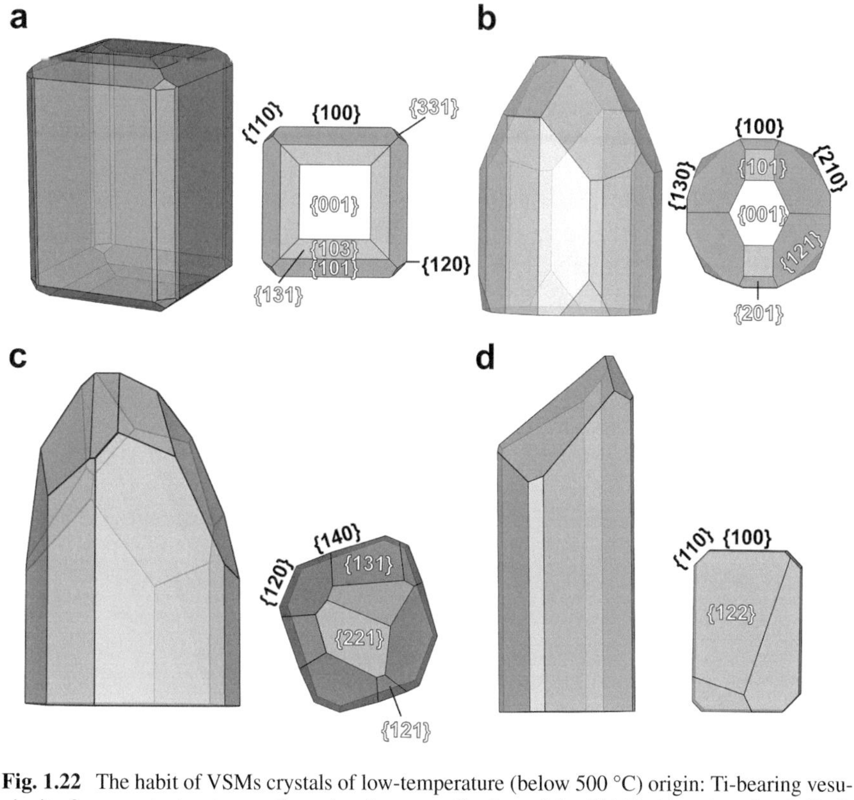

Fig. 1.22 The habit of VSMs crystals of low-temperature (below 500 °C) origin: Ti-bearing vesuvianite from contact-metasomatic rocks close to rodingites of the Shishimkiy mine, South Urals (**a**, after Kurbatov (1946); green vesuvianite from the Polyakovsky mine, South Urals (**b**, after Kurbatov (1946); vesuvianite from rodingite of the Bazhenovskiy deposit, Middle Urals (**c**, our data), and metasomatic rock uncovered by the Wessels mine, RSA (**d**, our data for cyprine). Prismatic forms are indicated by black color, white color indicates other forms. Investigation of simple forms of vesuvianite crystals from the Bazhenovskiy deposit and Wessels mine were performed by goniometer and camera of an Xcalibur EOS diffractometer

1.2.3 Color

The color of VSMs depends on impurities and varies widely from colorless, yellow, red and green in various shades to dark brown and black. The most important chromophore elements are iron, titanium, chromium, copper, manganese, iron and sulfur (Fig. 1.23).

Transparent varieties of VSMs are usually found in rodingites and similar metasomatic contact rocks. Among the chemical features, they are virtually free of Fe (up to 1 wt% of Fe_2O_3) and without other strong chromophores and are almost colorless. In particular, transparent VSMs specimens are found in rodingites from Jeffrey Mine, Canada (Fig. 1.24a, b) and Kosmet, Macedonia (Fig. 1.24c).

Causes of blue color of Cu-bearing vesuvianite are discussed in the papers by Dyrek et al. (1992), Hålenius et al. (2016), Panikorovskii et al. (2017f). Our conclusions confirmed that blue color of vesuvianite from Sauland, Norway is caused by divalent copper (Fig. 1.25).

Vesuvianite samples with Mn content reaching 3.8 wt% are known in metasomatic rocks of the Långban deposit, Sweden. Vesuvianite samples from the Franklin deposit, USA are also or rather rich in manganese. According to optical absorption spectra studies (Platonov et al. 1995; Czaja et al. 2018) trivalent manganese causes a vivid purple color of vesuvianite (Fig. 1.26a). As a result, not only "manganivesuvianite" (the old name of manganvesuvianite), but also cyrpine and Cu-bearing vesuvianite from the Wessels and N'chwaning mines, with the dominance of Cu^{2+} at the $Y1$ site have not blue, but lilac or maroon-red color (Fig. 1.26b).

Chromium is rarely present in VSMs in amounts exceeding 0.5 wt%, but even small admixtures of Cr^{3+} may have a significant influence on their color. Cr^{3+}-bearing vesuvianite samples are colored emerald green. Such vesuvianite varieties are known in rocks of several localities in Urals including Voikaro-Syn'insky massif (Fig. 1.27a), Bazhenovskoe deposit (Fig. 1.27b) and Monetnaya Dacha deposit (Krivovichev et al. 2013). Cr-containing vesuvianite was also noted in rodingites of Greece (Rogkala et al. 2022a). Cr-rich vesuvianite inclusions in diposide from rodingite veins cutting serpentinites in a chromium deposit in Xinjiang, China are characterized by high contents of both Cr_2O_3 up to 6.21 wt% and TiO_2 up to 7.43 wt% (Kobayashi and Kaneda 2010). Green vesuvianite with 1 wt% of Cr_2O_3 was found at Gerogetown, USA (Fig. 1.27c) and in Jeffrey Mine, Canada. The study of optical absorption spectrum of Cr-bearing vesuvianite (Platonov et al. 1993a; Hålenius 2000) showed that Cr^{3+} causes emerald green color, which is fully consistent with our data (Fig. 1.27).

Fe-bearing VSMs (Fig. 1.23), have different shades of green (rarely, brown) color (Manning 1968; Manning and Tricker 1975). Bottle-green and olive-freen color (Fig. 1.28) is caused by charge transfer between neighboring Fe^{2+}- and Fe^{3+}-centered octahedra (Calas et al. 2021). Isolated $Fe^{3+}O_6$ octahedra are the cause of brown color (Fig. 1.29).

Brownish red and orange color (Figs. 1.23 and 1.29) is typical for VSMs containing both Ti and Fe^{2+} in neighboring octahedra. This chromophore center occurs in various red minerals. In particular, the intense E||c-polarized band with

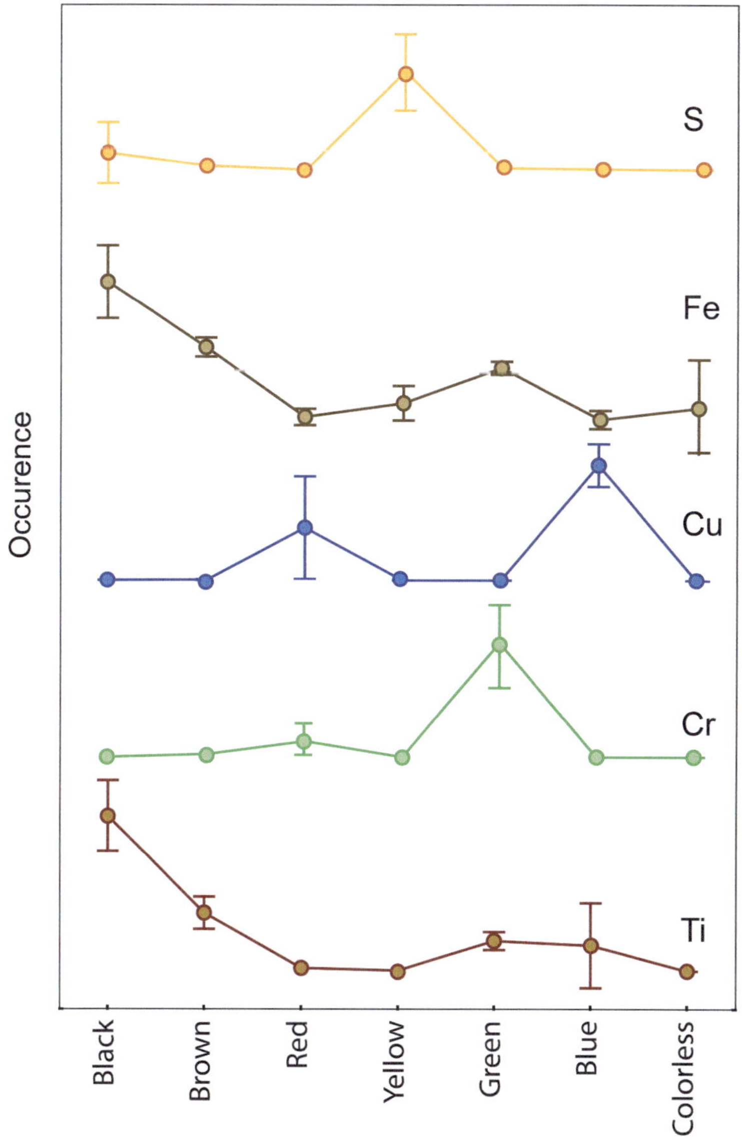

Fig. 1.23 Diagrams showing the dependence of vesuvianite color on the contents of Ti, Cr, Cu, Fe, S

the maximum at 20,500 cm^{-1} in the optical absorption spectra of red andalusite was assigned to the electronic Fe^{2+}/Ti^{4+} transition (Manning and Tricker 1975; Taran and Koch-Müller 2011). In red labuntsovite-Fe, Fe^{2+}- and Ti-centered octahedra share common edges (Aksenov et al. 2018).

Fig. 1.24 Colorless alumovesuvianite with Mn-bearing pink zones (**a**, Jeffrey Mine), druse of alumovesuvianite with 0.3 wt% of Fe_2O_3 from Jeffey Mine, Canada (**b**, sample 70,036 FM) and alumovesuvianite from rodingites of Kosmet, Macedonia (**c**, sample 55,211 FM); 1—alumovesuvianite, 2—diopside

Fig. 1.25 Blue Cu-bearing vesuvianite from Sauland, Norway (B, sample 15,377 FM); 1—vesuvianite, 2—Mn^{3+}-bearing zoisite, 3—quartz

Fig. 1.26 Mn^{3+}-bearing vesuvianite from Tuydo, Macedonia (**a**, sample 60,295 FM), Mn^{2+}, Mn^{3+}-bearing cyprine (Cu^{2+}-dominated at *Y*1 site) from Wessels, RSA (**b**, I.V. Pekovs sample); 1—vesuvianite, 2—calcite, 3—cyprine

Samples most rich in divalent iron and titanium have black or dark brownish red color (Figs. 1.23 and 1.28). In particular, the pair of neighboring Fe^{3+} and Ti-centered octahedra is the main cause of deep red (Fig. 1.29) to black color, as e.g. in Ti-rich modraite from Alchuri, Pakistan (see below) and in black Ti-rich VSM from Val d 'Aosta, Italy (Fig. 1.30a) as well as in Ti-rich modraite from Alchuri, Pakistan (Fig. 1.30b) and black Fe and Ti rich vesuvianite from Oyta, Japan (Fig. 1.30c).

VSMs with significant amounts of sulfide sulphur have a distinct yellow color (Figs. 1.23 and 1.31a, b). The most probable cause of yellow color is the $S_2^{\bullet -}$ radical anion (in more detail in Chap. 4).

1.2.4 Streak, Lustre, Hardness and Density

The VSMs streak varies from white (for light-colored samples) to brownish (for dark-colored samples). Most samples have vitreous (in crystals) or silky (in fibres) lustre, which becomes resinous in metamict varieties. The most common VSMs are translucent in thin sections.

The Mohs hardness is in the range from 5 to 6.5. The brittleness increases with the porosity and depends on the concentration of hydrogarnet-like defects in the VSMs structure (Galuskin et al. 2003b, 2007a; Ohkawa et al. 2009; Panikorovskii et al. 2016a). Density increases linearly from 3.30 to 4.05 g/cm^3 with increasing content of *REE*, Ti and Fe. The fracture is uneven to conchoidal. No cleavage is observed, but sometimes VSMs show parting on (100). Merohedral twins are typical

Fig. 1.27 Emerald-green Cr-bearing vesuvianite from chromitite of Voikaro-Syn'insky massif, Polar Urals, Russia (**a**), Crystal of Cr-bearing vesuvianite on calcite-clinochlore aggregate from a chromite lens. Bazhenovskoe asbestos deposit, Asbest, Middle Urals, Russia (**b**, specimen 6880 from the collection of B.I. Ermolenko); Cr-bearing alumovesuvianite from Gerogetown, USA (**c**, sample 51,073 FM); 1—vesuvianite, 2—clinochlore, 3—caclite

for manganivesuvianite (Armbruster et al. 2002) and magnesiovesuvianite (Panikorovskii et al. 2017e) and sometimes are observed for vesuvianite (Panikorovskii et al. 2016d). Vesuvianite dissolves slowly in concentrated hydrofluoric acid. Dehydroxylation of VSMs accompanied by amorphization or melting (in the case of boron-rich samples) takes place at ~1050 °C. Rapid cooling of vesuvianite melt results in the

Fig. 1.28 Fe^{2+}- and Fe^{3+}-bearing vesuvianite from rodingites of the Bazhenovskoe deposit, Middle Urals; 1—vesuvianite, 2—grossular, 3—caclite

formation of glass whereas in the course of slow cooling crystallization of grossular is observed.

1.2.5 Optical Properties

Optical properties of VSMs are shown in Table 1.1. In thin sections, these minerals are easily identified due to a high relief and very weak birefringence (pale gray tones of interference are common). Pleochroism is usually weak, although VSMs with high contents of Mn, Fe, and/or Ti are characterized by strong pleochroism from yellow to brown.

table

Despite tetragonal symmetry, VSMs often show anomalous biaxiality, which is especially characteristic of high-temperature phases (Groat et al. 1998), Prendel (1887) and Lyahovich (1955) paid attention to the zonal structure of wiluite crystals, in which the outer zones exhibit a thin oscillatory zonality and have $2V$ angle of 30–35°, regardless of the edge growths sector.

D.E. Arem noted sectorial zonation of optical properties of wiluite Fig. 1.32 and vesuvianite (Arem 1970). In their crystals different sectors demonstrate various

Fig. 1.29 Crystals of amber-yellow vesuvianite with clinochlore from Monte Somma, Italy (**a**, **b**) (sample 27,844 FM). Crystals of Fe^{2+}- and Ti-bearing vesuvianite 2 and 4 cm across. Fushian mine, Xintal City, Hubei, China (**c**)

birefringence along [001] with $2V$ angle in the range 0–65°. Abnormal biaxiality is also found in samples of F-rich vesuvianite from Pitkäranta, Karelia, Russia (Britvin et al. 2003) and Bellecombe, Italy (Tanaka et al. 2002) whose crystals have anomalous optical sectors with $2V$ angles of 10–45°.

All VSMs are characterized by a negative optical character until B_2O_3 content reaches ~2.60 wt% and then optical character changes to positive (Kurazhkovskaya et al. 2005). In particular, boron members of the vesuvianite supergroup, wiluite

Fig. 1.30 Ti-rich VSM from Bellecombe, Val d'Aosta, Italy (**a**), Modraite from Alchuri, Pakistan (**b**), black Fe- and Ti-rich vesuvianite from Oyta, Japan (**c**, sample 54,541 FM); 1—vesuvianite, 2—calcite, 3—chlorite

Fig. 1.31 S-bearing yellow vesuvianite crystal (**a**, sample 50–3057 SM) and zonal vesuvianite crystal with inclusion of pyrrhotite at the centre (the S content increases from centre to rim with increasing of color intensity) (**b**, sample 7359 MF), both from Monzoni, Predazzo, Italy. 1—vesuvianite, 2—pyrhotite

Table 1.1 Optical properties of VSMs after (Deer et al. 1982; Groat et al. 1998; Armbruster et al. 2002; Britvin et al. 2003; Panikorovskii et al. 2017e, a, f, b; Xu et al. 2019; Chukanov et al. 2019; Moiseev et al. 2020; Uher et al. 2024)

Mineral	ω	ε	δ	Optical sign
Vesuvianite	1.700–1.746	1.703–1.752	0.001–0.009	(–)
Wiluite	1.721	1.725	0.004	(+)
Fluorvesuvianite	1.702	1.699	0.003	(–)
Manganivesuvianite	1.727–1.735	1.716–1.724	0.012	(–)
Magnesiovesuvianite	1.725	1.721	0.04	(–)
Cyprine	1.744	1.732	0.012	(–)
Alumovesuvianite	1.725	1.722	0.003	(–)
Hongheite	1.720	1.725	0.005	(+)
Modraite*	1.749	1.740	0.009	(–)
Manaevite	1.760	1.750	0.010	(–)
Milanriederite	1.744	1.737	0.007	(–)

* Ti-rich variety, our data

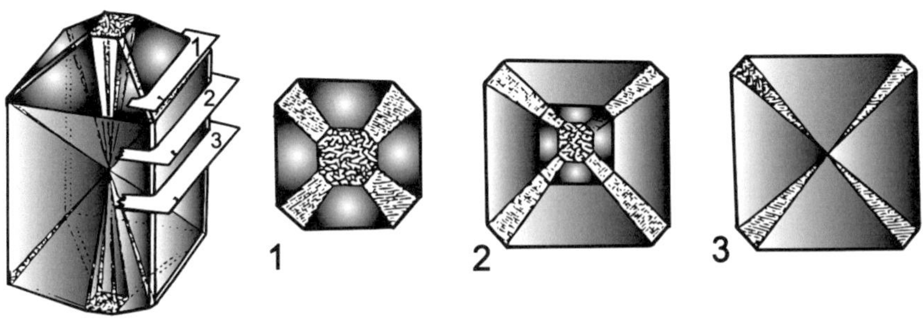

Fig. 1.32 Different sections of sectorial-zonal wiluite crystal (after Arem 1970) with modifications

and hongheite, are optically positive. The fact is that the boron filling of additional *T*1 and *T*2 sites, usually occupied by hydrogen atoms in the vesuvianite structure, simultaneously with an increase in the magnesium content (Fig. 1.33a) leads to a change in the ratio of the unit cell parameters (Fig. 1.33b). Presumably, this is the cause of the change of the optical sign. In Chap. 3, the presumed structural mechanism for changing the optical sign in the vesuvianite-wiluite series is considered in more detail.

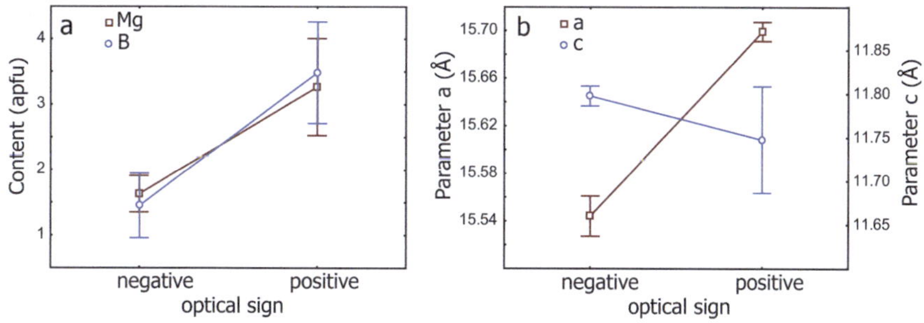

Fig. 1.33 Graphs of the dependence of the optical sign on the composition (**a**) and the parameters from VSM unite cell parameters (**b**)

1.3 Crystal Chemistry

1.3.1 General Formulae

According to the current mineralogical classification (Strunz and Nickel 2001), vesuvianite supergroup of minerals belongs to neso-sorosilicates (i.e. silicates containing both isolated SiO_4 and Si_2O_7 groups): 9.BG.35. The general formula of VSMs can be written as follows ($Z = 2$): $(X1_2X2_8X3_8X4)_{19}(Y1Y2_4Y3_8)_{13}T_{0\text{-}5}(Z_2O_7)_4[(ZO_4)_{10\text{-}x}(H_4O_4)_x](W)_9O_{1\text{-}3}$, where $x < 3$; among X sites populated by Ca^{2+} and subordinate amounts of Na^+, K^+, Fe^{2+}, REE^{3+}, and/or Bi^{3+}, the $X3$-polyhedron has a coordination number of 9 and the largest volume, the $X1$, $X2$ and $X4$ polyhedra are surrounded by 7 or 8 oxygen atoms; the $X1$ and $X4$ sites have square antiprism coordination and can be occupied by Ca^{2+}, REE^{3+} and Na^+.

The tetragonal $Y1$ pyramid can be occupied by the cations (in order of their abundance) Fe^{3+}, Mg^{2+}, Al^{3+}, Fe^{2+}, and Cu^{2+}; the $Y2$ and $Y3$ sites with octahedral coordination can be occupied by Al^{3+}, Mg^{2+}, Fe^{3+}, Ti^{4+}, Fe^{2+}, Mn^{3+}, Mn^{2+}, Cr^{3+} and Zn^{2+}.

The $Z1,2$ sites in the sorosilicate SiO_4 groups and the $Z3$ site of the nesosilicate Si_2O_7 group can be occupied by Si^{4+}, sometimes with subordinate vacancies coordinated by $(H_4O_4)^{4-}$ (in the case of isolated tetrahedra) and possible minor Al^{3+} admixture (discussed in Chap. 4).

The additional $T1$ and $T2$ sites with triangular and tetrahedral cordination can be occupied by B or be vacant whereas Al can occur only at the $T2$ site.

The anionic W sites can be populated by $(OH)^-$, F^-, Cl^- and O^{2-} (Machatschki 1932; Coda et al. 1970; Groat et al. 1992a; Dyrek et al. 1992; Galuskin et al. 2003b, 2005; Aksenov et al. 2016; Panikorovskii et al. 2016a, 2017d).

The problem of the correct writing the VSMs formulae remains actual, since the wide variations in the chemical composition do not always allow to reliably determine not only the specific type of the VSM, but also the belonging of the mineral to this group, having only the chemical composition data available (see Sect. 1.3.2). Table 1.2 shows the main historical stages of the refinement of the general crystal-chemical formula of the vesuvianite supergroup minerals.

The authors of the first structural description of vesuvianite (Warren and Modell 1931) suggested the formula $Ca_{10}Al_4(Mg,Fe)_2Si_9O_{34}(OH)_4$ derived without taking into account the half-occupied $X4$ and $Y1$ positions, the associated channel position O10, as well as some other oxygen positions. After 50 years, a new version of the vesuvianite formulae was proposed (Hoisch 1985; Valley et al. 1985): $Ca_{19}Mg_2Al_{11}Si_{18}O_{68}(O)(OH,F)$, which, with minor changes, was accepted until recent studies (Groat et al. 1992a; Galuskin et al. 2007a; Panikorovskii et al. 2017b). However, the discovery of new members of the VSMs necessitates the development of a full-fledged nomenclature for this group (Panikorovskii et al. 2017d), since the vesuvianite formula in its current form corresponds to several mineral species belonging to the vesuvianite supergroup.

Table 1.2 Evolution of the general vesuvianite and VSMs formulae

Formula	References
$Al_4(SiO_4)_3$	Clarke (1895)
$Ca_5(Mg,Fe,Mn,Na_2,K_2,H_2)Al_2[Al,Fe,Ti,B(OH,F)]Si_5O_{20}$	Weibull (1896)
$4[Ca_{10}Al_4(Mg,Fe)_2Si_9O_{34}(OH)_4]$	Warren and Modell (1931)
$2[X_{19}Y_{13}Z_{18}(O,OH,F)_{76}$; $X =$ Ca (Na, K, Mn); $Y =$ (Al, Fe^{3+}, Fe^{2+}, Mg, Ti, Zn, Mn), $Z =$ Si	Machatschki (1932)
$4[X_{10\text{-}u}Y_{6+u}Z_9(O,OH,F)_{38}$ $u \leq 1$; $X =$ Ca (Na, K, Mn); $Y =$ (Al, Fe^{3+}, Fe^{2+}, Mg, Ti, Zn, Mn), $Z =$ Si	Barth (1963)
$Ca_{16}(Ca_{2\text{-}x}Mg_x)Mg_2(Mg_{2\text{-}y}Al_y)Al_6(Al_ySi_{2\text{-}y})Si_{16}O_{68}(OH)_8$; $0 \leq x \leq 1.5$ $0 \leq y \leq 1.5$	Ito and Arem (1970)
$2[Ca_{19}(Mg,Fe,Al,Ti,Mn)_5Al_8(O,OH)_{10}(SiO_4)_{10}(Si_2O_7)_4]$	Coda et al. (1970)
$2[Ca_{19}Al_{11}Mg_2Si_{18}O_{69}(OH)_9$	Hoisch (1985)
$2[X_{19}Y_{13}Z_{18}T_{0\text{-}5}O_{68}W_{10}]$; $X =$ Ca, Na, REE^{3+}, Pb^{2+}, Sb^{3+}; $Y =$ Al, Mg, Fe^{3+}, Fe^{2+}, Ti^{4+}, Mn, Cu, Zn; Z = Si; W = (OH, F, O)	Groat et al. (1992a)
$2[X_{18}X'Y'Y_{12}Z_{18}O_{69}(OH,F)_9$; ${}^{7-9}X$, ${}^{7-9}X'$, ${}^{5}Y'$; ${}^{4}Z$	Armbruster and Gnos (2000b)
$Ca_{19}(Al,Mg,Fe,Mn)_{13}B_{<2.5}[(SiO_4)_{10-x}(OH)_{4x}](Si_2O_7)_4(OH,O,F,Cl)_{10}$; $x < 3$	Galuskin et al. (2007a)
$2[X_{16}X1_2X4Y1Y2_4Y3_8T_{0\text{-}5}(Z_2O_7)_4[(ZO_4)_{10\text{-}x}(H_4O_4)_x](W)_9O_{1\text{-}3}]$; $x < 3$; ${}^{7-9}X =$ Ca, Na, K, Fe^{2+}, *REE*, Bi; ${}^{8}X1,4 =$ Ca, Na; ${}^{5}Y1 = Fe^{3+}$, Mg, Al, Fe^{2+}, Mn^{3+}, Cu^{2+}; ${}^{6}Y2,3 =$ Al, Mg, Fe^{2+}, Fe^{3+}, Mn^{2+}, Mn^{3+}, Ti, Cr, Zn; $T =$ B, Al; $ZO_4 = (SiO_4, H_4O_4)$; $Z_2O_7 = (SiO_4)$ $W =$ OH, F, O, Cl	Panikorovskii et al. (2017b)

1.3.2 Chemical Variations and Formula Calculation

The wide variability of the chemical composition of VSMs, often observed even for the major components, is an obstacle to the identification of these minerals based on the data of chemical analysis. According to Kurbatov (1948), there is also the difficulty to express the composition of the vesuvianite with a single general formula. In a review on vesuvianite chemistry (Groat et al. 1992a), it was noted that deviations from the stoichiometry $Ca_{19}Si_{18}$ are most characteristic of samples with a significant boron content. Therefore, (Groat et al. 1994a) proposed a calculation of a boron-containing vesuvianite analogue (wiluite) formula with rationing for 19 calcium atoms and affinity elements for it, whereas for the rest of the group it is assumed that there are 50 cations in the formula.

The existence of the Si-deficient vesuvianite variety (Galuskin et al. 2003b, 2007a; Galuskin 2005; Ohkawa et al. 2009; Panikorovskii et al. 2016a) and its relatively widespread occurrence in nature preclude the possibility of calculating the VSMs chemical formulae based on 18 Si atoms. A possible way of calculating the formula based on 13 *Y* atoms (Y = Al, Mg, Fe, Ti, Mn, Cr, Cu, etc.) is not always satisfactory due to the presence of Al up to 0.6 atoms per formula unit (*apfu*) in the $T1$ position of many high-temperature VSMs samples (Groat et al. 1994a, b; Gnos and Armbruster 2006; Panikorovskii et al. 2017d). The calculation for 147 charges also can give an unsatisfactory result: for example, the total charge of cations in wiluite and VSMs close to it varies within the range 152^{+}–157^{+} (Panikorovskii et al. 2017d) Since in some cases the total content of B as well as Na, Sr, Bi, Pb, Sb, K, and *REE* is quite significant (Fig. 1.34a), the optimal method of formula calculation (used in this work) is based on 19 atoms at *X* sites.

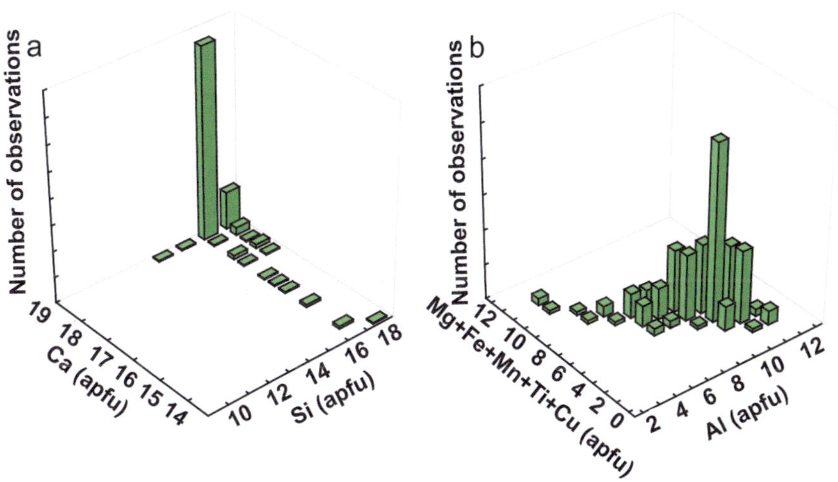

Fig. 1.34 Distribution of Si and Ca contents (**a**) and sum of major *Y*-components and Al (**b**) in VSMs

The VSMs can contain significant amounts of more than 30 elements of the periodic table, some of which are variable-valence elements. Information on their contents, is given below in accordance with their distribution between the groups of positions *X*, *Y*, *Z*, *T* and *W* in the structures of VSMs. According to their occurrence, the components are conventionally divided into the ***major*** and ***minor***.

X-sites

The main ***major*** component at *X* sites is calcium (Groat et al. 1992a). The minimal CaO content was found in *REE*-enriched VSMs from the skarnoids of the Kovdor massif (see Fig. 1.11c): 19.9 wt% or 14.2 *apfu*, which is slightly less than in the *REE*-containing vesuvianite from San Benito (USA) with 17 wt% of REE_2O_3 (Fitzgerald et al. 1987). The highest CaO content noted in the work (Kurbatov 1946) is 37.30 wt% or 19.3 *apfu*, which, taking into account the possible error of analytical measurements, corresponds to the normal calcium content of 19 *apfu* in vesuvianite sensu *strictu*.

There are large cations with effective ionic radii R_{ef} > 1.06 Å which can occur at the *X* sites (as a rule, in subordinate amounts with respect to Ca) and are considered here as ***minor*** components: REE^{3+}, Sr^{2+}, Na^+, K^+, Pb^{2+}, Sb^{3+}, Bi^{3+}, U^{4+}, Th^{4+}. The most significant contents were found for the following components: REE_2O_3—20.57 wt% (Crook and Oswald 1979), Sb_2O_3—21.21 wt% (Pan and Fleet 1992) and Bi_2O_3—20.42 wt% (Hålenius et al. 2013). The maximum content of Na_2O in VSMs is relatively low—1.49 wt% (Nyunt et al. 2009), which corresponds to 1.41 *apfu*. Taking into account the selective incorporation of Na into the *X*1 and *X*4 sites, in the case where Na content exceeds 1.50 *apfu*, the mineral would correspond to another species, "natrovesuvianite" (Panikorovskii et al. 2016b).

Y-sites

We did not find any vesuvianite in which all *Y*-sites (13 *apfu*) were populated exclusively by Al^{3+} atoms (Fig. 1.34b). The ***major*** components at the *Y* sites with the coordination numbers of 5–6 are Al, Mg, Fe, and Ti (Fig. 1.35). The content of Al_2O_3 varies from 4.33 wt%, or 3.11 Al *apfu*, for manaevite-(Ce) from the Kovdor massif, up to 21.62 wt% (Panikorovskii et al. 2017g; Moiseev et al. 2020) or 12.02 Al *apfu*, for alumovesuvianite from Asbestos, Canada (Fitzgerald et al. 1986a). Apparently, this is the limiting content of this element in the VSMs, as evidenced by the normal character of its distribution (Fig. 1.35a). According to our data, the average value of the aluminum content, is 9.6 *apfu* and coincides with that given by Groat et al. (1992a).

The most significant content of iron (up to 10.85 wt% FeO or 4.66 Fe *apfu*) is noted for As-containing vesuvianite occurring in skarn from Nedvědice, Czech Republic (Groat et al. 2013). The minimal values (0.04–0.40 wt% of FeO) were recorded for low-temperature VSMs, in particular, for alumovesuvianite and magnesiovesuvianite (Panikorovskii et al. 2017a, e). The Fe content in VSMs is distributed according to the normal law (Fig. 1.35b) with the average value of 1.80 *apfu*. The Fe^{2+}/Fe^{3+} ratio varies in a wide range and the total iron content is inversely proportional to the magnesium content (Rusakov et al. 2006). Divalent iron cations predominate in the reducing medium which results in the formation of Fe^{2+}-enriched VSMs whereas

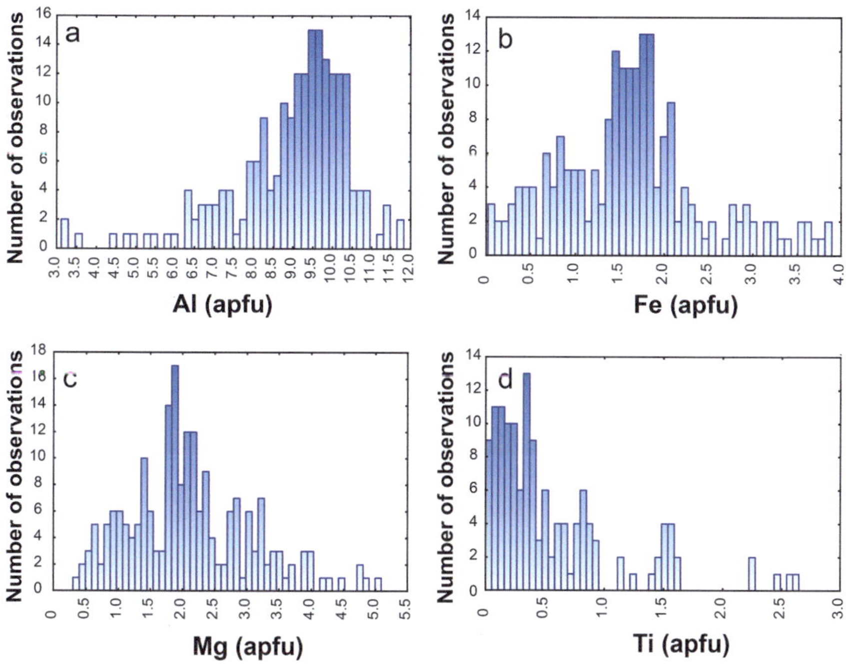

Fig. 1.35 Histograms of the Al (**a**), Fe (**b**), Mg (**c**) and Ti (**d**) contents in VSMs

in oxidizing environment ferric cations predominate and Mg-enriched VSMs with a high content of Fe^{3+} are formed. The possibility of entering a minor part of Fe^{2+} in the *X* position of the VSMs is assumed (Manning and Tricker 1975; Karwowski et al. 1996; Groat et al. 2013).

Magnesium is always present in the vesuvianite composition, in contrast to the similar grossular-andradite-uvarovite series, which usually allows to confidently distinguish them. The distribution of magnesium corresponds to a lognormal law with a minimum of 0.91 wt% of MgO or 0.68 Mg *apfu* in vesuvianite from the Franklin, USA (Groat et al. 1992a), maximum 6.65 wt% of MgO or 5.05 Mg *apfu* in wiluite from the bank of Wiluy River (Galuskin 2005) and an average content of 2.0 *apfu* (Fig. 1.35c).

The amount of titanium in VSMs depends on the type of host rocks: VSMs associated with granitic rocks practically do not contain titanium, and its maximum content is 5.85 wt% of TiO_2 or 2.2 Ti *apfu* (Fig. 1.35d) was recorded in vesuvianite from an Alpine type core in Alchuri, Pakistan. Unlike other ***major*** *Y*-group cations, titanium cannot occupy the *Y*1 tetragonal pyramid site (Aksenov et al. 2016).

The most common ***minor*** cations of the *Y*-group of sites are Cu^{2+}, Cr^{3+} and $Mn^{3+/2+}$, which content reaches the highest values 2.72 wt% of CuO, 2.31 wt% of Cr_2O_3 and 15.96 wt% of Mn_2O_3 in cyprine from Wessels, RSA (Panikorovskii et al. 2017f), vesuvianite from a hydrothermal vein in the chromites of the Monetnaya Dacha, Urals (Kurbatov 1946) and manganivesuvianite from the manganese deposit

N'chwaning II, RSA (Armbruster and Gnos 2000c), respectively. Vesuvianite from Franklin skarns, USA contains up to 2.35 wt% of ZnO (Groat et al. 1992a).

Z-sites

For a long time, only silicon was considered as the only ***major*** component in tetrahedral *Z* positions (Hoisch 1985; Allen and Burnham 1992; Groat et al. 1992a). Its content usually varies within narrow limits from 36.5 to 37.5 wt%, and the maximum value is 37.97 wt% (Weibull 1896). Recent works (Armbruster and Gnos 2000c; Galuskin et al. 2003b, 2007a; Ohkawa et al. 2009; Panikorovskii et al. 2016a) described samples with a significant Si deficiency (up to 32.02 wt% SiO_2 or 15.4 *apfu* Si), the charge disbalance being compensated by $(H_4O_4)^{4-}$ cluster (a ***minor*** component). According to the data by Groat et al. (2013), As^{3+} can also be incorporated into the *Z* sites (up to 1.78 wt% As_2O_3) and, according to our observations, Al can also be incorporated there (in Chap. 4, this matter is discussed in more detail).

The histogram of silicon content in the VSMs corresponds to the Cauchy distribution (Fig. 1.36a) with a sharp narrow maximum at 18 *apfu* on a fairly broad wings. In boron-containing VSMs, the average Si content is equal to 17.80 *apfu* (Fig. 1.36b), which is probably due to partial compensation of the excess positive charge resulting from the introduction of boron by replacing the corresponding part of Si^{4+} with Al^{3+} in tetrahedra (Groat et al. 1992a). In some analyses, the amount of silicon differs significantly from 18 *apfu* towards lower values. As a rule, such analyses correspond to altered (metamict or hydrated) samples.

T-sites

The ***major*** components at the additional *T* sites (*T*1—tetrahedral and *T*2—triangular) are B or Al, which replace H atoms of the hydroxyl group, otherwise these sites remain vacant. In wiluite (Galuskin 2005), the total boron content can reach 5.00 wt% of B_2O_3 or 4.2 B *apfu*. Chemical analysis of boron-containing VSMs (Fig. 1.37a) usually shows significantly less boron than single-crystal X-ray structural analysis (Fig. 1.37b) due to the partial incorporation of aluminum in the *T*1 position (Groat

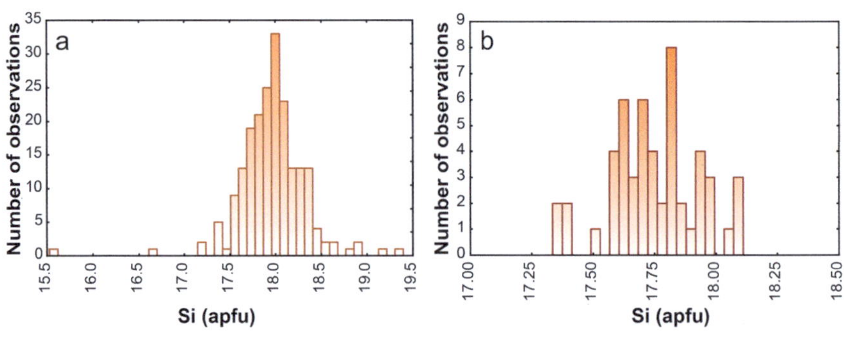

Fig. 1.36 Histogram of the Si content in all studied samples of VSMs (**a**) and in boron-containing VSMs (**b**)

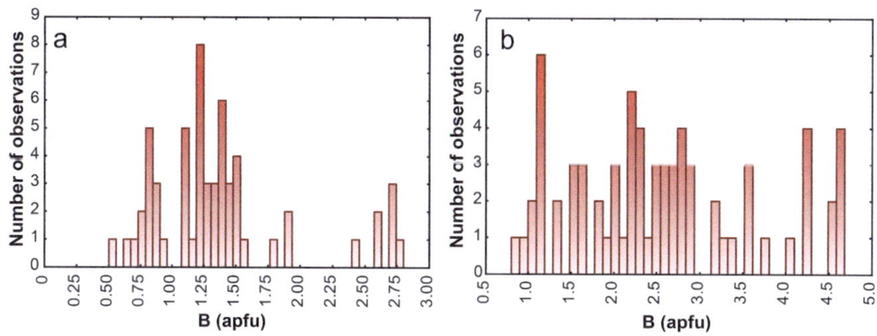

Fig. 1.37 Histograms of B content in VSMs according to chemical (**a**) and single-crystal XRD (**b**) data

et al. 1994b, 1996a; Panikorovskii et al. 2017d). Sulfur and carbonate anions are ***minor*** elements at the *T*2 site (the latter was found by us in a *REE*-Ti-enriched vesuvianite from the skarnoids of the Kovdor massif).

_W_-sites

In most VSMs samples, *W*-sites are occupied exclusively by hydroxyl ions (Fig. 1.38a). The maximum content of OH^- groups in the vesuvianite supergroup minerals corresponds to 5.69 wt% of H_2O or 18.2 OH^- groups per formula unit which is achieved by filling additional positions of the H atoms replacing silicon according to the substitution scheme $Si^{4+} \leftrightarrow 4H^+$ (Galuskin et al. 2003b). Besides the OH^- groups, another ***major*** component in VSMs is fluorine, the content of which in fluorvesuvianite from the Pitkäranta skarns, Karelia, Russia is up to 4.60 wt% or 7.2 *apfu*. This causes a decrease in the water content to 0.5 wt% or 1.7 OH groups per formula unit (Britvin et al. 2003).

In addition, a low H_2O content in the chemical composition of the VSMs (less than 1 wt%) can be associated with a significant boron content (Groat 1988a; Panikorovskii et al. 2017d). The ***minor*** component at W-sites is the chloride anion. The

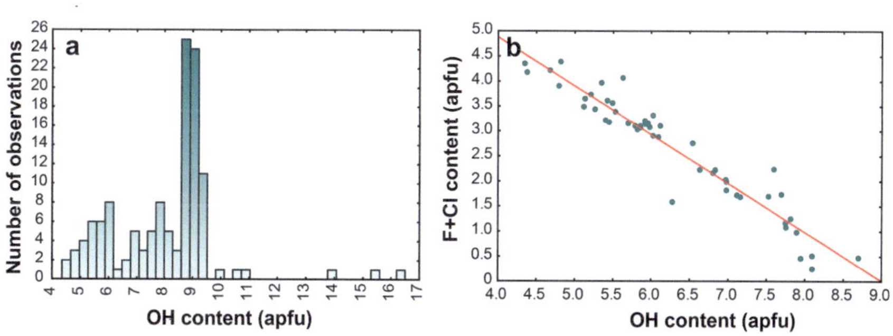

Fig. 1.38 Histogram of the OH content (**a**) and dependence of the total content of F and Cl on the OH content in VSMs without boron and silicon deficiency (**b**)

maximum chlorine content of 1.20 wt% or 1.0 *apfu* was recorded in the crusts of the late vesuvianite on wiluite crystals (Galuskin 2005). The average total amount of OH^-, F^-, and Cl^- in the VSMs, with the exception of their silica-deficient and boron-containing varieties, is 9.0 anions per formula unit (Fig. 1.38b).

The problems with the VSMs formula calculating arises from chemical composition variations (Table 1.3), which, in turn, are the result of complex substitution schemes playing a significant role in the crystal chemistry of this supergroup of minerals. Until recently, the role of beryllium and sulfur in the structures of VSMs remained unstudied (Groat et al. 1992a); also, the deficiency of silicon in its *REE*-variety was not explained. These issues are discussed in more detail in Chap. 4.

Table 1.3 Contents of different elements in VSMs (in oxides, except F, Cl and S)

Component	Content, wt%	References
SiO_2	32.02–37.97	Weibull (1896), Galuskin et al. (2003b)
Al_2O_3	4.33–21.62	Fitzgerald et al. (1986a)
TiO_2	Up to 5.85	Aksenov et al. (2016)
MgO	0.91–6.65	Groat et al. (1992a), Galuskin (2005)
MnO	Up to 14.34	Armbruster and Gnos (2000c)
FeO	0.04–10.85	Groat et al. (2013), Panikorovskii et al. (2017e)
Cr_2O_3	Up to 2.31	Kurbatov (1946)
CaO	19.90–37.30	Kurbatov (1946)
Na_2O	Up to 1.49	Nyunt et al. (2009)
BeO	Up to 9.20	Palache and Bauer (1930)
H_2O	0.50–5.69	Galuskin et al. (2003b), Britvin et al. (2003)
Bi_2O_3	Up to 20.42	Hålenius et al. (2013)
PbO	Up to 2.06	Groat et al. (1992a)
As_2O_5	Up to 1.78	Groat et al. (2013)
Sb_2O_3	Up to 21.21	Pan and Fleet (1992)
CuO	Up to 2.72	Panikorovskii et al. (2017f)
ZnO	Up to 2.35	Groat et al. (1992a)
F	Up to 4.60	Britvin et al. (2003)
Cl	Up to 2.40	Galuskin et al. (2005)
K_2O	Up to 0.66	Vogel (1887)
REE	Up to 20.57	Crook and Oswald (1979)
UO_2	Up to 0.84	Himmelberg and Miller (1980)
B_2O_3	Up to 5.00	Galuskin (2005)
ThO_2	Up to 2.90	(Our data)
CO_2	Up to 0.50	Groat et al. (1992a)
SO_3	Up to 0.35	Groat et al. (1992a)

Note Total Fe and Mn contents are conventionally presented as monoxides

1.3.3 Valence State of Iron in VSMs

Iron plays a significant role in the VSMs crystal chemistry and, according to various sources (Manning and Tricker 1975; Tricker and Manning 1979; Tricker et al. 1981; Karwowski et al. 1996; Kraczka and Żabiński 2003; Groat and Evans 2012), can occupy *X*3, *Y*1, *Y*2, *Y*3 and even the tetrahedral *Z* positions. Among the works listed above, it is worth mentioning the review by Rusakov et al. (2006), who investigated 17 Mössbauer spectra of the VSMs. The important result of this work is the occurrence of Fe ions exclusively in 5- and 6-coordinated positions. As a result of the reconsiderstion of these data, we have found that among VSMs with a magnesium content more than 2 *apfu*, the major part of iron is trivalent (Fig. 1.39). A significant predominance of Fe^{2+} ions over Fe^{3+} takes place only at Mg contents below 1 *apfu*. This confirms the general trend of an increase of the amount of trivalent iron together with an increase in the magnesium content of silicates.

The obtained observation may be useful for calculating the VSMs formulas (Panikorovskii 2025a). When the content of magnesium is more than 2 *apfu*, all iron can be considered as Fe^{3+}, and at lower magnesium contents the Fe^{2+}/Fe^{3+} ratio can be calculated using the empirical formula (Fig. 1.40).

$$Fe^{2+}/Fe^{3+} = 5.18 \times \exp(-1.0156 \times Mg(apfu))$$

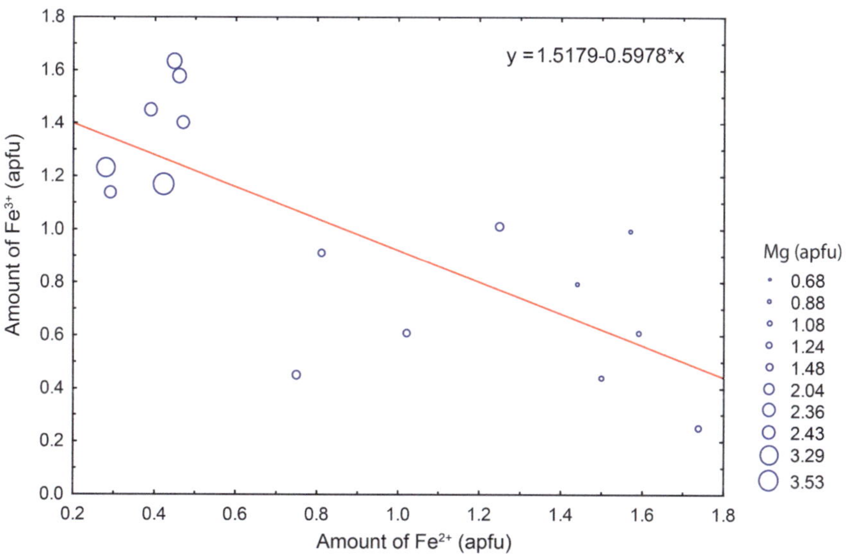

Fig. 1.39 The dependence of the Fe^{3+} content on the Fe^{2+} content (*apfu*) in the VSMs, according to Rusakov et al. (2006). Magnesium contents are shown by various circle sizes

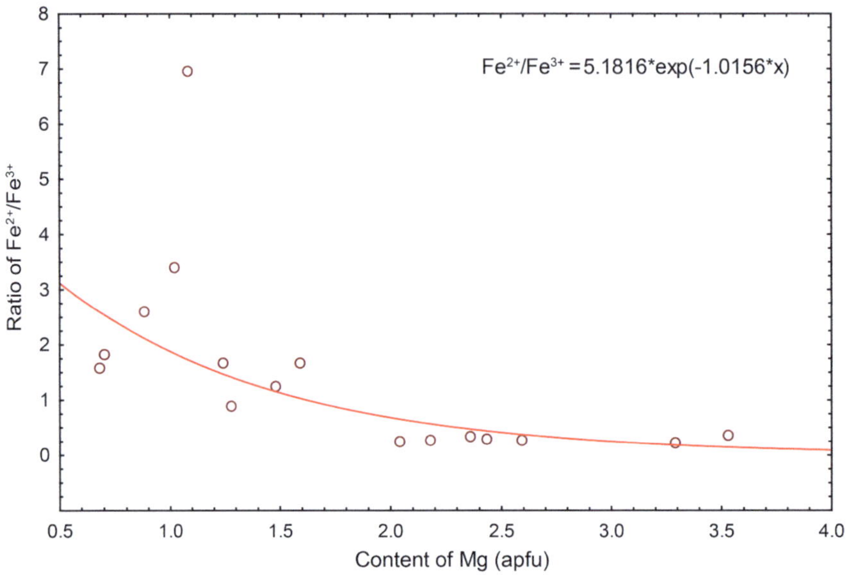

Fig. 1.40 Dependence of the Mg content (*apfu*) on the Fe^{2+}/Fe^{3+} ratio in VSMs, according to Rusakov et al. (2006)

It should be noted that this formula is convenient to be used for B-free (with <1 B *apfu*) VSMs, the chemical formula of which is calculated on 147 charge units; the total charge of *Y*-cations is + 37 which corresponds to $(M^{3+})_{11}(M^{2+})_2$.

1.3.4 Chemical Substitutions

In addition to the common schemes of homo- and heterovalent substitutions, some complex structural schemes of substitution are found in the VSMs. A striking example of complex substitution is the incorporation of boron into the wiluite structure. According to Belinskiy (1985), boron replaces silicon according to the scheme $[BO_4] \leftrightarrow [SiO_4]$. As it was later shown Groat et al. (1994a, 1996a, 1998), the amount of boron in chemical analyses does not depend on the amount of silicon, and boron atoms incorporate into the structure according to the following schemes: ${}^{T1}B + {}^{Y}Mg^2 + 2^{W}O \leftrightarrow {}^{T1}\square + {}^{Y}Al + 2^{W}(OH)$ and ${}^{T2}B + 2^{W}O \leftrightarrow {}^{T2}\square + {}^{W}(OH)$. Thus, boron atoms occupy additional $T1$ and $T2$ sites (vacant sites in B-free VSMs) in the triangular and tetrahedral coordination, respectively. These studies made it possible to distinguish wiluite as a separate mineral species (Groat et al. 1998).

Special attention to the occupancy of positions located in the channels of the structure and cation-ordering schemes is given in the works by Ohkawa et al. (1992), Ohkawa (1994), Armbruster and Gnos (2000b), Armbruster et al. (2002). Mainly, they showed a wide diversity of cations included in the square-pyramidal position.

Iron-free vesuvianite with Mn^{3+} at the *Y*1 site was approved as manganivesuvianite (Armbruster et al. 2002). A comprehensive study of iron-free VSMs (Panikorovskii et al. 2017a, e) resulted in the discovery of two other new members of the vesuvianite supergroup, alumovesuvianite and magnesiovesuvianite. Subsequently, Cu^{2+}-dominant (at the *Y*1 site) member of the vesuvianite supergroup was approved by the IMA CNMNC and described with the name cyprine (Panikorovskii et al. 2017f).

Data presented in a number of works (Henmi et al. 1994; Armbruster and Gnos 2000b; Galuskin et al. 2003b, 2007a; Ohkawa et al. 2009) confirmed the possibility of the substitution in accordance with the "hydrogatnet scheme": $SiO_4^{4-} \leftrightarrow H_4O_4^{4-}$. This substitution is common both in garnets and VSMs. For vesuvianite it was first considered by Allen and Burnham (1992).

The role of F and Cl in the vesuvianite formula was discussed in detail by Groat et al. (1992b), Britvin et al. (2003), Galuskin et al. (2005), which lead to the discovery of F-dominant VSM, fluorvesuvianite. Some other significant substitution schemes are described in detail in Sect. 1.4.1.

1.3.5 Crystal Structure

The first hypothesis about the vesuvianite crystal structure was proposed before the advent of single-crystal XRD analysis (Clarke 1895). In this model, vesuvianite was referred to the general type of sorosilicates with aluminum. Attempts to determine the structural formula of vesuvianite was continued in the works by Sjögren (1895), Jannasch and Weingarten (1896), Weibull (1896), long before the first successful description of the vesuvianite structure (Warren and Modell 1931) and subsequent works (Machatschki 1932; Kurbatov 1946). In this context, it should be noted that the mineralogist F. Machatschki calculated a more accurate formula of this mineral than the crystallographers B. Warren and D. Modell based on their structural data. However, the merit of the latter researchers is the construction of the structural model of vesuvianite (Fig. 1.41a) including the selection of one-dimensional grossular modules. These authors paid special attention to the role of the Si_2O_7 groups in the structure of the vesuvianite, which was previously considered as a sorosilicate (Jannasch and Weingarten 1896; Weibull 1896).

According to modern concepts, the crystal structure of the VSMs consists of the 1D modules with the grossular structure (Fig. 1.41b), running along [001] (the vesuvianite parameter *c* coincides with the edge cell parameter of grossular, ~11.80 Å) and the combined Si_2O_7 nesosilicate groups in the heteropolyhedral framework which form channels hosting half-populated (in the structural model for the space group *P*4/*nnc*) *Y*1 and *X*4 sites (Fig. 1.41c). The cationic positions located in the channels of the structure are in the *Y*1–*X*4–*X*4–*Y*1 sequence so that the interatomic distances *Y*1–*X*4, *Y*1–*X*4 and *X*4–*X*4 are less than 1.3, 1.3 and 2.5 Å, respectively (Fig. 1.41d).

The modular character of the vesuvianite structure largely determines properties of this mineral. For example, regular intergrowths of grossular and wiluite caused by the occurrence of garnet modules in the vesuvianite structure are described by

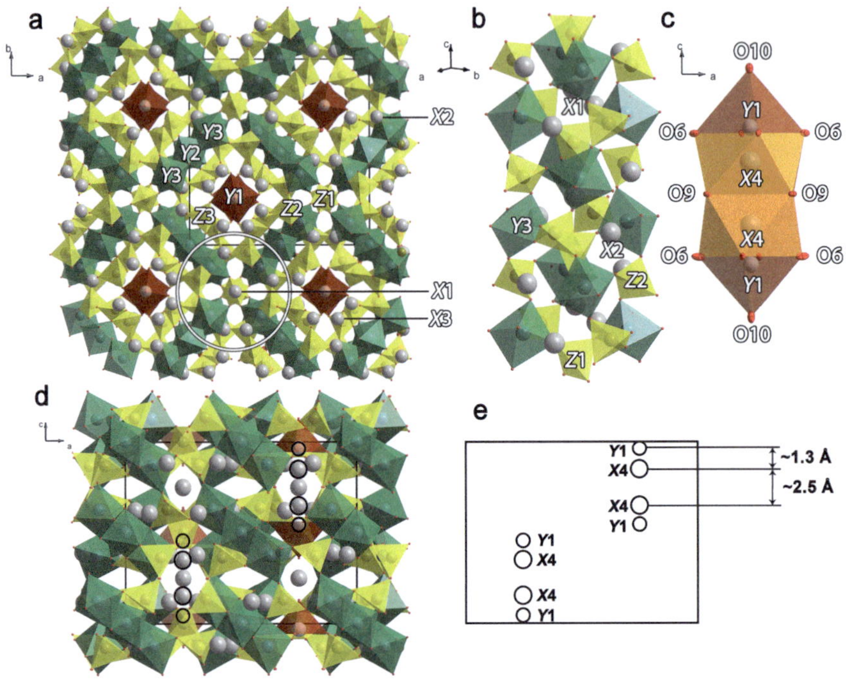

Fig. 1.41 The crystal structure of vesuvianite: a general view projected along the *c* axis (**a**); the grossular module projected along *c* axis marked by a circle in Fig. 1.41a (**b**); the *Y*1–*X*4–*X*4–*Y*1 sequence of positions with incomplete population (**c**); the projection along the *a* axis with half-populated *Y*1 and *X*4 sites (**d**) and the arrangement of the *X* and *Y* sites with incomplete occupancy in the unit cell (**e**)

Panikorovskii et al. (2014). Incorporation of Na^+ cations and $H_4O_4^{4-}$ groups into the vesuvianite structure (into the sites *X*1, *X*4 and *Z*1, *Z*2, respectively) is also due to the modular character of the structure: the $H_4O_4^{4-}$ groups occur in the garnet module and the Na^+ cations occupy the *X*-sites in the rest part of the crystal structure (Panikorovskii et al. 2016a, b).

1.3.6 Distortions of Coordination Polyhedra

The most significant variability of cation substitutions in the vesuvianite structure is observed at the *Y*1 square-pyramidal site and in the *Y*2,3 octahedra. Distortions of these polyhedra occur due to both lowering of the overall symmetry and owing to chemical substitutions. Structural distortions of vesuvianite with the *P*4*nc* space group from Bela Ophiolite, Pakistan were studied by Armbruster and Gnos (2000a). In this sample, the *Y*3A and *Y*3B octahedra, characterized by similar scattering factors, have significantly different mean bond lengths (<*Y*3A–O> = 1.942 Å and <*Y*3B–O> = 1.995 Å) due to different different contents of Mg and Al at the *Y*3A and *Y*3B sites [the refined populations are ($Al_{0.906}Fe_{0.094}$) for *Y*3A and ($Al_{0.63}Mg_{0.25}Fe_{0.12}$) for *Y*3B]. In the structure of the manganivesuvianite (space group *P*4/*n*), a similar ordering of bivalent cations (Mg^{2+} and Mn^{2+}) with their concentration in the *Y*3 octahedra is observed (Armbruster et al. 2002).

The distortion of the base of the tetragonal pyramid *Y*1 can occur due to the Jahn–Teller effect as a result of the incorporation of Mn^{3+} and Cu^{2+} atoms into the structure of manganvesuvianite and cyprine, respectively (Dyrek et al. 1992; Armbruster et al. 2002; Panikorovskii et al. 2017f). The observed bond lengths <Mn^{3+}– O> and <Cu^{2+}–O> in the bases of the *Y*1O_5 pyramids (distorted squares) are 1.975 (4) Å and 2.043 (3) Å, respectively.

The high degree of ordering of cations at the *Y*1 and *X*4 sites leads to the appearance of stresses in the structure, compensated by nonequivalent cation distribution of divalent cations in the *Y*3 octahedra (Panikorovskii et al. 2017f). This matter is discussed in detail in Sect. 1.4.5.

According to Hålenius et al. (2013), incorporation of Bi^{3+} into the largest (by volume) *X*3 polyhedra is accompanied by splitting this site into two sub-sites occupied by calcium (*X*3) and bismuth (*X*3′). Also vesuvianite structure can contain split five-coordinated *Y*1 site, which may be associated (regardless of symmetry) with the separation of Fe^{2+} and Fe^{3+} (Groat et al. 1992b, 1994a; Rusakov et al. 2006; Aksenov et al. 2016).

1.3.7 Hydrogen Positions and Hydrogen Bonding System

The presence of two different hydrogen positions in the vesuvianite structure was proposed based on the local bond-valence data (Yoshiasa and Matsumoto 1986; Groat et al. 1992a), as well as structural data which demonstrate that OH groups in the vesuvianite structure are associated with the O10 and O11 oxygen sites (Coda et al. 1970; Rucklidge et al. 1975). The study of polarized infrared (IR) spectra of various vesuvianite samples (Groat et al. 1995b) confirmed the presence of a hydrogen position on the fourfold axis in the structure channels and another hydrogen atom in the general position. The exact coordinates of hydrogen sites in the structure of low-symmetry (space group *P*4/*n*) and high-symmetry (space group *P*4/*nnc*) vesuvianite and its boron-containing variety were determined using neutron diffraction in the works by Groat et al. (1996a), Pavese et al. (1998a), Lager et al. (1999a).

The H11 site of the H1 hydrogen atom is located at the distance of 1.02 Å from the O11 oxygen atom, which coordinates the *Y*2 and *Y*3 octahedral sites, and at the distances of 2.42 and 2.62 Å from the O7 and O11 atoms, respectively (Lager et al. 1999a). This arrangement leads to a hydrogen bond bifurcation in the structure of a highly symmetric vesuvianite and non-equivalence of the H1*a* (H11A) and H1*b* (H11B) positions in the *P*4/*n* (Fig. 1.42) and *P*4*nc* structural models (Ohkawa et al. 1992; Pavese et al. 1998a; Armbruster and Gnos 2000c; Galuskin 2005; Ohkawa et al. 2009; Krivovichev et al. 2013; Panikorovskii et al. 2016c; Chukanov et al. 2018a).

The geometric parameters of hydrogen bonds in the structure of alumovesuvianite (*P*4/*n*) and vesuvianite from the Karmankul'skii cordon (*P*4*nc*) are given in Table 1.4.

The (H10) site of the H2 hydrogen atom is located on the fourfold axes in the structural channels, coordinates the O10 oxygen site and forms the hydrogen bond O10–H2···O10 (Fig. 1.43). In the structure of high-symmetry vesuvianite, the hydrogen position is located at a distance of 0.99 Å from oxygen at the O10 site and is half populated. In the interpretation of different authors, the position of H2 in the structure of low-symmetry VSMs is differently located. For example, in the work by Pavese et al. (1998a), three positions, H2*a*,*b*,*c* are given for H2 in *P*4/*n* vesuvianite, while in some other works (Armbruster and Gnos 2000b; Armbruster et al. 2002; Panikorovskii et al. 2017a, f) only one position is given for H2 in VSMs with the symmetry *P*4*nc* and *P*4/*n*.

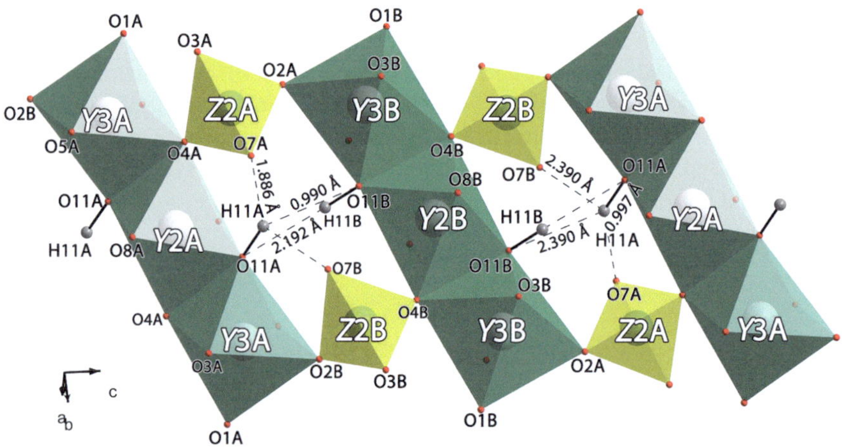

Fig. 1.42 Local arrangement of H1 atoms (H11A,B) and the scheme of hydrogen contacts and bonds in the crystal structure of alumovesuvianite

Table 1.4 Geometric parameters of the hydrogen-bonding system in the structure of alumovesuvianite (*P*4/*n*) and vesuvianite from the Karmankul'skii cordon (*P*4*nc*) according to Panikorovskii et al. (2016c, 2017a)

Mineral	Hydrogen bond	D–H(Å)	H···A(Å)	D···A(Å)	∠(DHA)
Alumovesuvianite	O10B–H10B…O10A	1.18(10)	1.55(10)	2.739(7)	180.00
Alumovesuvianite	O11A–H11A…O7A	0.99(2)	1.90(6)	2.747(3)	142(7)
Alumovesuvianite	O11B–H11B…O11A	0.99(2)	2.18(5)	3.059(3)	147(7)
Vesuvianite	O10B–H10B…O10A	1.04(5)	1.74(8)	2.784(7)	180.00
Vesuvianite	O11A–H11A…O11B	0.87(2)	2.25(4)	3.059(4)	153(6)
Vesuvianite	O11B–H11B…O7B	1.00(2)	1.83(5)	2.731(3)	148(6)
Vesuvianite	O11B–H11B…O7A	1.00(2)	2.37(5)	2.981(4)	118(6)

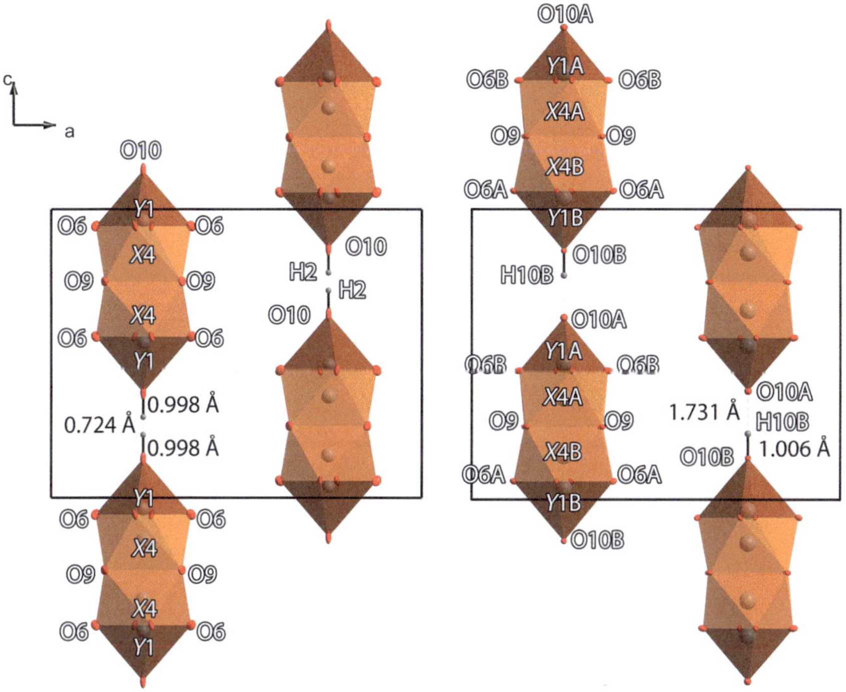

Fig. 1.43 The position of the hydrogen atom H2 (H10A,B) on the fourfold axis and the geometry of hydrogen bonds in the *P*4/*nnc* crystal structure of the vesuvianite with 50% occupancy of the H2 site, according to Lager et al. (1999a) (left) and *P*4/*n* alumovesuvianite, according to Panikorovskii et al. (2017a) (right)

Until recently, it was believed that in the vesuvianite structure there are only two hydrogen sites: H1 (the H11 site) in the general position and H2 (the H10 site) in the channels of the structure. However, it was shown (Galuskin et al. 2003b, 2007a; Galuskin 2005) that in Si-deficient varieties of vesuvianite, the substitution according to the scheme $SiO_4^{4-} \leftrightarrow H_4O_4^{4-}$ is realized and is accompanied by an increase of the <Z1–O> bond lengths up to 1.70 Å. Further studies using solid-state ^{1}H NMR, as well as differential scanning calorimetry, confirmed the presence of an additional hydrogen position in the structure of Si-deficient vesuvianite (Panikorovskii et al. 2016a).

1.3.8 *Structural Varietes*

Since the nineteenth century, the anomalous optical properties of vesuvianite caused by local deviations from tetragonal symmetry are known (Brauns 1883, 1891). Arem and Burnham (1969) showed the presence of reflexes that violate the extinction conditions for the space group *P*4/*nnc* in several vesuvianite samples from different localities. The cation ordering involvong the *X*4 and *Y*1 sites in the structure channels (Coda et al. 1970) leads to a change the overall symmetry of the mineral according to the following schemes: *P*4/*nnc* → *P*4/*n* (Giuseppetti and Mazzi 1983) or *P*4/*nnc* → *P*4*nc* (Ohkawa et al. 1992). The noncentrosymmetric (space group *P*4*nc*) vesuvianite variety of the is extremely rare, and reliable refinements of the structure in this space group are given in only a few papers (Ohkawa et al. 1992; Ohkawa 1994; Armbruster and Gnos 2000b, a; Panikorovskii et al. 2016c).

According to the data by Allen and Burnham (1992), the symmetry of vesuvianite is determined by its domain structure. A single domain is a group of unit cells with the same symmetry, *P*4/*n*(+), *P*4/*n*(–), *P*4*nc*(+) or *P*4*nc*(–) (Fig. 1.43). The overall symmetry of the crystal is determined by the size of the individual domains and their quantitative ratio. Table 1.5 shows possible cases of various combinations of domains with different sizes and symmetries in natural samples, according to data by Giuseppetti and Mazzi (1983), Allen and Burnham (1992), Ohkawa et al. (1992), Armbruster and Gnos (2000a) (Fig. 1.44).

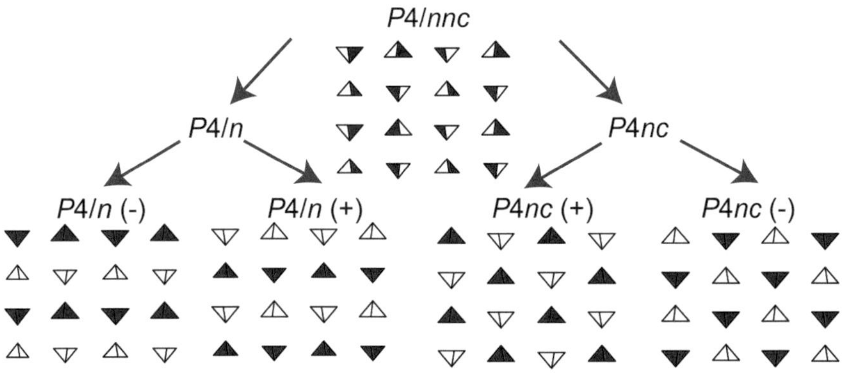

Fig. 1.44 Cation-ordering scheme of the *Y*1 (**a**, **b**) sites in the space groups *P*4/*nnc*, *P*4/*n* and *P*4*nc*

Table 1.5 General vesuvianite symmetry depending on the size and symmetry of domains encountered in nature

Domain size	Symmetry	Ratio	Space group
Large	*P*4/*n*(+), *P*4/*n*(–)	Any	*P*4/*n*
Large	*P*4*nc*(+), *P*4*nc*(–)	Any	*P*4*nc*
Large	*P*4/*n*(+), *P*4/*n*(–), *P*4*nc*(+), *P*4*nc*(–)	*P*4/*n* = *P*4*nc*	*P*4/*nnc**
Large	*P*4/*n*(+), *P*4/*n*(–), *P*4*nc*(+), *P*4*nc*(–)	*P*4/*n* < *P*4*nc*	*P*4*nc*
Large	*P*4/*n*(+), *P*4/*n*(–), *P*4*nc*(+), *P*4*nc*(–)	*P*4*nc* < *P*4/*n*	*P*4/*n*
Small	*P*4/*n*(+), *P*4/*n*(–), *P*4*nc*(+), *P*4*nc*(–)	*P*4/*n* ≈ *P*4*nc*	*P*4/*nnc*

* The overall *P*4/*nnc* symmetry observed instead expected $P\bar{4}$ (Allen 1985; Ohkawa 1994), because the $P\bar{4}$ model is not realized at the unit cell level (Galuskin 2005)

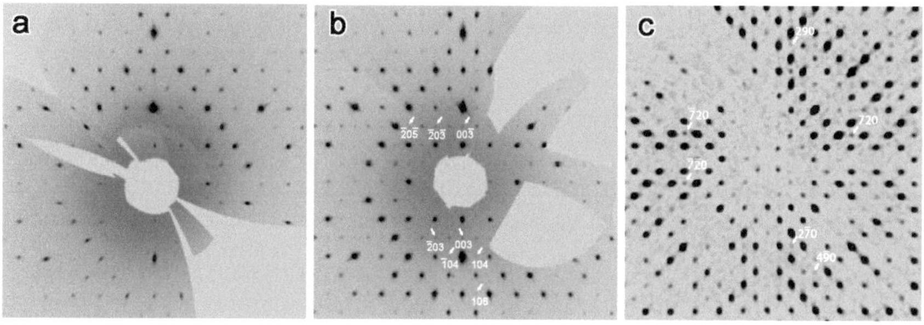

Fig. 1.45 Reconstruction of different sections of single-crystal X-ray diffraction patterns in the reciprocal space for Ewald Sphere VSMs with different symmetry: **a**—(*h*0*l*) section for wiluite (space group *P*4/*nnc*, no additional reflexes); **b**—(*h*0*l*) section for a low-symmetric boron-containing vesuvianite (space group *P*4/*n*, reflections $h + l \neq 2n$ are present); **c**—(*hk*0) section of vesuvianite from the Karmankul'skii cordon (space group *P*4*nc*, present reflections $h + k \neq 2n$). Violating reflections for the space group 4/*nnc* indicated by white font

If a crystal consists of small-sized domains randomly located relative to each other (a crystal with a low degree of order), it has the space group *P*4/*nnc*, and there is no additional glide plane violating reflections (Fig. 1.45a). Crystals with a high degree of ordering, which show additional violating diffraction reflections, can be described within the *P*4/*n* or *P*4*nc* space groups (Fig. 1.45b, c).

Additional reflections violating the *P*4/*nnc* space group are of three types: (1) *hk*0-type diffraction maxima where $h + k \neq 2n$; (2a) diffraction maxima of the 0*kl*-type where $k + l \neq 2n$ and (2b) diffraction maxima of the *hhl*-type where $l \neq 2n$. The first type of reflections refers to the *n* plane, perpendicular to the fourfold axis, type (2a) refers to the *n* plane lying in (100), and type (2b) corresponds to the *c* plane lying in (110). The presence of reflections of the first type excludes the space groups *P*4/*nnc* and *P*4/*n*. The presence of reflections of the second type (2a,b) excludes the space groups *P*4/*nnc* and *P*4*nc* (Armbruster and Gnos 2000a).

In the case when large-sized domains with *P*4/*n* and *P*4*nc* symmetry are present in equal amounts (see Table 1.5), reflections of the first and second types are observed. The larger are the domains, the stronger are violating reflections. In the case when the domains are described by a non-centrosymmetric group (for example, *P*4*nc*), vesuvianite crystals have nonlinear optical and piezoelectric properties (Allen and Burnham 1992; Panikorovskii et al. 2016c).

The transmission electron microscopy studies of optically anomalous (biaxial) zones in vesuvianite made it possible to assume that the common tetragonal symmetry of the mineral can be described as the result of pseudomerohedral twinning of *P*2/*n* symmetry domains (Veblen and Wiechmann 1991; Groat et al. 1995a). Based on the fact of the presence of reflections that violate tetragonal symmetry in the precision single-crystal XRD pattern, combined with the anomalous optical properties of the vesuvianite, it was concluded that the real symmetry of the mineral corresponds to the space group *P*2/*n* or *Pn* (Groat et al. 1993). The structure of the vesuvianite with triclinic growth sectors from Bellecombe, Aosta valley, Italy was refined in the *P*2/*n* space group (Tanaka et al. 2002).

1.3.9 Structural Complexity of Vesuvianite

A new approach to quantifying the structural complexity of a crystalline substance, as well as the availability of a function to calculate the required parameters in the *TOPOS* software package (Blatov et al. 2000), made it possible to widely use the complexity parameter (information in bits per unit cell or per atom) as a tool for studying the evolution of mineral formation in various systems (Avdontceva et al. 2015; Grew et al. 2016; Gurzhiy et al. 2016; Krivovichev et al. 2022). The presence of two or three various structural modifications of the VSMs (*P*4/*nnc*, *P*4/*n*, *P*4*nc*), in the same geological formation (for example, in the Wiluy River basin, Templeton skarns, and Kovdor massif), is of considerable interest for practical application of the complexity parameter.

The figurative points of structural complexity for various modifications of VSMs (Appendix 1) fall into two areas (Fig. 1.46), which correspond to the high-symmetry (*P*4/*nnc*) and the low-symmetry (*P*4/*n* and *P*4*nc*) varieties. All high-temperature *P*4/*nnc* modifications of the VSMs turned out to be structurally simpler than the low-temperature *P*4/*n* and *P*4*nc* forms. The similarity of the structural complexity for the *P*4/*n* and *P*4*nc* modifications of the VSMs emphasizes their structural proximity, which is observed in nature: in samples with the general *P*4/*n* symmetry, the *P*4*nc* domains are often present, and vice versa (see Sect. 1.3.5). As a rule, the structural evolution of vesuvianite occurs in the sequence *P*4/*nnc* → *P*4/*n* → *P*4*n* (Galuskin 2005), which corresponds to an increase in the structural complexity of the mineral structure with time (*i.e.* with decreasing of the crystallization temperature). Such observations are consistent with the general trend for inorganic crystalline compounds (Krivovichev 2013).

1.4 Vesuvianite as an Indicator of Crystallization Conditions

The relationship of crystal-chemical features of VSMs from different paragenetic associations with the stability of their structural modifications predetermines the possibility of practical use of the VSMs as a geothermometer (Panikorovskii et al. 2024a). The effectiveness of this approach can be demonstrated by the example of multi-stage (re)crystallization of the VSMs in the rocks of the Kovdor massif (see Chap. 4) and Wiluy deposits.

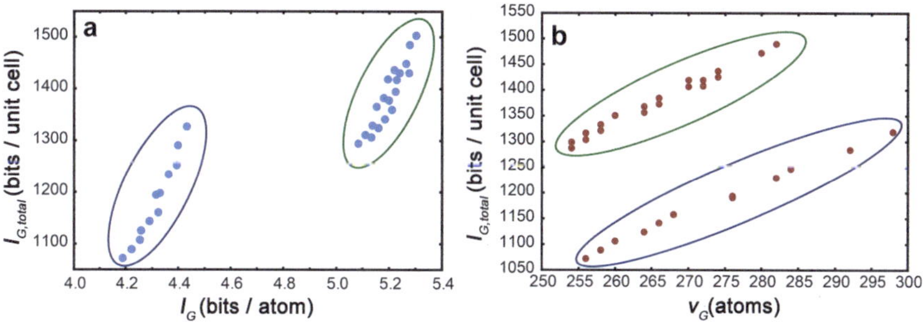

Fig. 1.46 Graphs showing the dependences of information per unit cell, $I_{G,\text{total}}$, on information per atom, I_G (**a**), and on the number of atoms in the unit cell (**b**). The points corresponding to samples with the *P*4/*nnc* symmetry and *P*4/*n* or *P*4*nc* symmetry are enclosed in the blue and green ovals, respectively

1.4.1 *Symmetry* Versus *Formation Temperature*

In the work by Allen and Burnham (1992), vesuvianite with *P*4/*n* symmetry (crystallized at temperature below 500 °C) was annealed at a temperature of 1000°C for 50 days, after which its symmetry did not undergo any changes. This fact led to the assumption that the symmetry of vesuvianite is associated only with crystallization conditions. In this regard, relationships between paragenesis and the space group were studied for a series of vesuvianite samples from Swiss and Italian Alps (Gnos and Armbruster 2006). It turned out that in the zone of granulite metamorphism, vesuvianite is replaced by melilite; *P*4/*nnc* modification of the mineral is associated with the amphibolite facies; in the rocks of the greenschist facies, vesuvianite with *P*4/*n* symmetry is widespread, and in the region of least metamorphism, *P*4*nc* modification is common.

The graph of the dependence of the symmetry of VSMs on the conditions of their crystallization (Fig. 1.47) clearly shows a trend of a decrease in the symmetry of the mineral with the decrease in the formation temperature of the rock whereas no dependence of symmetry on pressure (in the range of 1–6 kbar) was revealed. For high-symmetry VSMs samples (space group *P*4/*nnc*), the formation temperature is >500 °C, *P*4/*n* modification is formed in the range of 300–500 °C, and for rocks formed below 300 °C, VSMs with the space group *P*4*nc* are typical.

The correlation between the symmetry and the temperature of VSMs formation was completely confirmed by the statistical data on the number of glide plane violating reflections for the space group *P*4/*nnc*. For low-temperature VSMs, the number of violating reflections is larger than for high-temperature samples (Fig. 1.48a). Explanations of the relationship between the symmetry of the mineral and the number of glide plane violating reflections (and their types) are given in Sect. 1.3.5.

Hence, the correlation between the space group of VSMs and the temperature of their formation (Fig. 1.48b) confirms the correctness of the separation of the

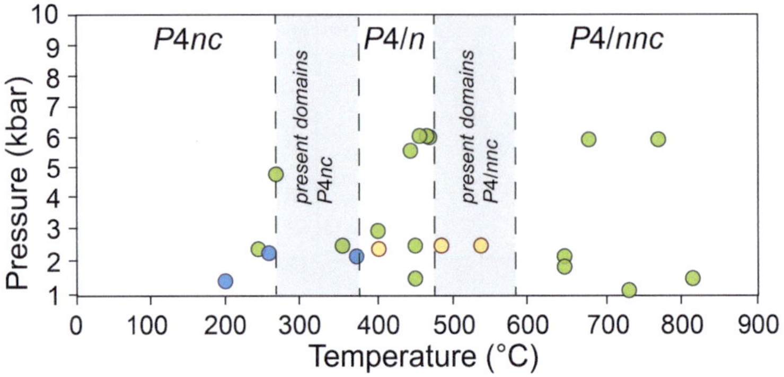

Fig. 1.47 The dependence of the vesuvianite space group on the P–T conditions of rock formation, according to Armbruster and Gnos (2000a, b) (blue circles), Frey et al. (1999), Gnos and Armbruster (2006) (green circles) and Justiniano et al. (2025b) (yellow circles). the shaded field corresponds to vesuvianite structures, where domains with different symmetry are present

VSMs into "high" (*P*4/*nnc*), with the temperatures of formation above 400 °C, and "low" (*P*4/*n* и *P*4*nc*), with the crystallization temperatures below 400 °C (Allen and Burnham 1992). The fact of the presence of the VSMs with *P*4/*n* symmetry only in the rocks formed at temperatures below 300 °C is noteworthy and allows distinguishing this structural variety as the lowest-temperature VSMs modification.

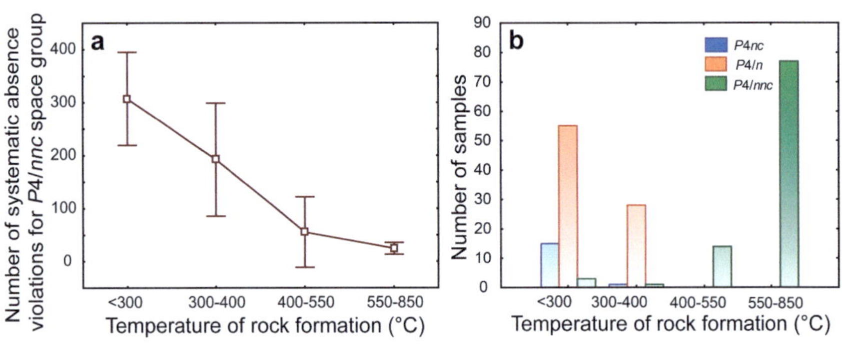

Fig. 1.48 The dependence of the number of glide plane violating reflections for the space group *P*4/*nnc*, on the temperature of formation of the rock (**a**) and the distributions of samples belonging to different space groups depending on the temperature of the rock formation (**b**). All measurements were carried out under identical conditions

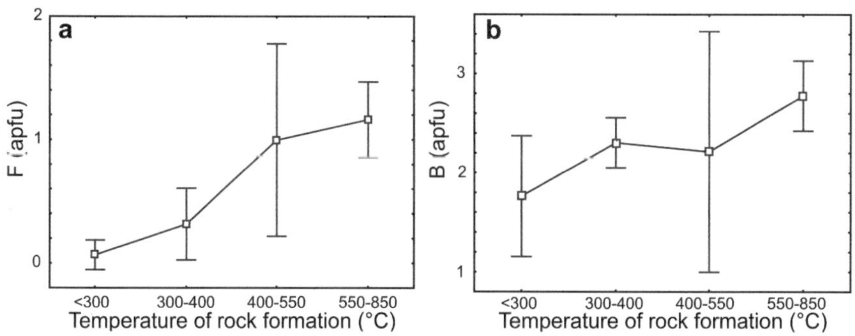

Fig. 1.49 The correlation between the temperature of rock formation and the mean contents of fluorine (**a**) and boron (**b**) in the VSMs composition

1.4.2 Correlation Between Chemical Composition and Temperature of Formation

The most important crystal-chemical marker of the formation conditions of VSMs is the amount of fluorine and/or boron in their composition. According to our data, the content of both components regularly increases with the temperature of formation of corresponding rock (Fig. 1.49), which confirms the assumptions put forward earlier (Groat et al. 1992b; Borovikova and Kurazhkovskaya 2006).

Indeed, the incorporation of these elements into the mineral structure is accompanied by the replacement of OH groups in hydroxyl positions (Groat et al. 1992b, 1994a), which is consistent with the general idea of a decrease in the amount of hydrogen in minerals with an increase in the crystallization temperature or the degree of metamorphism of the rock.

1.4.3 The Relationship Between the Chemical Composition and Symmetry of VSMs

In addition to the ***major*** components (Ca, Si, Al, Mg, Mn and Fe), containing in significant amounts in most VSMs samples, light elements (H, B and F), occurring in VSMs in variable amounts, and ***minor*** components, such as Ti, Na, S and Cr, are good indicators of the formation conditions (Fig. 1.50). In particular, relatively high-temperature (550–850 °C) wiluite and vesuvianite with a significant content of boron (see Fig. 1.50b) from skarn of the Titovskoe borate deposit, traps of Wiluy River basin (both in Russia), Ariccia in Italy and Templeton in Canada are characterized by the space group *P4/nnc* (Groat et al. 1992a; Bellatreccia et al. 2005a). VSMs with a smaller amount of boron crystallized at late stages of skarn formation often have *P4/n* or *P4nc* symmetry (Galuskin 2005).

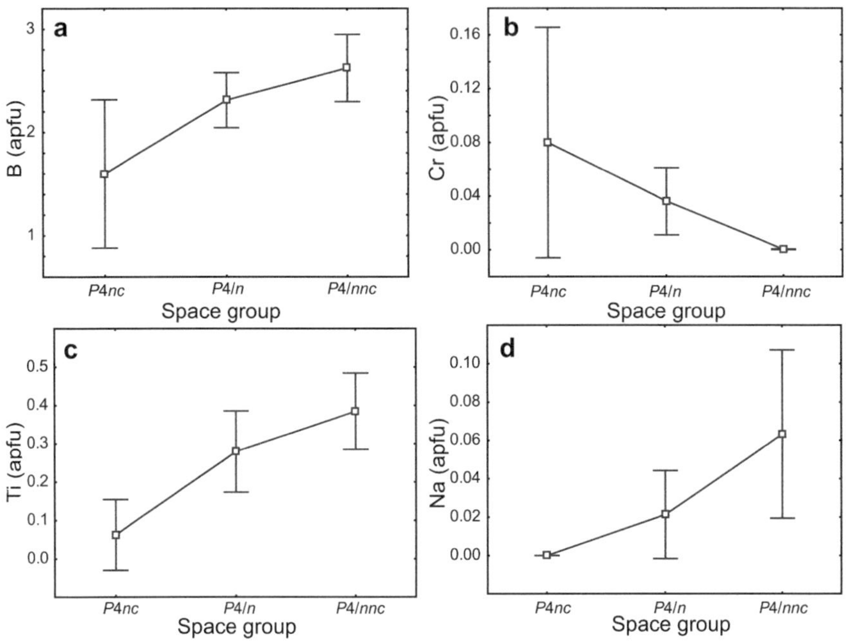

Fig. 1.50 The correlations between the space group of VSMs and the contents of boron (**a**), chromium (**b**), titanium (**c**) and sodium (**d**) in them

Most often, VSMs with significant Cr content originate from low-temperature hydrothermal veins inside chromium ores (Jeffrey mine in Canada; Bazhenovskoe, Klyuchevskoy mine and Monetnaya Dacha in Urals, Russia) and always have low symmetry, *P*4/*n* or *P*4*nc* (Krivovichev et al. 2013).

Titanium and sodium are characteristic ***minor*** components of high-temperature/high-symmetry VSMs (Fig. 1.28c, d). Decrease of the formation temperature is accompanied by the ordering of the general structure and ordering of cations in the crystal structures of VSMs which results in the self-cleaning from ***minor*** components, characteristic of most rock-forming minerals of various genesis (Ivanyuk et al. 2009).

1.4.4 Correlation Between Unit Cell Parameters and Chemical Composition

The first attempt to compare changes of chemical composition to changes of unit cell parameters (UCP) for VSMs was made by Groat et al. (1992a). Two trends are noted in the diagram of UCP values, one of which corresponds to boron-containing samples (the larger values of the *a* parameter correspond to the smaller values of the *c* parameter). The second trend is related to boron-free samples: for all such samples, the an increase in the Mg content is accompanied by an increase in the *a* parameter and the *a* value is linearly dependent on the Mg/(Mg + Al) ratio (Groat et al. 1992a).

Kurazhkovskaya and co-authors (2005) have linked the change of the optical sign, and the *a*/*c* ratio of wiluite and boron-containing vesuvianite with the degree of occupancy of the *T*1 site by boron.

Gnos and Armbruster (2006) revealed in the UCP diagram several different trends with their partial overlap (Fig. 1.51). Incorporation of titanium into the vesuvianite structure according to the scheme $^{Y}Ti^{4+} + {}^{Y}M^{2+} \leftrightarrow 2^{Y}M^{3}$ affects to the unit cell parameters in the same way as the "hydrogarnet" type substitution, $(SiO_4)^{4-} \leftrightarrow (H_4O_4)^{4-}$. This trend is observed because M^{2+} cations occurring in VSMs are larger than Al, and when vacancy occurs at the Si site, its polyhedral volume increases. To the previously noted trend associated with filling *T*1 and *T*2 positions (Groat et al. 1992a), a "hydro-manganese" trend is added, due to the incorporation of manganese into the vesuvianite structure with simultaneous substitution according to the scheme $^{Z1,2}(SiO_4)^{4-} \leftrightarrow {}^{Z1,2}(H_4O_4)^{4-}$.

The consideration of the relationship between the chemical composition of VSMs and the type of rocks that include them (Fig. 1.52a) results in the conclusion that there is a trend towards an increase of the contents of iron and titanium with an increase in the basicity of the rock, as well as the general trend of reducing the aluminum content from acid (granitic) to basic and, especially, alkaline rocks. A significant increase of unit cell parameters in the row from acid to ultrabasic rocks (Fig. 1.52b) is explained by the increase in the total content of iron and titanium, the ionic radii of which exceed that of aluminum (Shannon 1976): $^{VI}Al^{3+}(0.54) < {}^{VI}Ti^{4+}(0.61) < {}^{VI}Fe^{3+}(0.65) < {}^{VI}Fe^{2+}(0.78)$ Å.

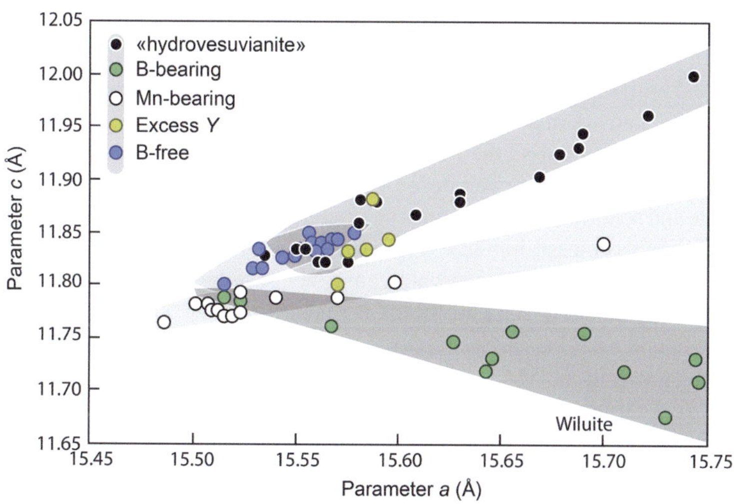

Fig. 1.51 Diagram of the dependence of unit cell parameters of VSMs on specific features of their chemical composition, according to Gnos and Armbruster (2006) with modifications

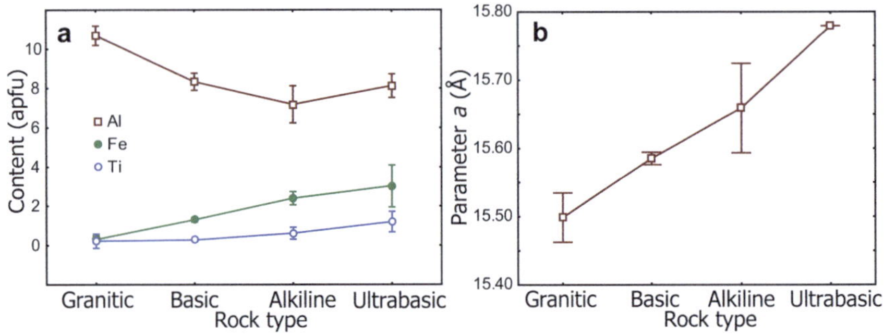

Fig. 1.52 The dependences of Al, Fe and Ti contents (**a**) and the *a* parameter (**b**) in VSMs on the type of rock

1.4.5 Metamict Vesuvianite

A metamict Th-bearing vesuvianite variety (with up to 2.00 wt% of ThO_2) was described in alkaline pegmatite from the foidolite massif Dahu-Unur, Tyva, Russia (Kononova 1960). Absolute age of these rocks (402–420 Ma) was determined by U–Pb isotopic dating of vesuvianite (Zhukov et al. 1961). The Th-bearing vesuvianite crystals have an oscillated zonality with significant chemical variations in different zones: the content of REE_2O_3 varies in the range 0–17 wt% and the content of FeO is in the range 0–15 wt%

The Dakhu-Unur massif is located in the core of a dome-shaped anticlinal valley composed of Proterozoic marbles. The massif, with an area of 1.6 km^2, is composed of ijolite-urtites, pyroxenites, and (in a small part) nepheline syenites. A satellite of the massif, 1.6×0.3 km^2 in size, is composed of ijolites. The marbles enclosing the massif are skarnified and partly transformed into calciphyres with diopside, nepheline and andradite, and are often strongly fluoritized.

In the Dakhu-Unur massif there are simple pegmatites forming small (up to 50 cm) schlieren in ijolites. They are composed of large-crystalline aggregates of nepheline, hastingsite and schorlomite. The largest pegmatite bodies have the form of oval stocks with steep contacts measuring 225×150 m^2. Their outer zones (up to 3 m thick) are composed of coarse-grained aggregate composed of nepheline, feldspar, andradite and hastingsite. The next zone 5-m-thick is mainly composed of the same minerals (only biotite and *REE*-bearing vesuvianite are added) which form larger crystals up to 20 cm in size (Fig. 1.53).

The *REE*-bearing vesuvianite is in a metamict state caused by the radioactive decay of uranium and thorium (ThO_2 2 wt%). Its metamict nature is confirmed by its appearance and properties including conchoidal fracture, resinous luster, reduced specific gravity, optical properties (isotropy, reduced refractive indices), and X-ray amorphic. Recrystallization of vesuvianite does not occur even after heating this mineral at a temperature of 700 °C.

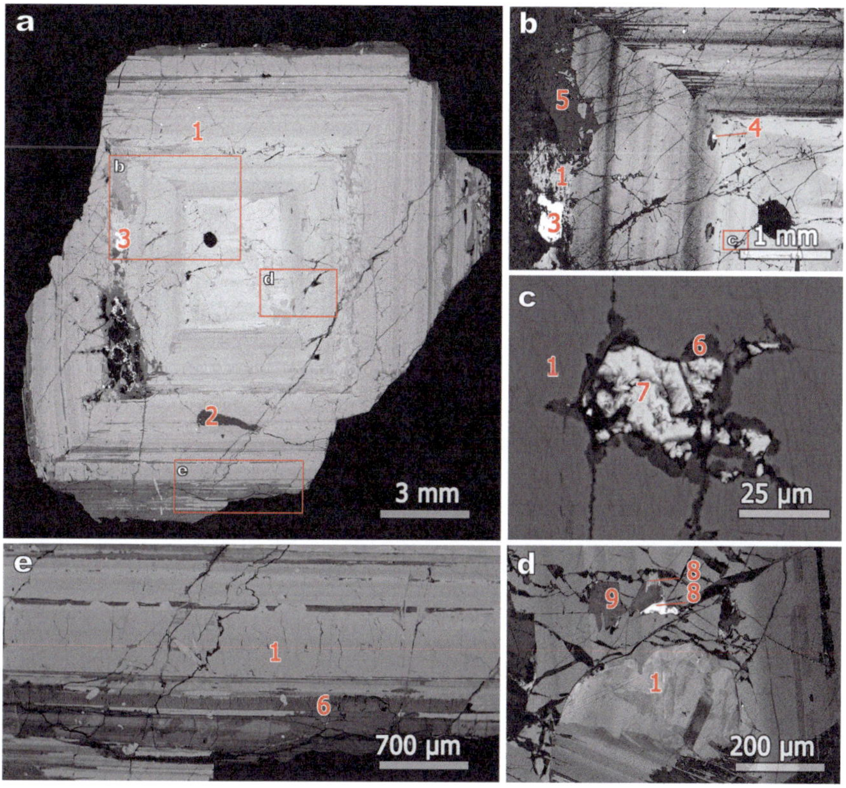

Fig. 1.53 Metamict Th-*REE*-bearing vesuvianite from the alkali massif Dahu-Unur, Tyva, Russia (sample 62,632 FM): BSE-image of the (001) section (**a**); inclusion of vesuvianite with ordinary composition in Th-*REE*-bearing vesuvianite (**b**); inclusion of an unnamed *REE*-bearing fluorcarbonate in Th-*REE*-bearing vesuvianite (**c**); several vesuvianite generations filling fracture in Th-*REE*-bearing vesuvianite (**d**); inclusions of diopside in Th-*REE*-bearing vesuvianite (**e**). 1—Th-*REE*-bearing vesuvianite, 2—albite, 3—goetite, 4—pyrite, 5—ordinary vesuvianite, 6—*REE*-bearing epidote, 7—unnamed *REE*-bearing fluorcarbonate, 8—Ca-deficient, Fe-enriched vesuvianite; 9—zoisite

1.4.6 Vesuvianite as Geochronometer

Vesuvianite group minerals are characterized by a high degree of stability during various superimposed processes. The results of the first U–Pb studies of vesuvianite group minerals revealed a number of limitations for its use as a potential source of geochronological information (Zhukov et al. 1961; Romer 1992a). Thus, a high degree of metamictization was established for vesuvianite from alkaline pegmatites of the Dakhu-Unur massif (Tuva, Russia) due to the significant content of actinides (Zhukov et al. 1961). Certain difficulties also arose for researchers due to the increased content of non-radiogenic lead in a number of vesuvianite samples (Romer

1992a) which required the use of additional correction methods when calculating the U–Pb age.

From the crystal chemistry point of view, the substitution scheme $Ca^{2+} \leftrightarrow Pb^{2+}$ in the structure channels along [001] is acceptable for the VSMs, which can explain the presence of a significant proportion of non-radiogenic Pb. It should be noted that vesuvianite is an important marker mineral of contact-metasomatic processes, which are often associated with various ore deposits. The high potential of VSMs as a source of geochronological information is obvious, but it is necessary to identify the factors that can affect the uranium and lead content in the structure, as well as the proportion of the inherited lead component. This determines the importance of conducting new studies of the vesuvianite group using modern methodological approaches. An important condition for the successful testing of new geochronometer minerals is the selection of "reference" objects for which reliable age information is available.

The age of 1550 ± 6 million years obtained for vesuvianite from the metasomatites of the Hopunvaara ore occurrence (Fig. 1.54) as a result of U–Pb (ID-TIMS) geochronological studies is consistent with the known age of skarn formation within the Pitkäranta ore region (1539 ± 9 million years: (Larin et al. 1991) and the age of formation of the granites of the Salmi batholith of 1547 ± 2 million years (Amelin et al. 1997). This demonstrates the reliability of the geochronological data obtained using vesuvianite and points to its high potential as a new geochronometer mineral for ore-bearing contact-metasomatic rocks.

According to the results of U–Pb geochronological studies, wiluite from contact-metamorphic rocks of the Wiluy River basin (Fig. 1.55) stands out noticeably with a lowered proportion of non-radiogenic lead ($Pb_c/Pb_t = 0.14$, identical for two measurements) compared to all currently analyzed VSMs (Wei et al. 2022; Zhang et al. 2022a; Stifeeva et al. 2024a; Han et al. 2025a). Vesuvianites from various contact-metasomatic rocks typically have a higher proportion of Pb_c with Pb_c/Pb_t from 0.30 to 0.80) (Wei et al. 2022; Zhang et al. 2022a).

The assumption that specific conditions of the mineral-forming environment may be the determining factor affecting the uranium and lead content in wiluite is not supported by the relatively average U content in wiluite (29.9–37.3 μg/g). For some vesuvianite, the uranium content is more than 500 μg/g (Han et al. 2025a), with an invariably high proportion of non-radiogenic lead. One of the main factors determining the low proportion of non-radiogenic lead in wiluite are its specific crystal-chemical characteristics.

The obtained results indicate the fundamental possibility of using vesuvianite for U–Pb geochronological studies. Together with the crystal-chemical characteristics obtained for wiluite from achtaragdite rocks of the Wiluy River, there is every reason to recommend this sample for further certification as a standard material when using local methods of geochronological studies.

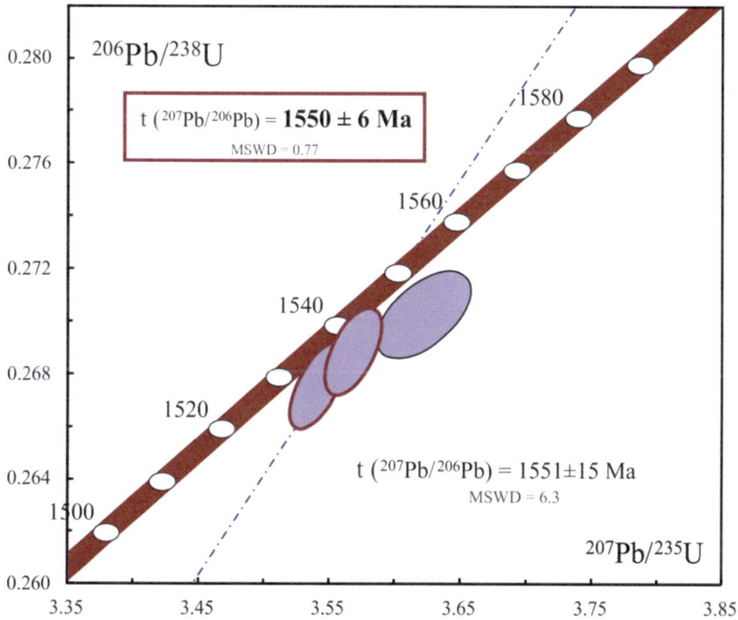

Fig. 1.54 Concordia diagram for vesuvianite from the metasomatites of the Hopunvaara ore occurrence

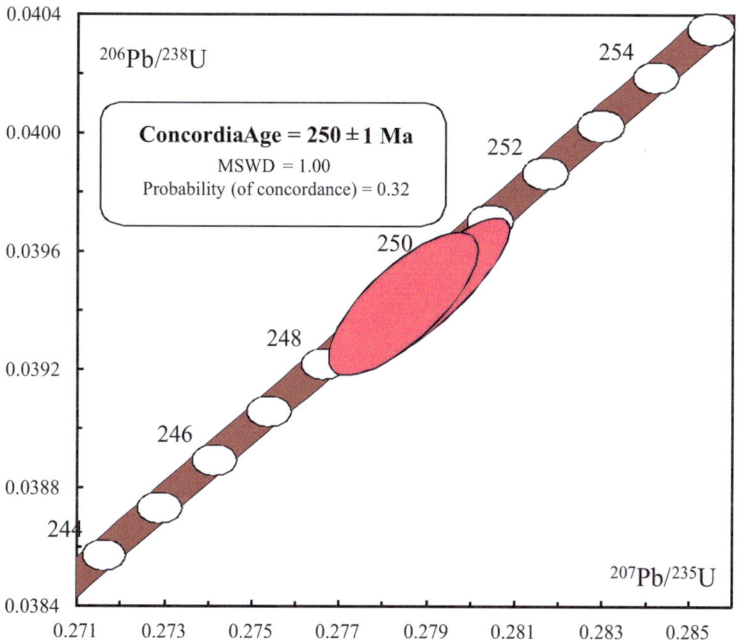

Fig. 1.55 Concordia diagram for wiluite from akhtarandite rocks of the Wilyui River

1.5 Synthesis of Vesuvianite

Several works on the synthesis of VSMs are known (Ito and Arem 1970; Hochella et al. 1982), but only recently single crystals have been obtained, which have clarified the structure. In addition to vesuvian proper, an analog with Ge instead of Si and Sr instead of Ca was obtained (Smart et al. 2023). We investigated the mechanism of *REE* incorporation into vesuvianite structure using the synthetic samples. The synthesis was carried out in a complex system imitating the formation of skarns in nature (methodic of synthesis given in Chap. 2).

As a result of the experiment, small vesuvianite crystals up to 200 μm in size with clinopyroxene inclusions were identified in the reaction zone with praseodymium oxide, in which a zonal distribution of praseodymium was observed (Fig. 1.56). The chemical composition of synthetic vesuvianite corresponds to the empirical formula $(Ca_{16.75}Pr_{2.18})_{18.93}(Al_{8.20}Mg_{2.97}Fe_{1.78}Ti_{0.20})_{13.15}Si_{17.90}O_{69}(OH)_9$.

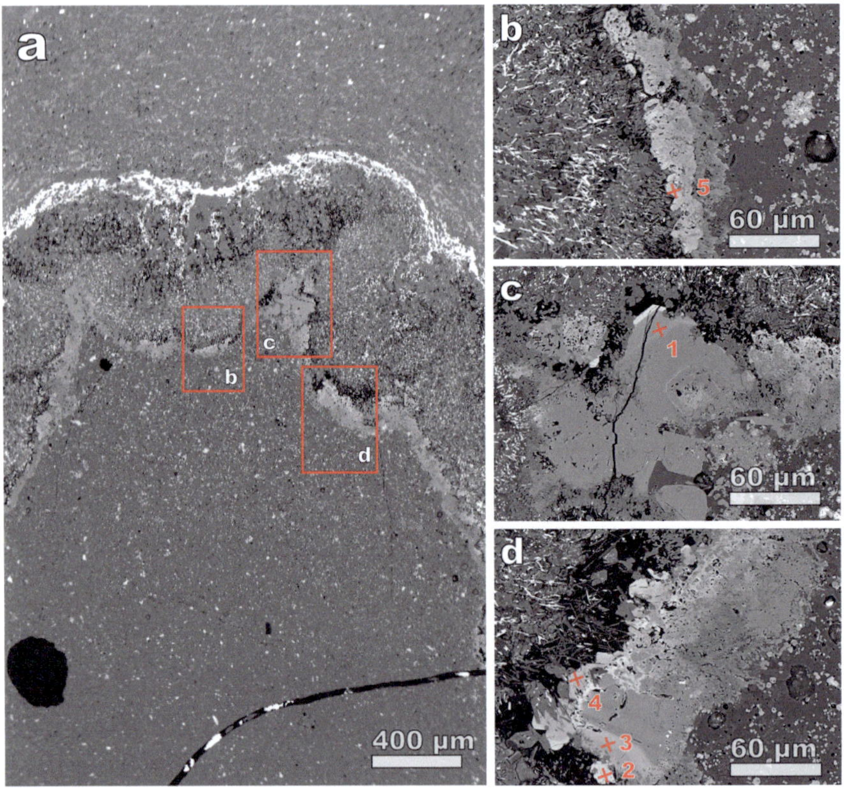

Fig. 1.56 SEM image of zoned synthetic *REE*-bearing vesuvianite. Vesuvianite analysis points are marked with numbers

Refinement of the crystal structure synthetic *REE*-bearing vesuvianite in the space group *P4/nnc* resulted in $R_1 = 0.051$ for 1672 independent reflections with intensities $Fo > 4\sigma(Fo)$. When refining the occupancy of Ca sites, it was noted that Pr incorporates at the *X*3 position with the coordination number of 9, which is completely consistent with the literature data on natural samples. In manaevite (Ce), *REE* also predominantly incorporated into the *X*3 position and partially occur at *X*4 (Moiseev et al. 2020). Thus, the possibility of selective incorporation of Pr at *X*3 position in the crystal structure of synthetic vesuvianite was demonstrated.

By now, many questions remain in the writing of formulas, names of minerals of the vesuvianite group and rules for the allocation of mineral species within this group. The data on the crystal chemistry of VSMs accumulated over the past 30 years make it possible to identify patterns in the incorporation of different cations into various positions of the structure, which makes it possible to generalize all the data in this work (Chap. 4), identify its main crystal chemical relationships and create the necessary basis for creating an up-to-date nomenclature.

Currently, the vesuvianite supergroup has 11 mineral species which differ from each other by the population of the cationic *X*3, *Y*1, *Y*3 *T*1,2 and anionic *W* sites (Table 1.6). Such cations as Fe^{3+}, Al^{3+}, Mn^{3+}, Mg^{2+}, Fe^{2+}, Cu^{2+}, Mn^{2+} can dominate at the five-coordinated *Y*1 site. The octahedral *Y*3 site can be Mg-, Fe^{3+}- or Al-dominant. The *T*1 and *T*2 sites in wiluite should be jointly more than half-occupied by boron atoms. Fluorvesuvianite is the F-dominant at *W* sites vesuvianite analogue.

Since the current classification of the VSMs has not yet been developed, the status of the group members with the dominance Mn^{2+} at the five-coordinated *Y*1 site remains unclear (Groat and Evans 2012). Also, according to our latest data, there are several varieties of wiluite with different populations of the *Y*1 site, which are described in the literature (Groat et al. 1998; Panikorovskii et al. 2017d) and meet modern criteria for distinguishing them as independent mineral species.

Some nomenclature questions still unresolves, for example, the question to be answered is whether the 25% admixture of 2+ cations in the *Y*3 position will be species-defining or not. Since the modraite formula is currently written as: $Ca_{19}Fe^{2+}Al_4(Al_6Fe^{2+}{}_2)(\square_4)\square[Si_2O_7]_4[(SiO_4)_{10}](OH)(OH)_9$. The described mechanisms of the incorporation into the structure of *REE* and Na indicate the existence of varieties and, possibly, mineral species differing in the occupancy of the *X* positions, described in more detail together with the other minerals of the vesuvianite supergroup in Chap. 3 and we made concluding remarks for forming general principles of VSM nomenclature in Chap. 4.

Table 1.6 Dominant components at cationic and anionic sites in the structures of VSMs*

Mineral	*Z*	*X*	*Y*1	*Y*2	*Y*3	*T*1	*T*2	O10	O11	O12	References
Vesuvianite	Si	Ca	Fe^{3+}	Al	Al	□	□	OH	OH	□	Giuseppetti and Mazzi (1983)
Fluorvesuvianite	Si	Ca	Fe^{2+}	Al	Al	□	□	F	F	□	Britvin et al. (2003)
Manganivesuvianite	Si	Ca	Mn^{3+}	Al	Al	□	□	OH	OH	□	Armbruster et al. (2002)
Cyprine	Si	Ca	Cu	Al	Al	□	□	OH	OH	□	Panikorovskii et al. (2017f)
Magnesiovesuvianite	Si	Ca	Mg	Al	Al	□	□	OH	OH	□	Panikorovskii et al. (2017e)
Alumovesuvianite	Si	Ca	Al	Al	Al	□	□	OH	OH	□	Panikorovskii et al. (2017a)
Wiluite	Si	Ca	Fe^{2+}	Al	Al	B	B	O	O	O	Groat et al. (1996a, 1998)
Hongheite	Si	Ca	Fe^{2+}	Al	Fe^{3+}	□	B	O	OH	O	Xu et al. (2017)
Milanriederite	Si	Ca	Fe^{3+}	Al	Mg	□	□	O	OH	□	Chukanov et al. (2019)
Manaevite-(Ce)	Si	Ca + *Ln*	Fe^{2+}	Al	Mg	B	□	O	O	□	Moiseev et al. (2020)
Modraite	Si	Ca	Fe^{2+}	Al	Al	□	□	OH	OH	□	Uher et al. (2024)

*VSMs can be characterized by different symmetry, including tetragonal (*P*4/*nnc*, *P*4/*n*, *P*4*nc*) or monoclinic (*P*2/*n*, *Pn*) space groups (Coda et al. 1970; Veblen and Wiechmann 1991; Allen and Burnham 1992; Ohkawa et al. 1992; Groat et al. 1993, 1995a; Ohkawa 1994)

The symbol □ denotes vacancy

Chapter 2
Experimental Methods

This chapter provides information on the methods of studying vesuvianite supergroup minerals (VSMs), the organization of the analytical work, as well as information on the processing and interpretation of the data obtained. The analysis of the crystal chemical characteristics of the VSMs was carried out according to the results of a comprehensive crystal chemical study of a representative collection of samples from the most known localities throughout the world (Table 2.1; see Appendix 2). Most of the research was conducted at the St. Petersburg State University (SPSU) and at the Kola Science Center (KSC RAS).

2.1 Work Organization

The actual basis of the investigations was a systematic collection of vesuvianite supergroup minerals, collected by the authors from various deposits in Russia and around the world (170 samples in total, Fig. 2.1), which includes samples from the collections of the Mineralogical Museum of St. Petersburg State University (St. Petersburg), Fersman Mineralogical Museum of the Russian Academy of Sciences (Moscow), Mineralogical Museum of Silesian University (Katowice) and I. V. Bel'kov Museum of Geology and Mineralogy of the Geological Instutute, Kola Science Center of Russian Academy of Sciences (Apatity), as well as the personal collections of A. A. Agakhanov (Fersman Mineralogical Museum of the Russian Academy of Sciences), S. N. Britvin, A. A. Antonov (St. Petersburg State University), E. V. Belogub (Institute of Mineralogy, Uralian Branch of Russian Academy of Sciences), E. V. Galuskin (Silesian University), Yu. V. Erokhin (Zavaritsky Institute of Geology and Geochemistry of Uralian Branch of Russian Academy of Sciences), V. Y. Karpenko, M. M. Moiseev (Fersman Mineralogical Museum of the Russian Academy of Sciences), I. V. Pekov (Moscow State University), N. V. Chukanov (Federal Research Center of Problems of Chemical Physics and Medicinal

T. L. Panikorovskii and N. V. Chukanov, *Vesuvianite-Supergroup Minerals*, Springer Mineralogy, https://doi.org/10.1007/978-3-032-06970-2_2

Table 2.1 Instrumental investigation methods

Method	Goal	Equipment
X-ray microanalysis/ electron microprobe	Precise determination of the chemical composition of VSMs and associated minerals	Scanning electron microscope (BSE) LEO-1450, EDX-analyzer Quantax; microprobe CAMECA MS46 (KSC RAS, Apatity); BSE Hitachi S-3400N, EDX-analyzer OXFORD, WDS-analyzer INCA (RC "Geomodel" SPSU, St. Petersburg); microprobe CAMECA SX100 (Warsaw University); BSE Philips XL30 ESEM, EDX-analyzer EDAX (Silesian University, Katowice), BSE CamScan MX2500S, EDX-analyzer OXFORD (A. P. Karpinsky Institute (VSEGEI), St. Petersburg)
Scanning electron microscopy	Investigation of morphological features, grain homogeneity, genetic relations, determination of the qualitative chemical composition of VSMs and associated minerals	
Inductively coupled plasma mass spectrometry	Determination of boron content	Mass spectrometer iCAP 6300 duo (Mekhanobr Engineering, St. Petersburg)
Differential scanning calorimetry/ thermogravimetry	Determination of phase stability temperatures, stages of weight loss and thermal effects of conversions, total weight loss	Differential scanning calorimetry spectrometer NETZSCH STA 449 F3 Jupiter (RC X-ray Diffraction Methods of Investigation SPSU, St. Petersburg)
Solid state MAS NMR	Determination of the coordination of the ^{5}B, ^{19}F, ^{23}Na, ^{27}Al, ^{29}Si and ^{139}La atoms in VSMs and their proportion	Spectrometer Bruker Avance III 400 WB (RC «Magnetic Resonance Research Methods» SPSU, St. Petersburg)
Infrared spectoscopy	Qualitative determination of the SO_3, BO_3, BO_4, H_2O, OH groups in VSMs, as well as estimation of the parameters of O–H···O hydrogen bonds	IR Fourier spectrometers Bruker Vertex 70 (RC X-ray diffraction methods of investigation SPSU, St. Petersburg), ALPHA FTIR (Institute of Problems of Chemical Physics, Russian Academy of Sciences, Chernogolovka)

(continued)

Table 2.1 (continued)

Method	Goal	Equipment
Photoelectron spectroscopy	Determination of local coordination and chemical state of Ca, Si, S, Al, Mg, B and Fe	Spectrometer Thermo Fisher Scientific Escalab 250Xi (RC Physical Methods for Surface Investigation, SPSU, St. Petersburg)
Elemental analysis	Quantitative determination of the contents of C, H	CNH-analizer Carlo Erba 1106 (Fersman Mineralogical Museum, Moscow)
Single-crystal XRD analysis	Determination and refinement of the crystal structure	Single-crystal XRD diffractometers STOE IPDS II, Agilent Technologies ExcaliburEos, Agilent Technologies SuperNova, Bruker APEX II и Bruker APEXDUO (RC X-ray Diffraction Methods of Investigation, SPSU, St. Petersburg)
Powder diffraction	Determination of the unit cell parameters, diagnostic of inclusions and associated minerals	Powder diffraction BrukerPhazerD2 difractometer, SC XRD difractometer Rigaku R-AXIS Rapid with Debye–Scherrer-Geometry (RC X-ray diffraction methods of investigation, SPSU, St. Petersburg)
Common optical microscopy	Selection of single crystals for research, obtaining color images of samples, obtaining optical spectra	Binocular Leica 60d, camera Levenhuk c800 (Crystallography Department, SPSU, St. Petersburg), camera Canon PowerShot G9 (GI FIC KSC RAS, Apatity), spectrometer AVASPEC-ULS2048X16 (National History Museum, Stockholm)
Optical microscopy in polarized light	Determination of optical sign, birefringence coefficients, 2*V* angle, pleochroism and other optical characteristics	Polarized microscope LOMO POLAM RP-1, set of immersion liquids (Crystallography Department, SPSU, St. Petersburg)
Goniometry	Determination of crystal forms	Four-circle goniometer on a diffractometer Agilent Technologies Xcalibur EOS (RC X-ray diffraction methods of investigation, SPSU, St. Petersburg)

Chemistry, Russian Academy of Sciences) and V. N. Yakovenchuk (Kola Science Center of Russian Academy of Sciences).

At the initial stage, grains for single-crystal XRD studies were selected. Thereafter these grains were used for electron microprobe measurements (25 preparations each with 10–20 grains, more than 3000 measurements in total). Along the way, material for the study of the petrological features of VSMs-containing rocks

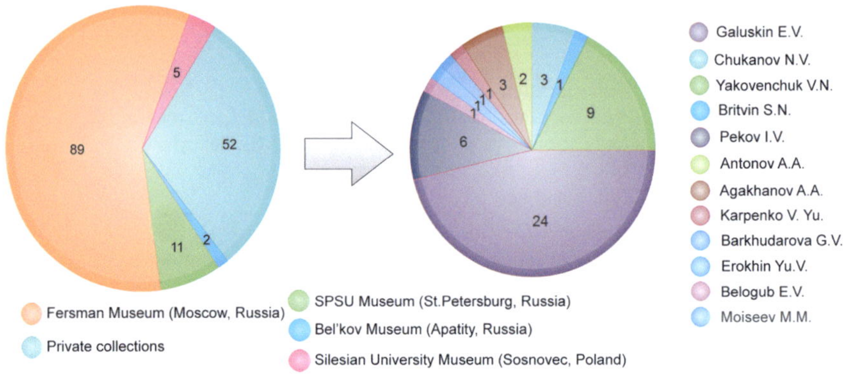

Fig. 2.1 Pie diagram of the number of obtained samples and their distribution over collections of two kinds (museum on the left and personal on the right). A part of samples from the collection of E. V. Galuskin was presented by I. O. Galuskina and E. D. Szełęg

was selected. For the most spectacular samples, color photographs were obtained together with G. Yu. Ivanyuk using Canon PowerShot G9 camera located in the Nanomaterials Research Center of KSC RAS, Apatity, Russia. The most interesting samples in respect with chemical composition or crystal structure were studied by additional spectroscopic methods: solid-state NMR, Mössbauer spectroscopy, differential scanning calorimetry/thermogravimetry, Raman, infrared and photoelectron spectroscopy.

2.2 Instrumental Methods

2.2.1 Powder and Single-Crystal XRD Diffraction

XRD studies were performed using equipment of the Crystallography Department and RC X-ray Diffraction Methods of Investigation, SPSU, St. Petersburg.

Powder X-ray diffraction was carried out using an automatic Bruker Phazer D2 diffractometer using Cu$K\alpha_{1+2}$ radiation (the wavelengths $\lambda_{CuK\alpha1} = 1.54059$ Å and $\lambda_{CuK\alpha2} = 1.54441$ Å), position-sensitive detector, Bragg–Brentano focusing scheme, at the sample rotation speed of 30 rpm, diffraction angle interval 2Θ from 10° to 80°, at T = 25 °C, in air atmosphere. The tube operated at 40 kV/30 mA. Shooting time was ~3.5 h. Samples were taken in a cuvette with a silicon monocrystal substrate, the diameter of the investigated area 2 × 2 mm.

Phase identification was performed using the PDXL2 v2.0.2 program and the PowderDiffractionFile powder diffraction database (PDF-2, 2011). The unit cell parameters were refined using the Rietveld method in the Topas program (Bruker AXS GmbH 2009).

Single-crystal XRD studies were performed on STOE IPDS II, Agilent Technologies Excalibur Eos, Agilent Technologies SuperNova, Bruker APEX II and Bruker APEX DUO single-crystal X-ray diffractometers equipped with Image-Plate (STOE IPDS II) and CCD-type detectors (others) on monochromatic Mo$K\alpha$. ($\lambda = 0.71069$ Å) radiation. Precision studies with the collection of full set of diffraction data were carried out with a scanning step of 0.5°, exposure time of 10–100 s per image, at 100 K in the 2θ range of 6.00–70.00°. Ordinary surveys were carried out with a scanning step of 1°, exposure time in the range of 10–100 s per image at room temperature in the range of 2Θ angles from 6.00° to 55.00°. The absorption correction was applied empirically using spherical harmonics, implemented in the SCALE ABSPACK calibration algorithm, in the CrysalysPro software package (Agilent Technologies 2014), or semi-empirically, using the APEX2 program algorithms (Bruker AXS 2014). The unit cell parameters were refined by the least squares method. The structures were refined using the SHELXL program (Sheldrick 2008) embedded in the OLEX2 software package (Dolomanov et al. 2009), or WinGX (Farrugia 2012). Visualization of crystal structures, electron density, calculation of polyhedral volumes, as well as the construction of theoretical X-ray diffraction patterns were performed using the Diamond 3.2f (Putz and Brandernburg 2014), VESTA 3 (Momma and Izumi 2011) and Mercury 3.10.1 (Macrae et al. 2008) programs, respectively.

2.2.2 *Electron Microscopy and Electron Microprobe Analysis*

The study of morphological features, grain homogeneity, genetic relationships and mineral associations, as well as the determination of the qualitative chemical composition of the VSMs, inclusions and associated minerals were performed in the RC "Geomodel" SPSU, St. Petersburg (the operators V. V. Shilovskikh and N. S. Vlasenko), Geological institute of KSC RAS (by A. V. Bazai), University of Silesia in Katowice (by E. V. Galuskin) using scanning electron microscopes Hitachi S-3400N (INCA 500), LEO-1450 (Quantax) and Philips XL30 ESEM (EDAX), respectively. The measurements were carried out at 20–30 nA, an accelerating voltage of 20 kV and an electron beam diameter of 1–5 μm.

A detailed processing of the elemental VSMs composition by electron microprobe (Table 2.2) was carried out on the following equipment: Hitachi S-3400N, INCA in Geomodel Resource Center at St. Petersburg State University, CamScan MX2500S, OXFORD in A. P. Karpinsky Russian Geological Research Institute, Cameca MS-46 in Geological Institute of the Kola Scientific Center of Russian Academy of Sciences and Cameca SX-100 in Warsaw State University (analyst P. Dzerzhanovsky). Chemical analyses of modraite from Shigar Valley were carried out by means of electron microprobe analyses using a digital scanning electron microscope Tescan VEGA-II XMU with an energy-dispersive spectrometer INCA Energy 450 at an accelerating voltage of 20 kV and electron beam current of 520 pA.

The main components, except elements with atomic numbers below 8, were determined at an accelerating voltage of 20 kV, a current of 20–30 nA, and an electron beam diameter of 5–20 μm. The boron content was determined using a specially elaborated method based on wave dispersion X-ray spectroscopy (Galuskin 2005) on a CAMECA SX100 microprobe with an accelerating voltage of 5 kV, a current of 100 nA and an exposure of 100 s at each point.

The empirical formulas were calculated using the MINAL module (Dolivo-Dobrovolsky 2016) for MS Excel spreadsheets.

Table 2.2 Parameters of electron microprobe analyses

Component	Detection limit (wt%)	Standart	
		SX-100	MS-46
B_2O_3	0.20	Ni/C-LSM	Datolite
F	0.15	Topaz	Fluorapatite
Na_2O	0.15	Albite	Lorenzenite
MgO	0.1	Forsterite	Pyrope
Al_2O_3	0.05	Sanidine	Pyrope
SiO_2	0.05	Sanidine	Diopside
SO_3	0.05	Barite	Barite
Cl	0.03	Vanadinite	Atakamite
K	0.03	Microcline	Microcline
CaO	0.03	Wollastonite	Diopside
TiO_2	0.01	Rutile	Lorenzenite
MnO	0.01	Spessartine	Rhodochrosite
FeO*	0.01	Hematite	Hematite
Fe_2O_3*	0.01		
CuO	0.01	Copper	Copper
Cr_2O_3	0.01	Corundum	Corundum
SrO	0.01	Celestine	Celestine
La_2O_3	0.01	$LaPO_4$	$LaCeS_2$
Ce_2O_3	0.01	$CePO_4$	$LaCeS_2$
Pr_2O_3	0.01	$PrPO_4$	$LiPr(WO_4)_2$
Nd_2O_3	0.01	$NdPO_4$	$LiNd(MoO_4)_2$
Sm_2O_3	0.01	$SmPO_4$	$LiSm(MoO_4)_2$
Gd_2O_3	0.01	$GdPO_4$	$LiGd(MoO_4)_2$
PbO	0.01	Vanadinite	Wulfenite
ThO_2	0.01	Cheralite	Thorite

* The FeO/Fe_2O_3 ratio was determined according to Mössbauer spectroscopy data

2.2.3 Differential Scanning Calorimetry

The study of VSMs samples by the differential scanning calorimetry method was performed in order to determine thermal effects at melting and weight loss steps on heating to 1200 °C using a synchronous thermal analyzer NETZSCH STA 449 F3 Jupiter at the research centrum of X-ray diffraction methods of investigation of the St. Petersburg State University. The measurements were carried out in an Ar atmosphere with a heating rate of 10 °C/min for samples with weights above 10 mg placed in platinum crucibles using corundum as a standard. Using synchronous thermal analysis, measurements of heat flows and measurements of the mass of the sample and corundum standard are carried out under completely identical conditions. Corundum was used as a standard of heat absorption in the working temperature range. The use of the differential method, which is based on the temperature comparison between the studied sample and the standard is due to a much greater sensitivity than traditional thermal analysis (Panikorovskii and Britvin 2013).

2.2.4 Elemental (CHN) Analysis

Direct determination of H_2O and CO_2 was carried out for the sample with an anomalous water content. In other cases, H_2O was determined by weight loss upon ignition in an inert atmosphere. According to structural data, the *T*2 position with three-fold coordination in the structure of manaevite-(Ce) is partially populated, but no BO_3 was detected in the chemical composition of this mineral. Based on these data, it was assumed that manaevite-(Ce) may contain carbonate groups. Measurment was carried out on the equipment of the Fersman Mineralogical Museum, Moscow by the operator V. Yu. Karpenko using a Carlo Erba 1106 analyzer. Combustion was carried out using a modified Dumas-Pregl method in a pewter capsule with oxygen added in a helium current of the chromatographic column. The initial temperature of the furnace was 1030 °C (during the flash, the temperature rised locally to 1800 °C). The chromatographic column filler was Porapak QS. The standards used for C and H were 2.55 mg samples of acetanilide, sulfanilamide and benzoic acid.

2.2.5 Infrared Spectroscopy

IR spectroscopic studies were carried out on a Bruker Vertex 70 Fourier spectrometer at the research centrum of X-ray diffraction methods of investigation of the St. Petersburg State University (operator T. L. Panikorovskii) and on an ALPHA FTIR Fourier spectrometer (Bruker Optics) at the Institute of Chemical Physics of the Russian Academy of Sciences (analyst N. V. Chukanov). Measurements were carried out at room temperature in the range 360–4000 см$^{-1}$ with a resolution of 4 см$^{-1}$ and the number of scans in the range 16–64. Samples were prepared according to the standard pelletization procedure by pressurizing a mixture prepared from 2 mg of VSM powder and 200 mg of KBr, predried at 120 °C, at 10 tons. The IR spectrum of a pellet of pure KBr was used as a reference blank. Data processing was performed using the OPUS software package.

2.2.6 *Inductively Coupled Plasma Mass Spectrometry*

Along with electron microprobe analyses, determination of boron content was performed by atomic emission method with inductively coupled plasma on the spectrometer iCAP 6300 duo (Mekhanobr engineering), at the wavelength 249.7 nm, with axial observation. The probe weight was in the range 200–250 mg.

2.2.7 *Solid-State NMR Spectroscopy*

Solid-state NMR analysis of the samples was carried out using a Bruker Avance III 400 WB spectrometer with an operating frequency of 400.23 MHz in the research centrum "Magnetic resonance research methods" St. Petersburg State University (operator A.S. Mazur). The measurements were carried out at room temperature. To record NMR spectra of ^{1}H, ^{11}B, ^{23}Na, ^{27}Al and ^{29}Si, a single-pulse mode was used with pulse frequency of 100, 66.4, 80, 55.5 and 25 kHz, a delay time of 20, 10, 5, 10 and 10 s, and 32, 3000, 1024, 4000 and 6000 scans, respectively. For ^{23}Na measurements, single-pulse mode with a pulse duration of 3 ms, delay time of 5 s and 1024 scans was applied. The ^{139}La spectrum was recorded using the spin-echo approach (the first pulse width was 0.8 μs, the delay time was 0.2 s, the sweep width was ±56.5 MHz, and the number of scans was 600,000). The spectra were visualized and processed in the MestReNova program (Willcott 2009).

2.2.8 *Optical Absorption Spectra in the Near Infrared, Visible and UV (NIR–Vis–UV) Range*

Polarized, room-temperature optical NIR–Vis–UV absorption spectra in the range 270–1100 nm (37,037–9091 cm^{-1}) were recorded on doubly sided polished single crystal sections at a spectral resolution of 1 nm using a AVASPEC-ULS2048X16 spectrometer attached via a 400 μm UV fiber cable to a Zeiss Axiotron UV-microscope. A 75 W Xenon arc was used as a light source and Zeiss Ultrafluar $10^{\times}$ lenses served as objective and condenser. The size of the circular measure aperture was 64 μm in diameter. A UV-quality Glan–Thompson prism with a working range from 250 to 2700 nm (40,000–3704 cm^{-1}) was used as a polarizer. The wavelength scale of the spectrometer was calibrated against Ho_2O_3 doped and Pr_2O_3/Nd_2O_3 doped standards (Hellma glass filters 666F1 and 666F7).

The thickness of the polished sections of crystal fragments was 46 μm. They were cut perpendicular to the crystallographic *a*-axis allowing measurements with the light polarization E||e*E* (i.e. parallel to the *c*-axis) and E||*O*. Spectra were measured on untreated crystal slabs as well as after subsequent heat treatment at 600 °C in H_2 atmosphere. The treated samples were placed in a gold container and pushed into a preheated horizontal tube furnace equipped with a quartz glass tube. The heating experiment lasted for 8 h and was terminated by pushing the samples out to the cold zone of the quartz tube.

2.2.9 Optical Characteristics

Optical characteristics (optical sign, birefringence coefficients, 2*V* angle, pleochroism, etc.) of the VSMs were determined at the Department of Crystallography of the St. Petersburg State University by T. L. Panikorovskii and E. Yu. Avdontseva with a LOMO POLAM RP-1 polarization microscope for suspensions in immersion liquids using a set of immersion liquids with high refractive indices.

2.2.10 Mössbauer Spectroscopy

Gamma-resonance studies were performed on a Wissel Mössbauer spectrometer (Physics Department of Moscow State University, analyst V. S. Rusakov) to determine the coordination and valence state of iron ions. The spectrometer is equipped with a standard ^{57}Co radioactive source in a rhodium matrix (Ritverc). Measurements in the absorption mode were performed at room temperature. Powdered samples were homogeneously distributed inside the punch holder and packaged as a Tablet with a density of 112 mg/cm^2 which corresponds to the optimal measuring conditions. The spectral fitting of the experimental Mössbauer spectra was carried out by the least squares method under the assumption of Lorentz spectral lines. The isomer shifts relative to metallic iron were measured. Mathematical analysis was performed by commercial software Mössbauer SpectRelax version 2. The detailed technique for measuring and interpreting spectra is described by Rusakov et al. (2006).

2.2.11 Photoelectron Spectroscopy

The photoelectron spectra were recorded on a Thermo Fisher Scientific Escalab 250Xi spectrometer at the Research Centrum Physical Methods for Surface Investigation of the St. Petersburg State University (by the operator Yu. M. Zhukov) on Al$K\alpha$ radiation (photon energy 1486.6 eV). The total energy resolution of the experiment was about 0.3 eV. The spectra were recorded in situ in the regime of a constant transmission rate at 50 eV for the analysis area with a diameter of 650 μm. The studies were conducted at room temperature in an ultrahigh vacuum, at a pressure of the order of 1×10^{-9} mbar. For the collection and analysis of data, the software package AVANTAGE was used.

2.2.12 Goniometry

Goniometric studies were performed during the coursework written by a first-year student of the Institute of Earth Sciences of St. Petersburg State University A. S. Rusina on a number of vesuvianite species from the Bazhenovskoe deposit, Middle Urals, Russia and cyprine from Wessels, Kalahari Manganese Fields, RSA. The crystal faces were indexed using the CrysalisPro software package and taking into account unit-cell orientation obtained with a diffractometer digital camera equipped with a four-circle goniometer (Agilent Technologies Xcalibur EOS).

2.2.13 Raman Spectroscopy

Raman spectra were obtained at the Dept. of Mineralogy, Faculty of Geology, Lomonosov Moscow State University (analyst N. V. Zubkova) for randomly oriented grains using an EnSpectr R532 spectrometer based on an OLYMPUS CX 41 microscope (Enhanced Spectrometry, San Jose, USA) coupled with a diode laser (λ = 532 nm) at room temperature. The spectra were recorded in the range of 100–4000 cm^{-1} with a diffraction grating (1800 gr mm^{-1}) and spectral resolution of about 6 cm^{-1}. The output power of the laser beam was in the range of 5–13 mW. The diameter of the focal spot on the sample was 5–10 μm. The backscattered Raman signal was collected with a 40× objective; signal acquisition time for a single scan of the spectral range was 1 s, and the signal was averaged over 50 scans. Crystalline silicon was used as a standard.

2.2.14 Photoluminescence Spectroscopy

Photoluminescence spectra were measured at the Vinogradov Institute of Geochemistry, Siberian Branch of Russian Academy of Sciences, Russia (analyst R. Yu. Shendrik) using a spectrometer based using an SDL-1 monochromator with a 600 lines per mm threaded diffraction grating (LOMO, St. Petersburg, Russia). The spectral slit width was 0.2 nm. Registration was carried out using a Hamamatsu H10721-04 photomodule (Hamamatsu, Sendai, Japan). Excitation was performed using a semiconductor laser with a wavelength of 405 nm and a power of 40 mW. The samples were fixed in a filling nitrogen cryostat, which was placed in a vacuum chamber.

2.3 Synthesis

Since materials doped with *REE* often have luminescent properties, such as synthetic analogs of apatite, the task of obtaining Pr-containing vesuvianite was set and implemented. Vesuvianite was synthesized on the reaction of liquid carbonate with solid silicate at high pressure and temperature (Anenburg and Mavrogenes 2018). The recipes and nominal compositions for both the silicate and carbonates are given in Table 2.3. The chemicals were mixed in an agate mortar and pestle using acetone until evaporation and dried in a 110 °C oven. The silicate composition was subsequently pelletized and fired at 1000 °C for decarbonation. A silver capsule was filled from below with 16.5 mg of powdered Re metal, 89.0 mg of carbonate mix, 66.6 mg of silicate mix, and 28.4 mg of additional Re metal. The capsule (Fig. 2.2) was cold-welded using the method of (Hack and Mavrogenes 2006).

The sealed capsule was inserted into a cylindrical MgO–graphite–talc–Teflon assembly, and loaded into a ¾ in. (19 mm) pressure vessel. Using an end-loaded piston cylinder apparatus, we simultaneously increased pressure and temperature to 600 MPa and 800 °C over ~10 min. Run duration was 87 h at experimental conditions,

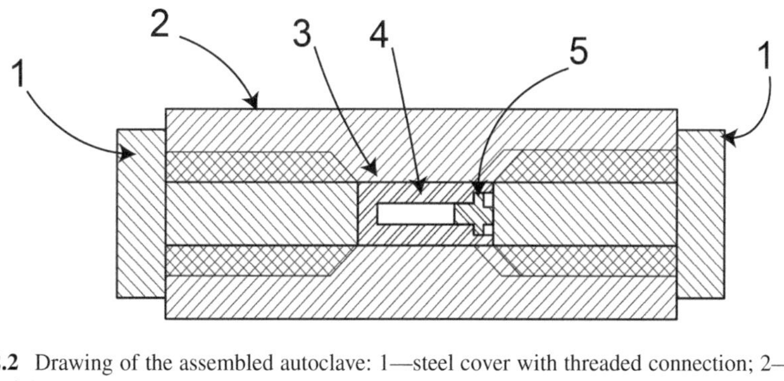

Fig. 2.2 Drawing of the assembled autoclave: 1—steel cover with threaded connection; 2—steel body of the autoclave; 3—silver capsule; 4—synthesis area; 5—silver capsule cover

Table 2.3 Composition of the mixtures used for the synthesis

Silicate				Carbonate			
	Recipe		Composition		Recipe		Composition
SiO_2	62.38	SiO_2	66.64	$CaCO_3$	69.69	CaO	48.09
Al_2O_3	14.94	Al_2O_3	15.97	$Ca(OH)_2$	6.76	MgO	7.34
Fe_2O_3	8.87	FeO	8.53	$Mg(OH)_2$	10.61	Pr_2O_3	2.91
$CaCO_3$	9.16	CaO	5.49	$Ca_3(PO_4)_2$	13.35	ThO_2	0.50
K_2CO_3	4.65	K_2O	3.38	Pr_6O_{11}	3.00	P_2O_5	6.12
				ThO_2	0.50	CaF_2	2.09
				CaF_2	2.09	H_2O	4.93
						CO_2	28.03

after which it ended by cutting power to the graphite heater causing temperature to drop below 100 °C in about 15 s. The capsule was then removed, mounted in epoxy resin and sectioned to expose the run products. For more details on the experimental procedure, see Anenburg and Mavrogenes (2018).

2.4 Data Processing and Interpretation

Data processing was carried out mainly with the MINAL software by D. V. Dolivo-Dobrovolsky (IGGD RAS), Statistica-12.0 (Statsoft) and Origin 8.1 (OriginLab Corporation).

The calculation of the structural complexity of various VSMs structural modifications were calculated using the modified Shannon formula, according to which the complexity of the crystal structure is estimated as the amount of information per atom (I_G) or per VSM unit-cell ($I_{G,\text{total}}$) (Krivovichev 2013):

$$I_G = -\sum_{i=1}^{k} p_i \log_2 p_i \tag{2.1}$$

$$I_{G,\text{total}} = -vI_G = -v\sum_{i=1}^{k} p_i \log_2 p_i, \tag{2.2}$$

where k is the number of regular systems of points in the structure under consideration, v is the number of atoms per unit cell, m_i is the multiplicity of the regular system of points occupied by atoms in the unit cell; $p_i = m_i/v$. The calculation of structural complexity was carried out by *TOPOS* software package (Blatov et al. 2000).

The distortion D index was calculated by the formula proposed by Baur (1974):

$$D = \frac{1}{n} \sum_{i=1}^{n} \frac{[l_i - l_{av}]}{l_{av}},$$

where l_i is the distance between the central atom of the i-coordinated atom and l_{av} is the average bond length.

Chapter 3
Minerals of the Vesuvianite Supergroup

This chapter presents results of chemical analysis, IR spectroscopy, thermogravimetry, Mössbauer spectroscopy, solid-state NMR, powder diffraction, occurrence, physical properties and specific features of the crystal structures of VSMs from various parageneses. Below, the description of minerals is given in the order of the reducing number of their samples in the collection (Fig. 3.1). For assignment of the VSMs to the mineral species, at least, structural and chemical composition data were used.

The data presented in this chapter are given in the context of the development of the current VSMs nomenclature, in accordance with the VSMs crystal chemistry and current rules of the CNMNC (Bosi et al. 2019a, b). The main purpose of the elaboration of the vesuvianite supergroup nomenclature is a maximum simplicity of formula calculating in the most common cases when only chemical data are available.

3.1 Vesuvianite Sensu Stricto

Vesuvianite was first found by M.A. Kappeler at the Somma-Vesuvius volcanic complex, Campania, Italy (Kappeler 1723). This mineral was was firstly confused with garnets, schorl, and even obsidian. Later vesuvianite was described as a mineral species with its present name (Werner 1795). Currently, according to www.mindat.com, the number of known vesuvianite localities is 1.462, and according to the SciFinder database, there are hundreds of publications about the minerals of this supergroup. It is worth noting that until 1998 all minerals of this group were described under the name vesuvianite (Groat et al. 1998).

T. L. Panikorovskii and N. V. Chukanov, *Vesuvianite-Supergroup Minerals*, Springer Mineralogy, https://doi.org/10.1007/978-3-032-06970-2_3

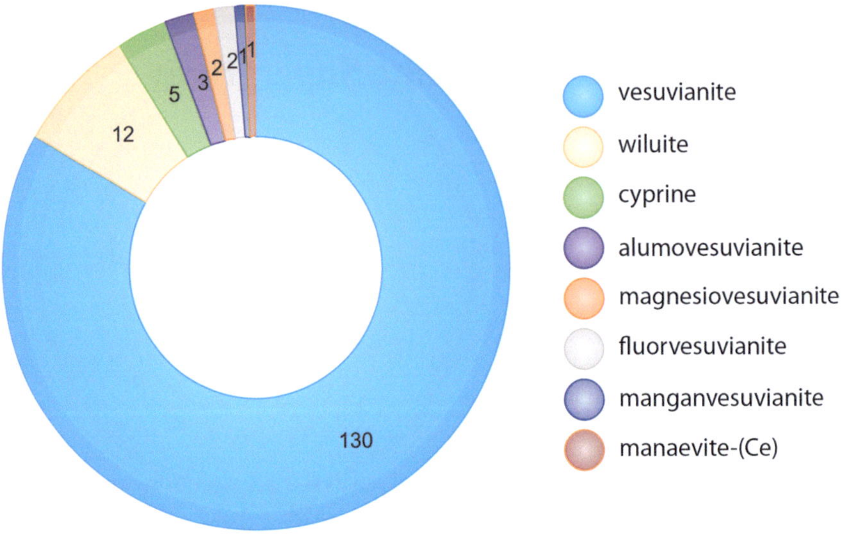

Fig. 3.1 Pie diagram of the distribution the studied VSMs samples belonging to different mineral species

3.1.1 General Information

According to the www.mindat.com database and our data, more than 95% of VSMs samples belong to vesuvianite and wiluite. Most data provided in Sect. 3.1 are related to vesuvianite sensu strictu. Data on other VSMs are given for comparison. Vesuvianite has many synonyms, most of which are not currently used. In addition to the eleven names of valid minerals approved by the IMA CNMNC, the following names can be found in the literature of the XIX–XX centuries: idokrase, egeran, gahnite genevite, pyramidal garnet, volcanic chrysotile, loboite, hekumit, xanthite, frugardite, jevrenovite, duparcite, as well as cebolite (a mixture of melilite and Si-deficient vesuvianite. In the modern literature (from the last quarter of the twentieth century to the present day), the following names were applied to different chemical varieties of the VSMs: Be-vesuvianite, chromvesuvianite, Ce-vesuvianite (synonymous with manaevite-Ce), and hydrovesuvianite.

The list of deposits and the number of studied VSMs samples together with the information about their mineral associations and the geological environment is given in Table A.3.

3.1.2 Appearance and Physical Properties

For vesuvianite sensu stricto from its type locality in the vicinity of Vesuvius, which formed during the eruption of 79 and was described by A. Werner (Gilg et al. 2001); skarn assemblage including clinopyroxene, phlogopite, wollastonite, leucite, grossular-andradite garnet series, spinel, and scapolite is characteristic (Fig. 3.2a). Vesuvianite-containing skarns are highly variable in the mineral composition and vary widely in alkalinity from the more common basic skarns with leucite, olivine and clinohumite to rocks with syenite composition containing sanidine, clinopyroxenes and amphiboles as the major components (Balassone et al. 2011). The color of a mineral from a type locality (Ciriotti et al. 2009) ranges from dark brown and reddish brown to brownish yellow and golden yellow. Vesuvianite from the type locality is characterized by smooth color transitions from different shades of brown to green (Fig. 3.2b) due to different Fe^{2+}/Fe^{3+} ratios (see below).

Most VSMs samples slightly differ in their physical properties (Table 3.1), however, significant variations in their optical characteristics (Table 1.1), density and hardness are observed as a result of significant variations in chemical composition. In Table 1.9, data for milanriederite and hongheite are given for comparison.

Fig. 3.2 Vesuvianite samples from the Vesuvius volcano: greenish-brown vesuvianite with yellow grossular and clinohumite (sample 51061 FM); 1—vesuvianite, 2—grossular, 3—clinohumite

Table 3.1 Physical properties of VSMs

Mineral	Vesuvianite	Wiluite	Manganivesuvianite	Fluorvesuvianite	Milanriederite	Modraite
Color	Green, brown	Marsh green, brown	Red, lilac	Colorless, blue	Dark brownish red	Brown
Crystal forms	{100}, {001}, {110}, {111}, {131}, {011}	{100}, {001}, {110}, {110}, {011}, {101}	{100}, {001}, {110}, {131}, {011}, {120}	{100}, {110}, {101}, {111}	{111}, {001}, {110}, {100}	{111}, {001}, {110}, {100}
Density (g/cm^3)	3.23–3.42	3.36	3.40	3.46	3.53	3.44
Hardness	5.5–6.5	6	6–7	6	6	6
Optical sign	–	+	–	–	–	–
ω	1.703–1.752	1.721	1.735	1.702	1.744	1.749
ε	1.700–1.746	1.725	1.724	1.699	1.737	1.740
Crystal habit	Prismatic	Short-prismatic	Prismatic	Acicular	Dipyramidal	Prismatic
Mineral	Cyprine	Magnesio-vesuvianite	Alumovesuvianite	Manaevite-(Ce)	Hongheite	
Color	Blue, lilac	Colorless, pink	Colorless, pink	Orange, brown	Green, brown	
Crystal forms	{100}, {110}, {331}, {111}, {120}, {101}	{100}, {110}, {111}, {101}, {001}	{100}, {110}, {210}, {111}, {101}, {001}, {112}, {221}	{100}, {001}, {110}, {110}, {011}, {101}	No data	
Density (g/cm^3)	3.40	3.30	3.31	3.85	3.45	
Hardness	6.5	6	6.5	4.5	6–7	
Optical sign	–	–	–	–	+	
ω	1.744	1.731	1.725	1.760	1.720	
ε	1.732	1.725	1.722	1.750	1.725	
Crystal habit	Prismatic	Acicular	Prismatic	Epitaxial crusts	Prismatic	

3.1.3 Chemical Composition

All VSMs samples whose crystal structures were studied by the first author of this book T. L. Panikorovskii (156 samples in total) were characterized by electron microprobe analyses including a method specially elaborated for determining boron content (see Sect. 2.2.2) for B-bearing samples. Table 3.2 presents electron microprobe data for a representative vesuvianite from the type locality (Balassone et al. 2011). Statistical data on the chemical composition of all VSMs studied in this work are given above, in Sect. 1.3.2. The formula coefficients calculations were based on 19 atoms of the *X*-group (Ca, Na, K, Sr, *REE*, U, Th, etc.), because other calculation methods show less satisfactory results. As an example, formulae of a Si-deficient, B-containing, "hydrovesuvianite" (analysis 45) from Yakutia (Fig. 3.3d) calculated by different ways are given below.

The calculation formula based on 50 $X + Y + Z + T$ cations (1), 18 Si atoms (2), 13 *Y* atoms (3), and 19 *X* atoms (4) give the following results:

1. ${}^{X}(Ca)_{\Sigma 19.32}{}^{Y}(Al_{7.20}Mg_{3.85}Fe^{2+}{}_{1.98}Ti_{0.28}Mn_{0.05})_{\Sigma 13.36}{}^{T}(B)_{0.76}{}^{Z}(Si_{16.51}[H_4O_4]_{1.49})_{\Sigma 18}$ $O_{69}{}^{W}([OH]_{10.16}Cl_{0.04})_{\Sigma 10.20}$ (charge disbalance −0.8, excess cations and anions at *X*, *Y*, *W* sites).
2. ${}^{X}(Ca)_{20.99}{}^{Y}(Al_{7.83}Mg_{4.18}Fe^{2+}{}_{2.16}Ti_{0.31}Mn_{0.06})_{\Sigma 14.54}{}^{T}(B)_{0.83}{}^{Z}(Si)_{18}O_{69}{}^{W}([OH]_{17.51}$ $Cl_{0.05})_{\Sigma 17.56}$ (charge disbalance −1.56, significant excess cations and anions at *X*, *Y*, *W* sites).
3. ${}^{X}(Ca)_{18.79}{}^{Y}(Al_{7.00}Mg_{3.74}Fe^{2+}{}_{1.93}Ti_{0.28}Mn_{0.05})_{13}{}^{T}(B)_{0.74}$ ${}^{Z}(Si_{16.11}[H_4O_4]_{1.89})_{\Sigma 18}$ $O_{69}([OH]_{8.12}Cl_{0.04})_{\Sigma 8.16}$ (charge disbalance −1.56, lack cations and anions at *X*, *W* sites).
4. ${}^{X}(Ca)_{19}{}^{Y}(Al_{7.08}Mg_{3.78}Fe^{2+}{}_{1.95}Ti_{0.28}Mn_{0.05})_{\Sigma 13.14}{}^{T}(B)_{0.75}$ ${}^{Z}(Si_{16.29}[H_4O_4]_{1.71})_{\Sigma 18}$ $O_{69}([OH]_{8.84}Cl_{0.04})_{\Sigma 8.88}$ (charge disbalance +0.71).

For most VSMs (~75% of all samples), the way of formula calculation (the calculation basis) is not important: the formulae obtained by any method agree well with each other. However, based on the above data, the most universal calculation formula method is based on the normalizing on 19 cations of the *X*-group. Due to the significant boron content, which occupies additional structural positions, the calculation on 50 cations turns out to be incorrect. The calculation on 18 silicon atoms is also useless because of the deficiency of Si in "hydrovesuvianite". Different Fe^{2+}/Fe^{3+} ratios and/or the incorporation of aluminum in the *T* and *Z* positions makes the calculation based on 13 cations of the *Y*-group wrong.

According to the results of chemical analyses, the contents of the most variable components range as follows: REE_2O_3 0–20.57 wt%, FeO 0.18–6.23 wt%, MnO 0–14.34 wt%, MgO 0.57–5.21 wt%, SiO_2 33.45–37.43 wt%, TiO_2 0–5.85, and F 0–2.66 wt%. Variations in the VSMs composition and the description of isomorphic substitutions schemes are given in detail in Chaps. 1 and 4. Like the minerals of the garnet supergroup, the VSMs are good indicators of crystallization conditions (Gnos and Armbruster 2006).

Vesuvianite samples often possess concentric (in particular, oscillatory) or sectorial zonality in relation to such components as Cr, Fe, Al, Mn, Cu, Ti and Mg.

Table 3.2 The chemical composition of vesuvianite samples (major components) in comparison with the composition of the vesuvianite from Vesuvius, after (Balassone et al. 2011)

Sample	1	2	4	5	7	8
SiO_2	36.50	36.33	35.95	36.20	36.18	35.94
TiO_2	–	2.36	4.12	1.18	–	1.35
Al_2O_3	18.17	16.94	14.02	17.23	17.15	17.05
FeO	1.06	4.80	4.25	3.34	6.71	4.41
MnO	0.64	0.21	0.19	0.07	–	–
MgO	2.34	1.33	2.47	1.79	0.57	1.34
CaO	35.77	34.59	35.65	35.91	35.35	35.81
Na_2O	–	0.67	–	–	0.24	–
B_2O_3	–	–	–	–	–	–
Cr_2O_3	–	–	–	–	–	–
CuO	0.78	–	–	–	–	–
Ce_2O_3	–	–	–	–	–	–
K_2O	–	–	–	–	–	–
SO_3	–	–	–	–	–	–
BeO	–	–	–	–	–	–
H_2O*	1.46	1.80	2.71	1.80	1.66	1.78
F	2.66	1.95	–	1.97	2.16	1.98
Cl	–	–	–	–	0.13	–
–O≡F	1.12	0.82		0.83	0.91	0.84
–O≡Cl					0.03	
Total	98.26	100.16	99.36	98.66	99.21	98.82
apfu						
Si	18.10	18.00	17.88	17.88	17.93	17.87
Ti		0.88	1.54	0.44		0.50
Al	10.62	9.89	8.22	10.03	10.02	9.95
Fe	0.44	1.99	1.73	1.38	2.78	1.83
Mn	0.27	0.09	0.08	0.03		
Mg	1.73	0.98	1.83	1.32	0.42	0.99
Ca	19.00	18.36	19.00	19.00	18.77	19.00
Na		0.64			0.23	
Cu	0.32					
H	4.83	5.94	9.00	5.93	5.50	5.90

(continued)

Table 3.2 (continued)

Sample	1	2	4	5	7	8
Cl					0.11	
F	4.17	3.06		3.07	3.39	3.10

1—Ostebo, Norway (sample 67940 FM), 2—Sakharjok, Keivy (sample 50578 FM), 4—Akhmatovskaya pit, Russia (sample 28975 FM), 5—Rila Mt., Bulgaria (sample 73599 FM), 7—Ulug-tanzek, Tuva (V. Yu. Karpenko's sample), 8—Eger, Czech (sample 15978 FM)

* H_2O content calculated based on F + Cl + OH = 9 for samples without Si-deficite and takes into account of $SiO_4 \leftrightarrow H_4O_4$ substitution scheme for samples with significant degree (>0.5 *apfu*) of Si-deficite

** H_2O content measured by TG with heating up to 1200 °C

Sample	9	11	12	13	25	26
SiO_2	36.54	36.72	36.81	35.86	36.94	36.37
TiO_2	–	–	–	–	0.68	1.77
Al_2O_3	16.80	15.84	18.09	18.13	16.81	16.93
FeO	3.48	4.63	3.84	4.77	3.40	3.52
MnO	–	–	–	0.05	–	0.38
MgO	2.74	2.74	1.18	0.71	2.25	1.53
CaO	36.38	36.70	35.81	36.26	34.71	35.07
Na_2O	–	–	–	–	0.69	0.37
B_2O_3	–	–	–	–	–	–
Cr_2O_3	–	–	–	–	–	–
CuO	–	–	–	–	–	–
Ce_2O_3	–	–	–	–	–	–
K_2O	–	–	–	–	0.26	–
SO_3	–	–	–	–	–	–
BeO	–	–	–	–	–	–
H_2O*	2.04	2.79	1.75	2.76	1.56	1.80
F	1.34	–	2.05	–	2.42	1.95
Cl	0.38	–	–	–	0.18	–
–O≡F	0.57		0.86		1.02	0.82
–O≡Cl	0.09				0.04	
Sum	99.04	99.42	98.67	98.54	98.84	98.87
apfu						
Si	17.81	17.74	18.23	17.54	18.07	18.04
Ti					0.25	0.66
Al	9.65	9.02	10.56	10.45	9.69	9.90
Fe	1.42	1.87	1.59	1.95	1.39	1.46
Mn				0.02		0.16

(continued)

Table 3.2 (continued)

Sample	9	11	12	13	25	26
Mg	1.99	1.97	0.87	0.71	1.64	1.13
Ca	19.00	19.00	19.00	19.00	18.19	18.64
Na					0.65	0.36
K					0.16	
H	6.62	9.00	5.79	9.00	5.90	5.94
F	2.07		3.21		3.75	3.06
Cl	0.31				0.15	

9—Ak-Chagyl, Kazakhstan (sample 38476 FM), 11—Khewra, Pakistan (sample 65273 FM), 12—Frag el Ma, Marocco (sample 70024 FM), 13—Barbotova Yama, Russia (sample. 23700 FM), 25—Templeton, Canada (sample 31413 FM), 26—Koytezek, Tajikistan A. A. Agakhanov's sample)

* H_2O content calculated based on F + Cl + OH = 9 for samples without Si-deficite and takes into account of $SiO_4 \leftrightarrow H_4O_4$ substitution scheme for samples with significant degree (>0.5 *apfu*) of Si-deficite

** H_2O content measured by TG with heating up to 1200 °C

Dash—content below detection limit

Sample	28	29	30	31	32	33
SiO_2	36.46	37.43	34.60	36.34	35.65	34.82
TiO_2	1.28	–	5.77	1.33	2.32	0.14
Al_2O_3	17.35	17.51	13.31	16.66	16.60	14.59
FeO	3.46	1.67	5.84	4.99	4.72	3.08
MnO	–	0.05	–	0.31	0.21	0.15
MgO	1.23	3.29	0.83	1.43	1.29	4.77
CaO	35.68	35.32	33.18	35.00	34.10	36.79
Na_2O	–	–	0.90	0.35	0.41	
B_2O_3	–	–	–	–	–	1.30
CuO	–	–	–	–	–	–
SO_3	–	1.38	–	–	–	–
BeO	–	–	–	–	–	–
H_2O*	1.88	2.69	2.26	1.47	1.74	4.30**
F	1.75	–	0.83	2.61	1.91	
Cl	–	–	–	–	–	0.26
–O≡F	0.74		0.35	1.10	0.81	
–O≡Cl						0.06
Sum	98.35	99.34	97.17	99.39	98.14	100.14
apfu						
Si	18.12	18.79	17.63	18.08	18.15	16.78
Ti	0.48		2.21	0.50	0.89	0.05

(continued)

Table 3.2 (continued)

Sample	28	29	30	31	32	33
Al	10.16	10.36	7.99	9.77	9.96	8.29
Fe	1.44	0.70	2.49	2.08	2.01	1.24
Mn		0.02		0.13	0.09	0.06
Mg	0.91	2.46	0.63	1.06	0.98	3.43
Ca	19.00	19.00	18.11	18.66	18.60	19.00
Na			0.89	0.34	0.40	
B						1.08
S		0.52				
H	6.25	9.00	7.67	4.89	5.92	13.80
F	2.75		1.33	4.11	3.08	
Cl						0.17

28—Hazlov, Czech (sample 61902 FM), 29—Manzoni, Italy (sample 7359 FM), 30—Alchuri, Pakistan (N. V. Chukanov's sample), 31—West Keyvi (V.N. Yakovenchuk's sample), 32—Sakharjok, Keivy (sample 976 BM), 33—epitaxial crust of vesuvianite from Wiluy, Russia (E. V. Galuskin's sample)

* H_2O content calculated based on F + Cl + OH = 9 for samples without Si-deficite and takes into account of $SiO_4 \leftrightarrow H_4O_4$ substitution scheme for samples with significant degree (>0.5 *apfu*) of Si-deficite

** H_2O content measured by TG with heating up to 1200 °C

Dash—content below detection limit

Sample	34	35	36	38	39	40
SiO_2	36.37	36.56	37.13	36.67	36.70	36.89
TiO_2	0.19	2.22	–	0.45	–	–
Al_2O_3	16.31	16.49	16.15	15.99	17.54	16.03
FeO	4.44	2.48	2.10	4.00	3.19	4.50
MnO	0.54	0.29	–	0.38	–	0.55
MgO	2.50	2.45	3.12	3.24	2.21	2.54
CaO	35.55	36.04	37.10	35.55	36.66	36.06
Na_2O	–	–	–	–	–	–
B_2O_3	–	–	–	–	–	–
Cr_2O_3	–	–	1.38	–	–	–
CuO	–	–	–	–	–	–
Ce_2O_3	–	–	–	–	–	–
K_2O	–	–	–	–	–	–
SO_3	–	–	–	–	–	–
BeO	–	–	–	–	–	–
H_2O*	1.72	2.74	2.82	2.11	2.79	2.74
F	2.02	–	–	1.24	–	–

(continued)

Table 3.2 (continued)

Sample	34	35	36	38	39	40
Cl	–	–	–	0.05	–	–
–O≡F	0.85			0.52		
–O≡Cl				0.01		
Sum	98.79	99.27	99.8	99.15	99.09	99.31
apfu						
Si	18.14	17.99	17.75	18.29	17.75	18.14
Ti	0.07	0.82		0.17		
Al	9.59	9.56	9.10	9.40	10.00	9.59
Fe	1.85	1.02	0.84	1.67	1.29	1.85
Mn	0.23	0.12		0.16		0.23
Mg	1.86	1.80	2.22	2.41	1.59	1.86
Ca	19.00	19.00	19.00	19.00	19.00	19.00
Cr			0.52			
H	5.71	9.00	9.00	7.01	9.00	9.00
F	3.19			1.95		
Cl				0.04		

34—Vesuvius, Italy (sample 5228 FM), 35—Stara Kamienica, Poland (E. Szełeg's sample), 36—Kluchevskii mine, Russia (sample 27952 FM), 38—Vesuvius, Italy (sample 27844 FM), 39—Karmankul'skii cordon, Russia (sample 21155 FM), 40—Bazhenovskoe, Russia (sample 79479 FM)

* H_2O content calculated based on F + Cl + OH = 9 for samples without Si-deficite and takes into account of $SiO_4 \leftrightarrow H_4O_4$ substitution scheme for samples with significant degree (>0.5 *apfu*) of Si-deficite

** H_2O content measured by TG with heating up to 1200 °C

Dash—content below detection limit

Sample	41	42	43	44	45
SiO_2	36.54	36.67	36.13	34.32	33.45
TiO_2	2.21	–	0.40	0.71	0.76
Al_2O_3	16.48	13.83	16.14	12.19	12.34
FeO	2.48	6.23	4.21	4.63	4.79
MnO	0.29	0.15	0.45	0.05	0.12
MgO	2.45	3.37	2.50	4.82	5.21
CaO	36.02	36.63	35.48	36.29	36.41
Na_2O	–	–	0.09	–	–
B_2O_3	–	–	0.12	1.03	0.89
Cr_2O_3	–	–	–	–	–
CuO	–	–	–	–	–
Ce_2O_3	–	–	0.11	–	–

(continued)

Table 3.2 (continued)

Sample	41	42	43	44	45
SO_3	–	–	–	–	–
BeO	–	–	0.06	–	–
H_2O*	1.81	2.79	2.00	4.39	4.88
F	1.73	–	1.71	–	–
Cl	0.44	–	–	0.06	0.05
–O≡F	0.73	–	–	–	–
–O≡Cl	0.10	–	0.72	0.01	0.01
Sum	99.62	99.67	96.68	98.48	98.89
apfu					
Si	17.99	17.75	17.95	16.77	16.29
Ti	0.82		0.15	0.26	0.28
Al	9.56	7.89	9.45	7.02	7.08
Fe	1.02	2.52	1.75	1.89	1.95
Mn	0.12	0.06	0.19	0.05	0.05
Mg	1.80	2.43	1.85	3.51	3.78
Ca	19.00	19.00	18.89	19.00	19.00
Na			0.09		
B			0.10	0.87	0.75
Ce			0.02		
Be			0.07		
H	5.94	9.00	6.32	14.32	15.84
F	2.70		2.68		
Cl	0.36			0.05	0.09

41—Hessen, Germany (sample 51056 FM), 42—Kovdor (V. N. Yakovenchuk's sample), 43—Vesuvius, Italy after (Balassone et al. 2011), 44, 45—epitaxial crusts, Wiluy River basin, Russia (G. V. Barkhudarova's sample)

* H_2O content calculated based on F + Cl + OH = 9 for samples without Si-deficite and takes into account of $SiO_4 \leftrightarrow H_4O_4$ substitution scheme for samples with significant degree (>0.5 *apfu*) of Si-deficite

** H_2O content measured by TG with heating up to 1200 °C

Concentric zonality is usually due to changes in the concentration of these elements during crystallization, but crystal growth by the chemical wave mechanism may be an alternative cause of oscillatory zonality (Kapral and Showalter 1995). The main cause of the sectorial zonality is different ability of different faces of the growing crystal to selective incorporation of impurities. The epitaxial Si-deficient vesuvianite is characterized by the sponge morphology of agregates (Fig. 3.3d). Sometimes, clinochlore overgrowth on vesuvianite is observed and vice versa (Fig. 3.3f). The chemical composition of vesuvianite from its type locality is close to the average

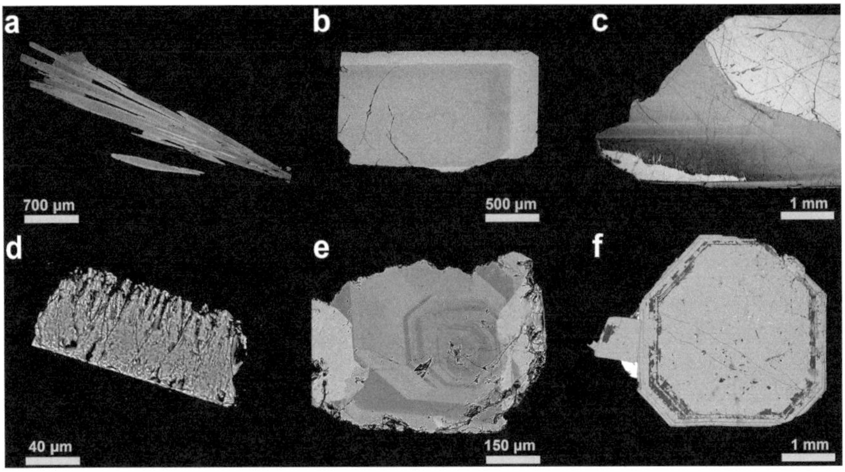

Fig. 3.3 BSE images of a split zonal VSM crystal from Tas–Khayakhtakh, Russia (**a**, E. V. Galuskin's sample), vesuvianite crystal from the Akhmatovskaya pit, Russia (**b**, E. V. Galuskin's sample), Sludyanka, Russa (**c**, L. Z. Reznitskiy's sample) sponge-like Si-deficient vesuvianite crystals from Wiluy, Russia (**d**, E. V. Galuskin's sample), zonal vesuvianite crystal from Kluchevskiy mine, Russia (**e**, sample 27952 FM) and clinochlore overgrowth on vesuvianite prism faces from Gukunvara, Finland (**f**, sample 16908 FM)

vesuvianite composition. Sometimes, mechanical impurities can distort analytical data. In this reference, the CuO content close to 1 wt% in the sample from an unusual association of the Ak-Chagyl deposit, Kazakhstan (Fig. 3.4) should be noted: the measured Cu impurity originates from thin films of chrysocolla and malachite along the cracks between individuals of the radial vesuvianite aggregates. However, vesuvianite sectors without cracks didn't contain Cu admixture.

3.1.4 Thermogravimetry and Differential Scanning Calorimetry

The data of thermogravimetry (TG) and differential scanning calorimetry (DSC) of vesuvianite samples from Somma-Vesuvius, Italy (Fig. 3.5b) and Stara Kamienica, Poland (Fig. 3.5c) and Si-deficient "hydrovesuvianite" sample from the Wiluy River basin, Russia (Fig. 3.5a) demonstrate significant differences in the H_2O content (1.60, 2.74 and 4.30 wt%, respectively) associated with additional hydrogen sites in the "hydrovesuvianite" samples. These hydrogen atoms incorporate into the vesuvianite structure in accordance with the substitution scheme $SiO_4^{4-} \leftrightarrow H_4O_4^{4-}$ (Galuskin et al. 2003b). The slightly reduced H_2O content for vesuvianite from Somma-Vesuvius may be due to a partial substitution of OH^- for F^- (Groat et al. 1992a).

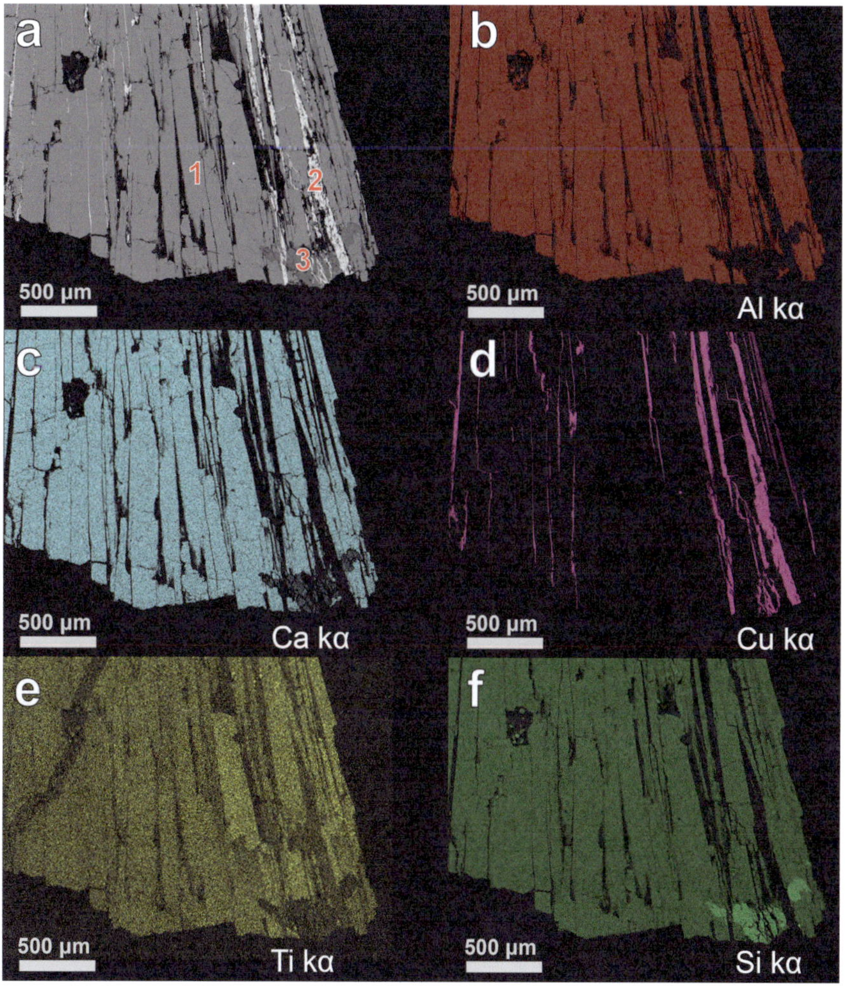

Fig. 3.4 Images of radial aggregates of vesuvianite from the Ak-Chagyl skarn deposit, Kazakhstan (sample 38476 MΦ) in the BSE mode (**a**) and characteristic X-ray radiation of Al *K*α (**b**), Ca *K*α (**c**), Cu *K*α (**d**), Ti *K*α (**e**) and Si *K*α (F). 1—vesuvianite, 2—malachite, 3—diopside

According to Żabiński et al. (1996) and Földvári (2011), single-step dehydroxylation of the low-temperature and high-temperature VSMs accompanied by water evolution occurs in the temperature ranges of 820–1090 °C and 990–1150 °C, respectively. The total mass loss for vesuvianite reported in the literature does not exceed the theoretical content of 9 OH^- groups per formula unit.

Among the samples studied in this work, Si-deficient B-bearing vesuvianite forming epitaxy on wiluite (Fig. 3.3d) demonstrated anomalous two-stage mass loss. The first stage of mass loss (2.08 wt%, Fig. 3.5a) observed in the temperature range

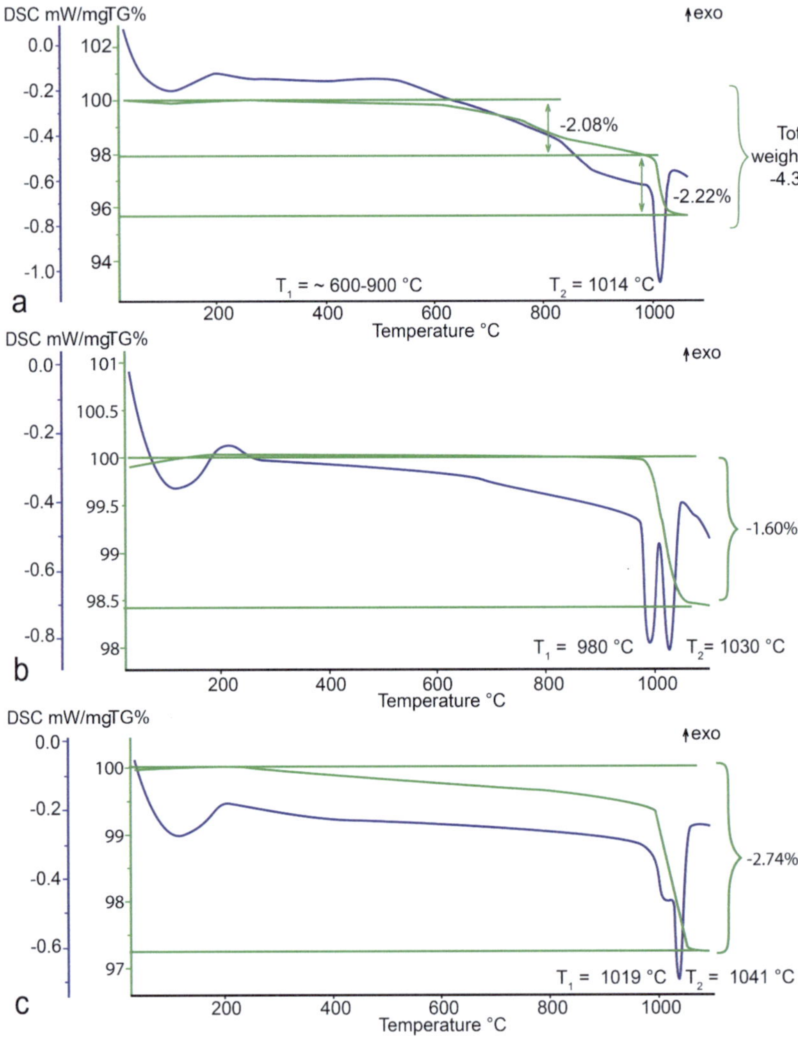

Fig. 3.5 DSC and TG curves (shown in blue and green, respectively) for "hydrovesuvianite" from the Wiluy River basin (**a**, E. V. Galuskin's sample), vesuvianite from Vesuvius (**b**, sample 27844 FM) and vesuvianite from Stara Kamienica (**c**, E. Szełeg's sample). Unlike ordinary vesuvianite (mass loss ~2.5%), "hydrovesuvianite" demonstrates two weight loss steps. The endothermic peak in the range of 20–150 °C is due to the fact gas density changes nonlinearly with heating

600–900 °C corresponds to the loss of the $(H_4O_4)^{4-}$ clusters similarly to that previously described for synthetic katoite (Rivas-Mercury et al. 2008). The second stage of mass loss (2.22 wt%) and endothermic peak with a minimum at the temperature of 1014 °C correspond to the loss of OH^- groups from the OH11 and OH10 sites (Żabiński et al. 1996). The subsequent melting of the mineral (after preliminary cooling to room temperature) leads to the formation of a mixture of grossular, wollastonite, quartz and a glassy phase.

Single-step weight loss (in the temperature range 980–1050 °C) was observed for vesuvianite samples with low H_2O contents of 1.60 and 2.74 wt% (Fig. 3.5b, c) which correspond to the OH^- groups (at the sites OH11 and OH10) whose total content does not exceed 9 groups per formula unit. The presence of two endothermic peaks in this area is likely due to the separate dehydroxylation processes involving the OH10 and OH11 sites (Panikorovskii et al. 2017e).

3.1.5 Infrared Spectroscopy

Infrared (IR) spectra of VSMs (Fig. 3.6) contain bands of O–H, Si–O and B–O stretching vibrations, Si–O–Si, O–Si–O, O–B–O and *M*···O–H bending modes as well as lattice modes involving vibrations of the *M*···O and *M*···F ionic bonds (where *M* are *Y*-cations) and translations and rotations of silicate and borate groups as a whole (Chukanov et al. 2018a; Chukanov and Vigasina 2020).

Bands in the ranges 1200–1570 and 1060–1200 cm^{-1} correspond to stretching modes of the BO_3^{3-} and BO_4^{5-} groups, respectively. The ranges of 740–840 and 840–1040 cm^{-1} correspond to *M*···O–H bending and Si–O stretching vibrations, respectively. In the region with wave numbers below 740 cm^{-1}, bending vibrations of silicate and borate groups, as well as (below 500 cm^{-1}) *M*···O stretching vibrations, appear. Much stronger bands of silicate groups mask the weak bands of bending vibrations of the BO_3 group in the range of 650–900 cm^{-1}.

The following empirical correlations between O–H stretching frequencies in IR spectra of minerals and O···O and H···O distances (from structural data) were established by Libowitzky (1999):

$$\nu\left(\mathrm{cm}^{-1}\right) = 3592 - 304 \cdot 10^{9} \cdot \exp[-\mathrm{d}(\mathrm{O}\cdots\mathrm{O})/0.1321]$$

$$\nu\left(\mathrm{cm}^{-1}\right) = 3632 - 1.79 \cdot 10^{6} \cdot \exp[-\mathrm{d}(\mathrm{H}\cdots\mathrm{O})/0.2146]$$

During two decades, this publication was of a great importance because it emphasized the existence such correlations as a general trend. However, later it became obvious that the Libowitzky's equations are a very rough approximation and have a restricted applicability. First, it is to be noted that at high frequencies (above 3500 cm^{-1}) substantial deviations from the Libowitzky correlations are common because O–H stretching frequencies depend not only on O···O and H···O distances, but also on the nature of cations coordinating O–H groups and H_2O molecules, as well as on the O–H···O angle, and the influence of these factors becomes most evident in case of weak hydrogen bonds. The Libowitzky equations predict that maximum possible values of O–H stretching frequencies for minerals are 3592 and

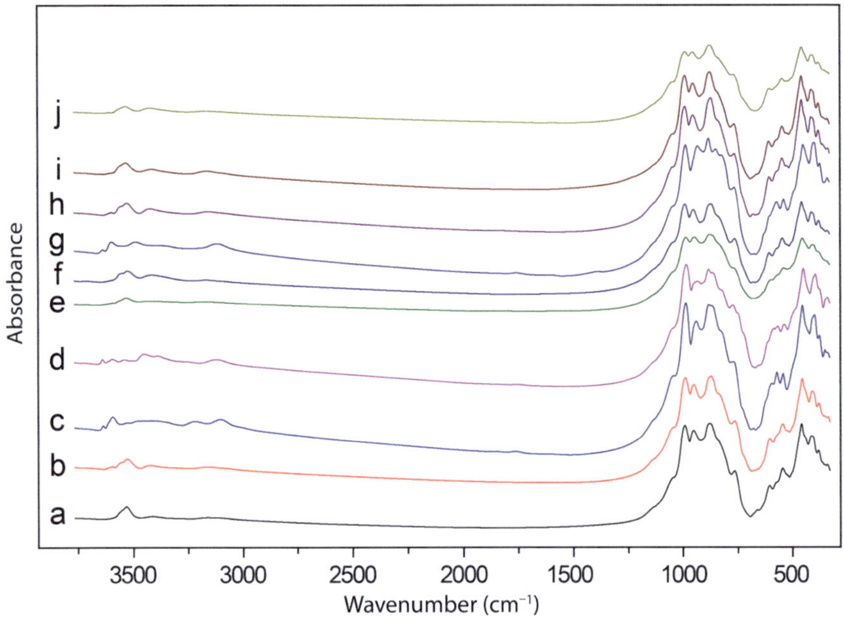

Fig. 3.6 Infrared spectra of vesuvianite from Østebø, Vest-Agder, Norway (**a**, sample 67940 FM), Eger, Karlovy Vary Region, Czech Republic (**b**, sample 15978 FM), Barbotova Yama, South Urals, Russia (**c**, sample 23700 FM), Akhmatovskaya pit, South Urals, Russia (**d**, sample 28975 FM), Templeton, Québec, Canada (**e**, sample 31413 FM), Hazlov, Karlovy Vary Region, Czech Republic (**f**, sample 61902 FM), Khewra, Punjab Region, Pakistan (**g**, sample 65273 FM), Frag el Ma, Marrakech-Safi Region, Morocco (**h**, sample 70024 FM), Rila, Urdini Lakes, Bulgaria (**i**, sample 73599 FM), Western Keivy, Kola Peninsula, Russia (J, sample 976 BM)

3632 cm^{-1} respectively. However, in many minerals including for magnesium serpentines, brucite, kaolinite, amphiboles etc. observed frequencies are much higher and even can exceed 3700 cm^{-1}, and the VSMs are not an exception.

The VSMs infrared spectra in the O–H stretching vibration region 13 characteristic absorption bands (denoted by letters from A to M) can be distinguished. Their wavenumbers are given in the Table 3.3. These designations were used in many publications on the VSMs. Bands A–H have predominant polarization along the fourfold axis ($E \wedge c < 35°$), I band is polarized perpendicular to the *c* axis, and J–M bands are polarized strictly along *c* (Groat et al. 1995b; Bellatreccia et al. 2005b). The A–H bands can be attributed to the stretching vibrations of the O11–H1 bond, and the J–M bands are related to the vibrations of the O10–H2 bond. There is no satisfactory assignment for the I band within the framework of the available structural data. In fact, in the IR spectra of the samples studied in this work, the position of the similar band varies from 3217 to 3300 cm^{-1}, and this band is not observed in most spectra. One possible explanation for the anomalous polarization of the I band is its assignment to the SiO–H bands. The silanol group usually appears as a series of bands at

Table 3.3 Absorption bands corresponding to O–H vibrations (Groat et al. 1995b)

Band designation	Wavenumber (cm^{-1})	Assignment
A	3670	O(11)H1
B	3635	(O11)H1
C	3596	(O11)H1
D	3567	(O11)H1
E	3524	(O11)H1
F	3487	(O11)H1
G	3430	(O11)H1
H	3383	(O11)H1
I	3240	OH
J	3210	$O(10)H_2$
K	3156	$O(10)H_2$
L	3120	$O(10)H_2$
M	3054	$O(10)H_2$

lower frequencies. Another explanation involves variants with $(OH)_4$, $B(OH)_3$ and $B(OH)_4$ groups.

Assignment of the A–M bands and their attachment to certain wavenumbers is an idealization: in fact, the positions of the bands observed in the IR spectra substantially deviate from corresponding "ideal" values. Moreover, in the spectra of many VSMs samples there are bands that fall out of this classification. In particular, in the IR spectra of more than a half of the samples studied, there is a band in the range of 3440–3470 cm^{-1} which occupies an intermediate position between the F and G bands and can be assigned to adsorbed water.

Comparing the IR spectra with the data on chemical composition shows that there is a distinct relation between the intensity of this band and the content of titanium in the mineral (Fig. 3.7). Thus, with a high probability, the bands at 3440–3470 cm^{-1} are related to vibrations of the O11–H1 bond coordinated by Ti. From this pattern falls a high-titanium sample 28975, in the IR spectrum of which, instead of one band in the interval 3440–3470 cm^{-1}, there is a strong F band, according to the nomenclature by Groat et al. (1995b), with a maximum at 3488 cm^{-1} and a shoulder at 3460 cm^{-1} as well as a weaker band at 3424 cm^{-1}.

According to Groat et al. (1995b), the F band is present in the IR spectra of all samples with the titanium content higher than 0.35 *apfu* (the samples V4, V5, V6, V28, V30, V31, V38, and V45), whereas in the spectra of most low-titanium samples this band is not observed or has a low intensity.

Sample V49 is an exception to this regularity: a strong F-band is present in its spectrum, whereas the content of titanium is below its detection limit. The possible explanation of this anomaly is as follows. Among VSMs studied in this work, the sample 28975 FM is the only titanium-containing vesuvianite with space group $P4/n$ (Table A.4). Besides ordering of cations in the structural channels, cation ordering is also observed in the octahedral *Y*3A,B sites coordinated by the O11–H1(11) hydroxyl

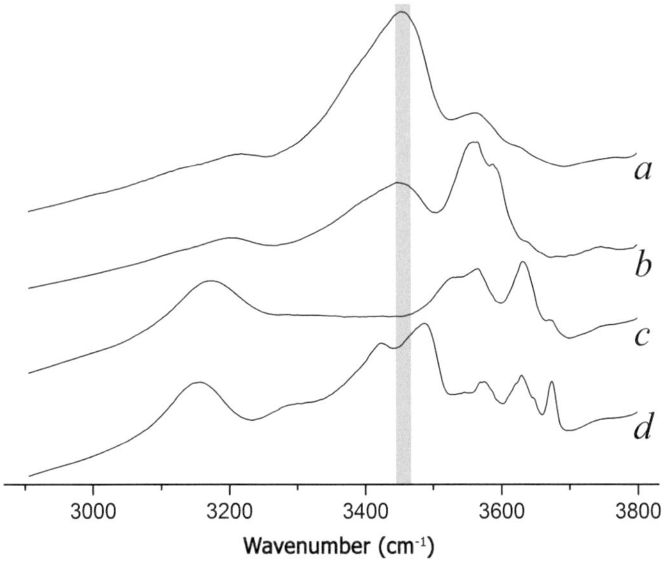

Fig. 3.7 The O–H stretching region of VSMs IR spectra with various Ti content: **a**, sample 7086 FM with $Ti_{2.21}$; **b**, sample 61902 FM with $Ti_{0.48}$; **c**, sample 60211 FM with $Ti_{0.00}$, and **d**, sample 28975 FM with $Ti_{1.54}$ (anomalous high-titanium sample with *P*4/*n* symmetry). The area of the F band is highlighted in gray

groups (Fig. 3.8). The high degree of ordering is confirmed by different refined occupancies of the *Y*3A and *Y*3B octahedral sites, which are $[(Al, Mg)_{0.87}(Ti, Fe)_{0.13}]_{1.00}$ and $[(Al, Mg)_{0.78}(Ti, Fe)_{0.22}]_{1.00}$, respectively. One more confirmation of the cation ordering is the difference in the average lengths of metal–oxygen bonds in the *Y*3 octahedra, which are equal to 1.936 and 2.005 Å for the <*Y*3A–O > and < *Y*3B–O> bonds, respectively. The nonequivalence of the sizes of the *Y*3A and *Y*3B octahedra, which are coordinated by hydroxyl groups, leads to differences in the relative positions of the protons (angles for the strongest hydrogen bonds O11A–H11A···O11B and O11B–H11B···O11A differ by 15°). This fact ultimately, causes splitting of bands related to the stretching vibrations of these groups, including those in the region of 3440–3470 cm^{-1}. Probably, the appearance of the F band in the IR spectra of high-titanium vesuvianite given in Groat et al. (1995b) is caused by changes in the hydrogen-bonding system in the presence of titanium at the *Y*3 sites.

Most of the low-titanium VSMs studied in this work have *P*4/*n* symmetry (Table A.4). Among them, there are samples of cyprine with a high content of Cu and Mn^{3+}. In their IR spectra (and only in these spectra), there is a band at 3375–3380 cm^{-1}, close to the H band. In the spectrum of the sample from N'Chwaning, RSA this band is stronger than in the spectrum of the sample from Wessels, RSA with a relatively lower content of Mn^{3+} and a higher content of Cu. Considering these data, the band at 3375–3380 cm^{-1} is logically attributed to vibrations of the O11–H1 group, coordinated by Mn^{3+}.

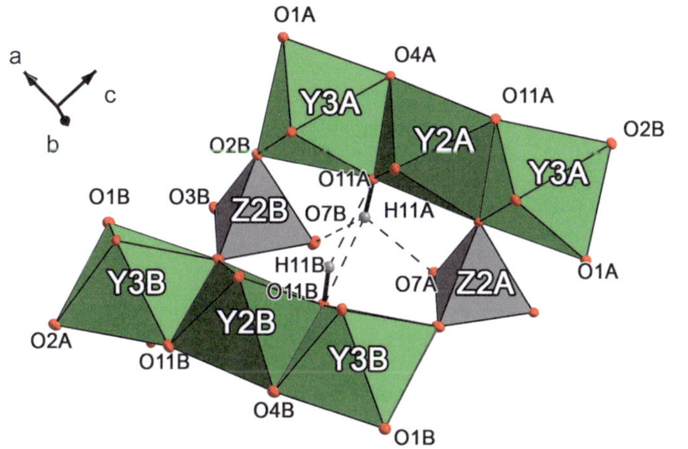

Fig. 3.8 The configuration of hydrogen bonds formed by the O11–H11 group in the crystal structure of the sample 28975 FM. Thermal ellipsoids are drawn at the 50% probability level

The position of the D band varies from 3560 to 3567 cm^{-1}, and its intensity increases with increasing fluorine content. This band has the highest intensity in the IR spectra of fluorvesuvianite and F-enriched vesuvianite (the samples 73599, 15978, 27844, 70024, 61902, 67940, and 50578: see Galuskin et al. (2003a), Britvin et al. (2003); Fig. 3.9). At the same time, in the IR spectra of samples that practically do not contain fluorine (the sample 65273 and holotype samples of alumovesuvianite and magnesiovesuvianite), the D band is absent. Thus, taking into account the polarization, this band can be attributed to the of the O11–H1 bond vibrations in the presence of fluorine in the adjacent position O11.

It was suggested that the J band (at 3203 cm^{-1}) is also characteristic of high-fluorine vesuvianite and refers to OH groups at the O10 site coordinated by Fe in *Y*1 (Galuskin et al. 2003b). However, this assumption was not supported by our observations. The J band is quite distinct and often has a relatively high intensity in the IR spectra of most VSMs, including a number of F-free, low-F and Fe-deficient samples (for example, the samples 41304 and 27728, as well as alumovesuvianite and magnesiovesuvianite), and its wavenumber varies significantly from sample to sample (in the range from 3190 to 3225 cm^{-1}). The position of the J band corresponds to a strong hydrogen bond, which may be in disagreement with the weak H-bond acceptor, which is the F^- ion. However, with due regard of the polarization (Groat et al. 1995b), the assignment of the J band to OH groups at the O10 site is beyond doubt. Variations in the position of this band, as well as the presence of several K–M bands in the IR spectra, are most likely associated with different sets of cations that coordinate the O10 position.

The BO_3 groups in VSMs are connected with the $Y1O6_4O10$ polyhedra via O10 oxygen atom to form the cluster $T2Y1O_7$ (Fig. 3.10) where $T2 = B$ and $Y1 = Fe^{3+}$, Fe^{2+}, Mn^{3+}, Cu^{2+}, Al, or Mg. As a result, four degrees of freedom corresponding

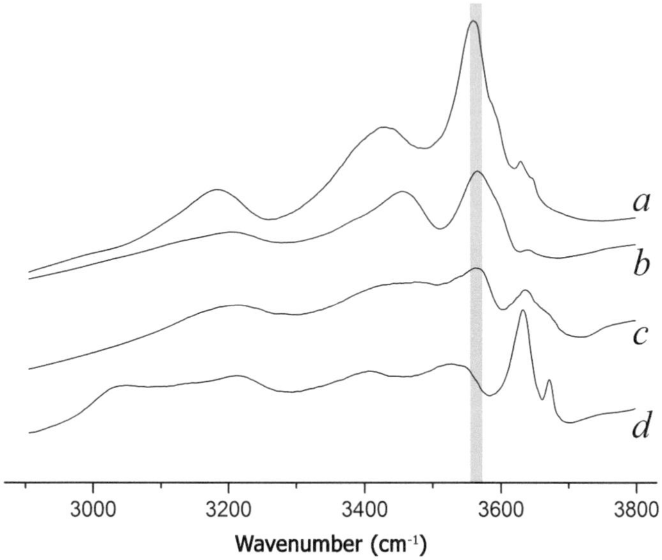

Fig. 3.9 IR spectra in the region of O–H stretching vibrations of VSM samples with different fluorine content: fluorvesuvianite (holotype) N. V. Chukanov's sample (**a**), sample 50578 FM with $F_{7.16}$ (**b**), $F_{3.06}$; sample 43180 FM with $F_{3.06}$ (**c**), $F_{0.24}$; alumovesuvianite holotype with $F_{0.024}$ (N. V. Chukanov's sample). The area of the D band is highlighted in gray

to the bond lengths *T*2–O12 (×2), *T*2–O10, and *Y*1–O10 are involved in stretching vibrations of BO_3^{3-}. This results in four nondegenerate modes and, consequently, the expected number of absorption bands in the region of stretching vibrations of BO_3 groups (i.e. 1200–1570 cm^{-1}) is equal to 4. However, in the IR spectrum of wiluite only three bands are observed in this region: the peaks at 1267 and 1373 cm^{-1} and the shoulder at 1415 cm^{-1} ((Panikorovskii et al. 2017c), see Fig. 3.11). The fourth band corresponding to symmetric vibrations of BO_3^{3-} is forbidden for a regular BO_3 triangle and is weak in case of weak-distorted BO_3 triangle. The latter case takes place in wiluite: the bond lengths *T*2–O10 and *T*2–O12 are 1.39–1.40 и 1.32 Å, respectively (Panikorovskii et al. 2017c). Weak absorption between 1267 and 1373 cm^{-1} may correspond to the symmetric stretching mode of BO_3^{3-} groups (Fig. 3.11).

Unlike wiluite, most boron-rich VSMs contain significantly distorted BO_3 triangle. For example, in a sample from Gulshad, Kazakhstan the bond lengths *T*2–O10 and *T*2–O12 are equal to 1.384 and 1.20 Å, respectively. As a result, four distinct IR bands (at 1557, 1467, 1419, and 1365 cm^{-1}) are observed in the range 1200–1570 cm^{-1} (curve *b* in Fig. 3.11). As compared to wiluite, these bands are substantially shifted towards high frequencies because of shorter B–O bonds and shortened *Y*1–O bond (2.044 Å for the mineral from Gulshad and 2.15 Å for wiluite: (Panikorovskii et al. 2017c; Chukanov et al. 2018a). These differences may be due to different predominant cations in the *Y*1 site: Mg in wiluite and Fe^{3+} in the sample from Gulshad.

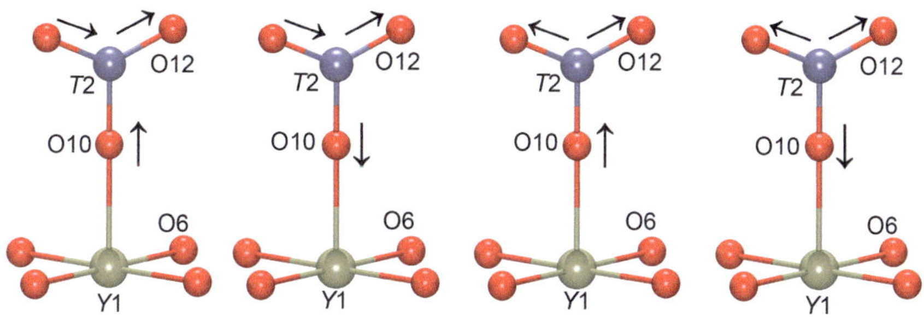

Fig. 3.10 Diagram illustrating shifts of O atoms for different stretching modes of the BO_3 triangle in VSMs

Fig. 3.11 IR spectra of vesuvianite-group minerals in the region of B–O-stretching vibrations: wiluite from its type locality (**a**), B-rich VGM from Gulshad (**b**), a typical B-bearing vesuvianite from Somma-Vesuvius complex, Italy (**c**), and anomalous B-rich VSM from Titivskoe (**d**)

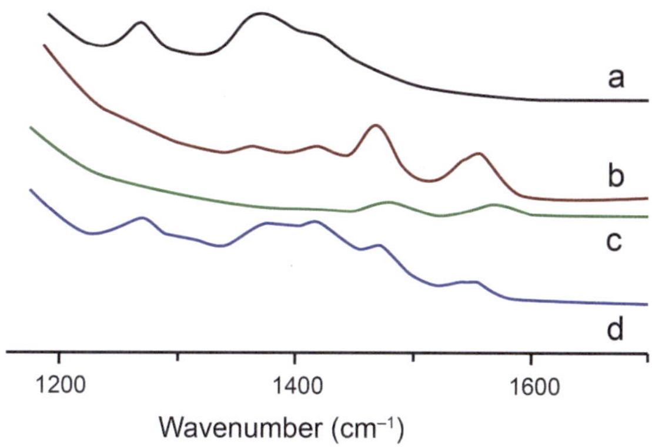

An anomalous IR spectrum with 6 absorption bands in the range 1200–1570 cm^{-1} shows boron-rich VSM from Titivskoe boron deposit, Yakutia, Russia (curve *d* in Fig. 3.11). Additional bands may be related to incorporation of CO_3 into vesuvianite structure, which way incorporation remains unexplored (Groat et al. 1992a).

Earlier, based on the analysis of the IR spectra of a limited VSMs collection, it was concluded that "in the spectra of low vesuvianite, the high-frequency band of ν_3 vibrations is shifted by 10–15 cm^{-1} to lower frequencies compared to the spectra of high-vesuvianite and is either broadened or split intp two closely located peaks with the maxima at 980 and 970 cm^{-1} (Kurazhkovskaya et al. 2005). Our data do not fully support this conclusion. Indeed, in the IR spectra of most low-symmetry vesuvianite with a low boron content, there is an intense band in the interval of 962–968 cm^{-1}. An exception is the sample 23700, for which a similar band is observed at 973 cm^{-1}. In the IR spectra of most highly symmetric vesuvianites, this band shifts to 976–986 cm^{-1}, but there are some exceptions (namely, the samples 9578, 51109, 21370, 5136 and 43180, whose IR spectra contain a strong band at 962–968 cm^{-1}).

Another conclusion that was made in the work by (Kurazhkovskaya et al. 2005) is that the "splitting of the low frequency band of ν_3 vibrations at 915–930 cm^{-1} is ~30 cm^{-1}". In the spectra of "low vesuvianite", a doublet of 925 + 895 cm^{-1} is given, but it is not explained why this doublet is associated with the splitting of the band at 920 cm^{-1}, and not with the appearance of a new band at 895 cm^{-1}. According to the data of the present work, the presence of two bands in this region is possible in the spectra of both high-symmetric and low-symmetric VSMs, and there is no relationship between the positions of these bands and symmetry.

In the works by Kurazhkovskaya et al. (2005), Borovikova and Kurazhkovskaya (2006), the splitting of the ν_3 band with a shoulder at about 900 cm^{-1} and the presence of the doublet of bending vibrations (~575 + 615 cm^{-1}) are the specific features of low-symmetry VSMs. As mentioned above, the first of these proposed regularities is not confirmed on a large group of samples. The second regularity is confirmed by our data as a general trend, but there are some exceptions: a doublet of approximately equal intensity in the region of ~575–615 cm^{-1} is present in the spectra of the high-symmetry VSMs samples 51109, 5136 and 43180.

3.1.6 Solid-State Nuclear Magnetic Resonance

^{1}H solid-state nuclear magnetic resonance. The proton NMR spectra of most of the studied VSMs samples (Fig. 3.12b–e) contain two peak areas in the intervals of 7.60–6.50 and 1.60–0.95 ppm (Table 3.4). The Si-deficient vesuvianite sample studied by Panikorovskii et al. (2016a) (Fig. 3.12a) contains an additional band at 3.38 ppm. The calculation of the O···O distances *d* for the O–H···O hydrogen bonds using the formula δ_{iso}(ppm) = 79.05–0.255·d(Å), proposed by Yesinowski et al. (1988), gives the values of 2.80–2.85 and 3.03–3.06 Å, respectively, and 2.96 Å for the additional band at 3.38 ppm (Fig. 3.12a) in the spectrum of a sample of Si-deficient vesuvianite (Table 3.4).

The chemical shift values for the strongest peak in the range of 7.60–6.50 ppm correspond to the O(11)–H(1*a*)···O(7) fragments with the O···O distances of 2.784(4) Å determined using neutron diffraction in the work by Lager et al. (1999a). The O(11)–H(1*a*)···O(7) hydrogen bond has the O···O distances in the range of 2.74–2.82 Å (Ohkawa et al. 1992; Pavese et al. 1998a; Armbruster and Gnos 2000c; Galuskin 2005; Ohkawa et al. 2009), which may refer to the peak of the same nature.

The band at 3.38 ppm is absent in the spectra of samples with normal silicon contents (~18.0 *apfu*) and can correspond to the hydrogen bonds of the additional H3 position, resulting from the substitution of the hydrogarnet type. A similar peak at 3.40 ppm was noted in the ^{1}H NMR spectrum of synthetic hydrogarnet with the composition $Li_{3.8}H_{3.2}La_3Zr_2O_{12}$ (Larraz et al. 2015). Peaks near 1.60–0.95 (ppm) correspond to the O(11)–H(1)···O(11) links with the O···O distances of 3.01–3.04 Å (Pavese et al. 1998a; Lager et al. 1999a).

^{27}Al solid-state nuclear magnetic resonance. The most clearly distinguished intense peak in the region of 0–8 ppm in the ^{27}Al NMR spectra of VSMs (Fig. 3.13a–e) corresponds to aluminum in octahedral coordination (in the *Y*2A,B and *Y*3A,B

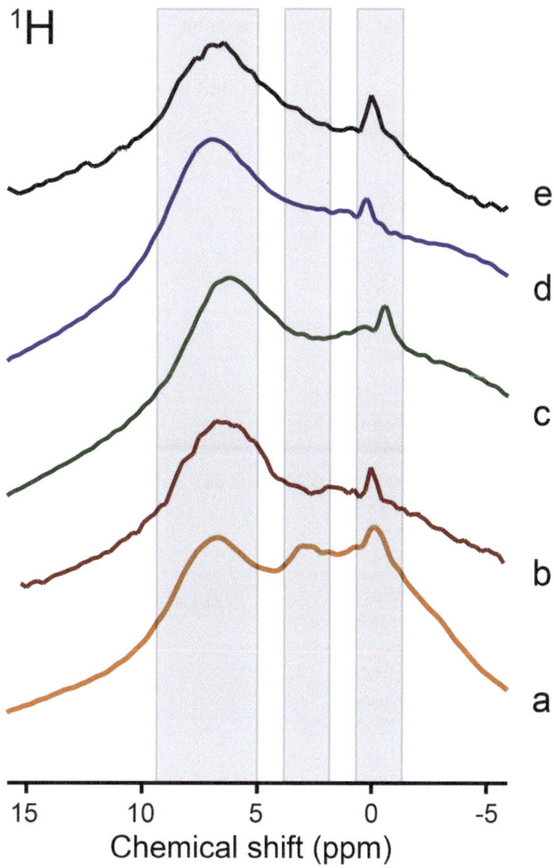

Fig. 3.12 ^{1}H solid-state NMR spectra of epitaxial "hydrovesuvianite" crusts on wiluite from the Wiluy River basin, Russia (**a**, E. V. Galuskin's sample), Somma-Vesuvius, Italy (**b**, sample 27844 FM), and Lupikko, Karelia, Russia (**c**, sample 189/12409 SM), and vesuvianite from Koytezek pass, Tajikistan (**d**, A. A. Agakhanov's sample) and Stara Kamienica, Poland (**e**, E. Szełeg's sample)

positions) (Phillips et al. 1987). The half-width of this asymmetric peak varies considerably, increasing with an increase in the content of other components (Fe, Ti, Mg, Cr) at this site.

According to the works by Olejniczak and Żabiński (1996), Panikorovskii et al. (2017a, d), a weak shoulder at ~40 ppm (Fig. 3.13a) corresponds to Al at the five-coordinated *Y*1A,B sites. Aluminum in tetrahedral coordination is characterized by a signal at 60 ppm (Panikorovskii et al. 2017c) in the ^{27}Al NMR spectrum of a VSM which found in a skarn sample from Pitkäranta, Karelia, Russia (Fig. 3.13c). It should be noted that the presence of five-coordinated Al detected based on NMR data is a characteristic feature of vesuvianite with low-temperature parageneses (primarily, rodingites), while the presence of Al with four-fold coordination is characteristic of a vesuvianite from skarns and metamorphic rocks. Weak symmetrical peaks at ±175 ppm correspond to spinning sidebands.

2D multiquantum study of the ^{27}Al NMR spectrum of vesuvianite (Fig. 3.14) showed the absence of additional peaks which would correspond to nonequivalent octahedral aluminum positions (*Y*2A,B and *Y*3A,B). In the resulting spectrum,

Table 3.4 Chemical shifts of bands in the ^{1}H NMR spectra, Si contents and O···O distances in VSMs

Sample	Chemical shift (ppm)			Content of Si (*apfu*)	O···O distances (Å)		
Stara Kamienica, Poland (E.V. Galuskin's sample)	7.55		1.57	17.99	2.80		3.04
Koytezek pass, Tajikistan (A. A. Agakhanov's sample)	6.55		1.16	18.04	2.84		3.05
Lupikko, Karelia (sample 189/12409 SM)	6.56		1.11	17.98	2.84		3.06
Somma-Vesuvius, Italy (sample 27844 FM)	6.88		1.48	17.81	2.83		3.04
"Hydrovesuvianite", Wiluy, Russia (E. V. Galuskin's sample)	6.74	3.38	0.95	16.79	2.83	2.96	3.06

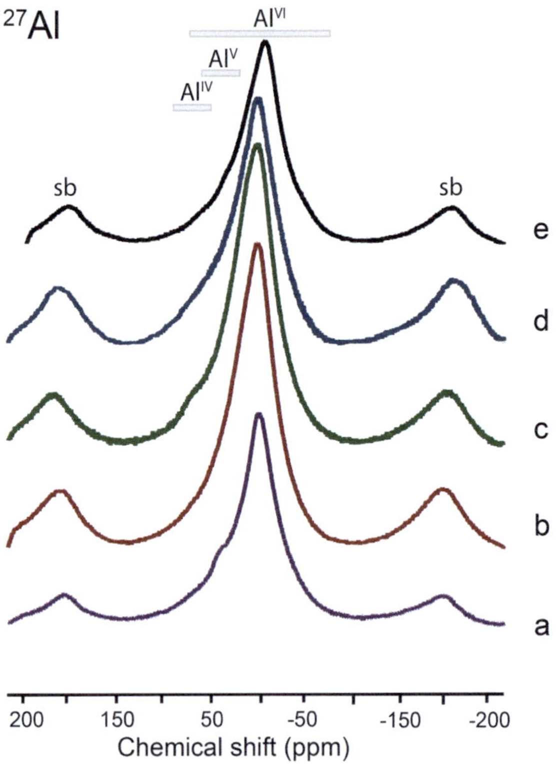

Fig. 3.13 ^{27}Al solid-state NMR spectra of vesuvianite samples from Stara Kamienica, Poland (**a**, E. Szełeg's sample), from Vesuvius volcano (**b**, sample 27844 FM), from Lupikko, Karelia, Russia (**c**, sample 189/12409 SM), from Koytezek pass, Tajikistan (**d**, A. A. Agakhanov's sample), and from the Bazhenovskoe deposit, Urals, Russia (**e**, A. A. Agakhanov's sample); sb are the spinning sidebands

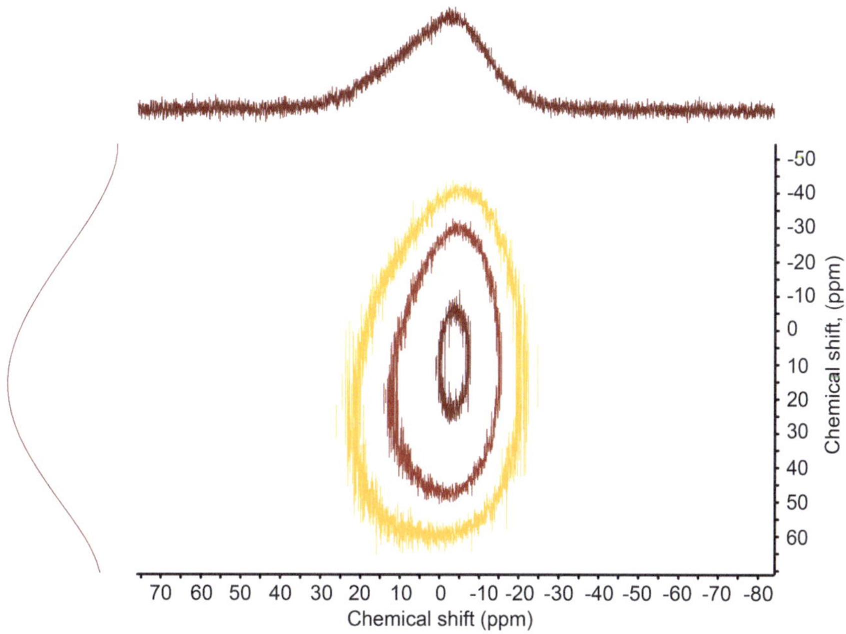

Fig. 3.14 ^{27}Al 2D MAS NMR spectrum of vesuvianite from the Koytezek pass, Tajikistan (D in Fig. 3.13, A. A. Agakhanov's sample)

there is an asymmetrical intense peak with the weighted mean position of 0 ppm, corresponding to aluminum with six-fold coordination.

^{23}Na solid-state nuclear magnetic resonance. The NMR spectrum of the sodium-rich vesuvianite sample 979 BM (Fig. 3.15a) is presented by an asymmetric quadrupole doublet centered at –15.8 ppm, which, according to published data (Xue and Stebbins 1993; Fechtelkord et al. 2001), corresponds to Na in the 8- and ninefold coordination and this is fully consistent with the single-crystal X-ray structural data (Panikorovskii et al. 2016b). In contrast to our case, symmetric Gaussian peaks with a chemical shift of 0.0–2.0 ppm (Zhou et al. 2007) are characteristic of 6-coordinated sodium, whereas sodium in tenfold coordination corresponds to a maximum with the chemical shift of about –25 ppm (Xue and Stebbins 1993).

^{19}F solid-state nuclear magnetic resonance. The ^{19}F NMR spectrum of the F-enriched sample from Lupikko, Karelia, Russia (Fig. 3.15b) is presented by an intense peak with a chemical shift of –109.02 ppm with sidebands at –63 and –165 ppm. Its location is shifted relative to the fluorine peak in the octahedral environment of Al in micas (in the range from –130 to –180 ppm) and can correspond to both fluorine in the FO_4 coordination of the square-pyramidal *Y*1 site and fluorine in octahedral coordination (Huve et al. 1992). Our NMR data are in agreement with the structural data of the works by Groat et al. (1992b, 1996a).

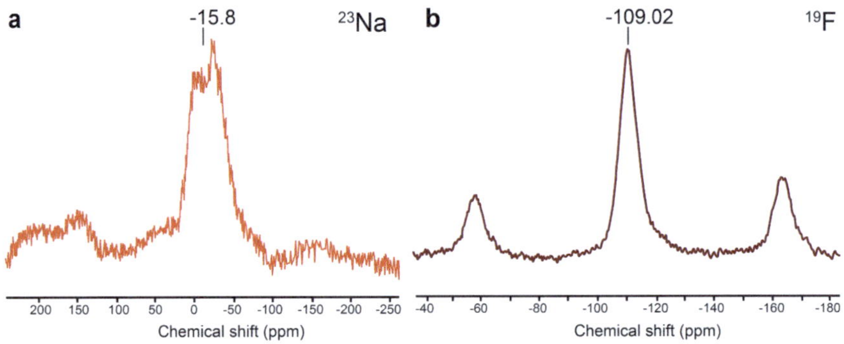

Fig. 3.15 MAS NMR ^{23}Na spectrum of Na-enriched vesuvianite (**a**, sample 97BM) from Western Keivy, Kola Peninsula, Russia presented by an asymmetric quadrupole doublet with a center at –15.8 ppm and MAS ЯМР ^{19}F spectrum for F-enriched vesuvianite from Lupikko, Karelia, Russia (**b**, sample 189/12409 SM) presented by a single peak at –109.02 ppm

3.1.7 Powder X-Ray Diffraction

Powder X-ray patterns (PXRD) of some vesuvianite samples are shown in Fig. 3.16a–f. For most of them, observed reflections are identical, but the *hhl* reflections in the spectrum of the vesuvianite from the Kluchevskii mine (Fig. 3.16d) has preferential orientation along the [110] direction. It should be noted that a slight shift of the *hhl* and *h*00 lines for Si-deficient B-bearing vesuvianite from Yakutia towards smaller 2Θ angles is characteristic of boron-containing vesuvianite (Panikorovskii et al. 2017c). The presence of additional reflections on the powder diagrams related to a decrease in the symmetry of *P*4/*nnc* → *P*4/*n* and *P*4/*nnc* → *P*4*nc* for these samples was not observed. Symmetry-reducing reflections are clearly visible only in single-crystal diffraction patterns. Differences in the intensities of some lines, for example, reflection 440 in the spectrum (Fig. 3.3.16f), are associated with the preferential orientation of the grains (elongation along the *hh*0 prisms). The data on the intensities, the positions of the reflections in PXRD and their *hkl* indices are given in Tables A.5, A.6, A.7, A.8, A.9 and A.10.

3.1.8 Mössbauer Spectroscopy

To clarify the role of the the valence state of iron in the VSMs crystal chemistry, the Mössbaur spectra of two samples of green (Fig. 3.17a) and brown (Fig. 3.17b) vesuvianite from volcanogenic skarns of the Somma-Vesuvius complex were studied (see photos of the samples in Figs. 1.29a and 3.2). These spectra are most satisfactorily described by four or five quadrupole doublets with parameters given in Table 3.5. It should be noted that the interpretation of doublets is unambiguous: it is consistent with X-ray structural data (Panikorovskii et al. 2017b) and corresponds to that in the

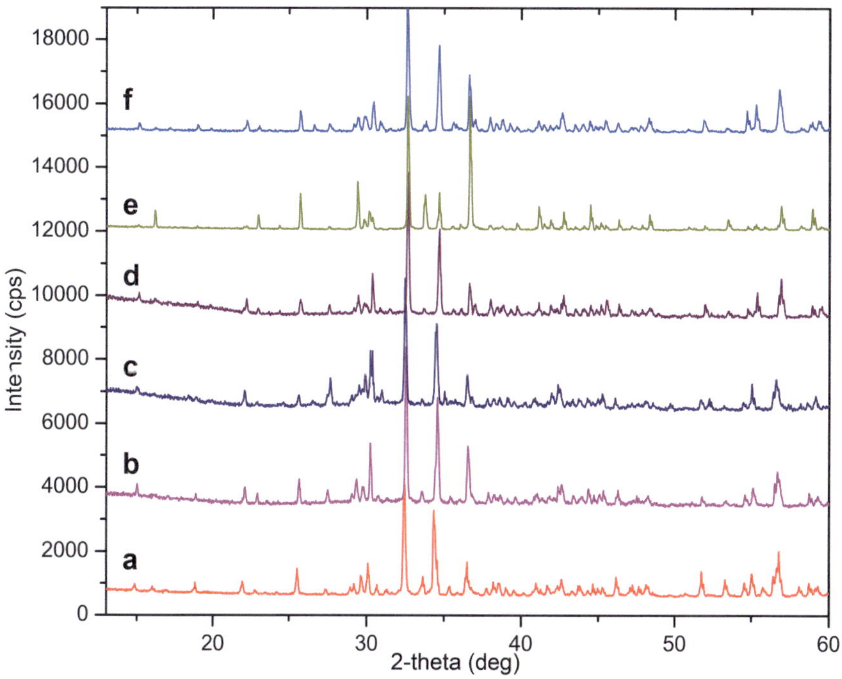

Fig. 3.16 Powder diffraction patterns of Si-deficient B-bearing vesuvianite from Yakutia (**a**, E.V. Galuskin's sample) and vesuvianite from Bazhenovskii deposit (**b**, A. A. Agakhanov's sample), from Kovsor (**c**, A. A. Agakhanov's sample), from Koytezek pass, Tajikistan (**d**, Agakhanov's sample), from Kluchevskiy mine, Middle Urals, Russia (**e**, sample 27962 FM) and from Vesuvius, Italy (**f**, sample 27844 FM)

works by Rusakov et al. (2006), Groat and Evans (2012). The names of the doublets are taken from Rusakov et al. (2006) and correspond to all possible combinations of Fe^{2+} and Fe^{3+} coordination in VSMs structure. Doublet *a* is usually superimposed by doublet *b* and they both correspond to Fe^{2+} at the *Y*1 site.

The most intense doublet *c* with the isomer shift (IS) of ~0.38 mm/s and quadrupole splitting (QS) of ~0.50 mm/s corresponds to Fe^{3+} in the octahedral *Y*3 position (Kraczka and Żabiński 2003; Rusakov et al. 2006). The weak doublet *e* with the parameters IS = 0.40–0.52 mm/s and QS = 1.00–1.30 mm/s can be associated with Fe^{3+} in the *Y*1 position. The strong doublet *d* with the parameters IS = 0.75–0.85 mm/s and QS = 0.40–0.50 mm/s can be associated with Fe^{2+} in the octahedral *Y*3 position. The doublet *b*, like other doublets with similar ultrafine IS parameters of 1.03–1.10 mm/s and QS of 2.80–2.92 mm/s, corresponds to Fe^{2+} in the *Y*1 position (Manning and Tricker 1975; Rusakov et al. 2006; Groat et al. 2013). The doublet *g* with the parameters IS = 0.92 mm/s and QS = 2.40 mm/s, first noted for the VSMs by Panikorovskii et al. (2017c), according to the data by Burns (1994) and Geiger et al. (2000) can be attributed to octahedral Fe^{2+} (i.e. Fe^{2+} in the *Y*2 position).

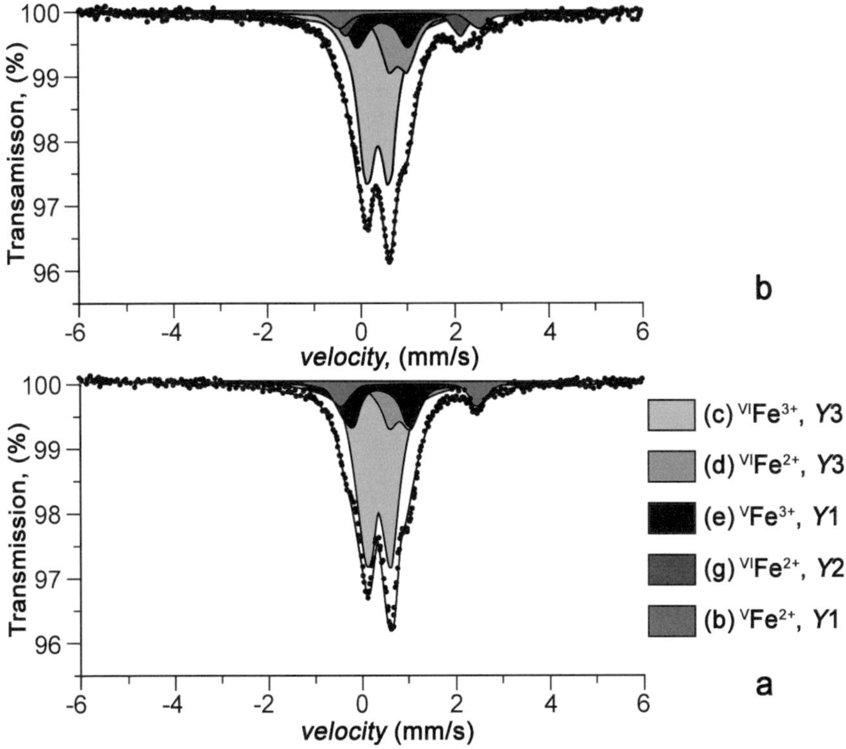

Fig. 3.17 Mössbaur spectra of vesuvianite from Vesuvius, Italy (**a**, sample 51062 FM), (**b**, sample 27844 FM)

Table 3.5 The values of the hyperfine parameters of the individual components of the ^{57}Fe nuclear gamma resonance spectrum of the vesuvianite samples 27844 FM and 51062 FM from Vesuvius, Italy

Doublet	Isomer shift, (mm/s)	Quadrupole splitting (mm/s)	Line width, (mm/s)	Area (%)	Assignment
Sample 27844 FM					
c	0.377 ± 0.002	0.492 ± 0.004	0.460 ± 0.004	54.4 ± 0.6	$^{VI}Fe^{3+}$ at *Y*3
e	0.481 ± 0.006	1.076 ± 0.011		12.8 ± 0.7	$^{V}Fe^{3+}$ at *Y*1
d	0.821 ± 0.003	0.396 ± 0.004		17.9 ± 0.5	$^{VI}Fe^{2+}$ at *Y*3
b	1.039 ± 0.012	2.920 ± 0.031		6.1 ± 0.5	$^{V}Fe^{2+}$ at *Y*1
g	0.921 ± 0.009	2.400 ± 0.022		8.8 ± 0.5	$^{VI}Fe^{2+}$ at *Y*2
Sample 51062 FM					
c	0.377 ± 0.007	0.528 ± 0.014	0.432 ± 0.004	59.9 ± 0.4	$^{VI}Fe^{3+}$ at *Y*3
e	0.405 ± 0.004	1.312 ± 0.010		16.3 ± 0.5	$^{V}Fe^{3+}$ at *Y*1
d	0.842 ± 0.030	0.438 ± 0.060		14.8 ± 0.4	$^{VI}Fe^{2+}$ at *Y*3
b	1.030 ± 0.007	2.920 ± 0.014		9.0 ± 0.3	$^{V}Fe^{2+}$ at *Y*1

The studied samples demonstrate the tendency of color to change from dark reddish-brown to green with an increasing fraction of iron in the trivalent state from 57 to 76%. At the same time, the dominant role of Fe^{3+} in the $Y1$ position for the vesuvianite from its type locality should be noted, which can be useful in determining the vesuvianite end-member formula.

3.1.9 Single-Crystal X-Ray Structural Analysis

The refinements of the vesuvianite crystal structures were performed in three space groups (*P*4/*nnc*, *P*4/*n*, *P*4*nc*). The final choice of symmetry is based on the number and intensities of glide-plane violating reflections (if present) for a particular space group (see Sect. 1.3.7). Attempts of refinement in low-symmetric groups, such as $P\bar{4}$, *P*2/*n* and $P\bar{1}$ (Veblen and Wiechmann 1991; Tanaka et al. 2002) for the samples with abnormal optical properties (i.e. biaxial, with the 2 *V* angle > 5°) showed no deviation of the general vesuvianite symmetry from tetragonal. For low-symmetry samples, the symbol A or B was added to the site names. The list of studied samples, unit cell parameters (UCP), boron content (according to X-ray structural data), as well as the number of violating reflections for the *P*4/*nnc* and *P*4/*n* space groups, channel population positions in the *P*4/*nnc* and *P*4/*n* structures and final R_1- factors are given in Table A.11.

As noted in Chap. 1, in the crystal structure of vesuvianite, three groups of sites differing in the size of cations and their coordination can be distinguished: *X*, *Y* and *Z* (Fig. 3.18). The *Y* positions (Fig. 3.18a) have a tetragonal pyramid coordination ($Y1$ site) or an octahedral coordination ($Y2$ and $Y3$ sites). The *Z* positions are in a tetrahedral coordination and belong to isolated tetrahedra ($Z1$ and $Z2$) and nesosilicate Z_2O_7 groups ($Z3$). The *X* sites have coordination numbers from 7 to 9 and, as a rule, are populated exclusively with calcium cations; the $X3$-polyhedron has a coordination number of 9 and the largest volume, the $X1$, $X2$ and $X4$ polyhedra are surrounded by 7 or 8 oxygen atoms. Due to the close distances between the sites $Y1$ and $X4$ (see details in Sect. 1.3.3), these sites cannot be occupied simultaneously. The total occupancy of $Y1$ and $X4$ is 1. In high-symmetry VSMs the occupancy factor of each of these sites is ½.

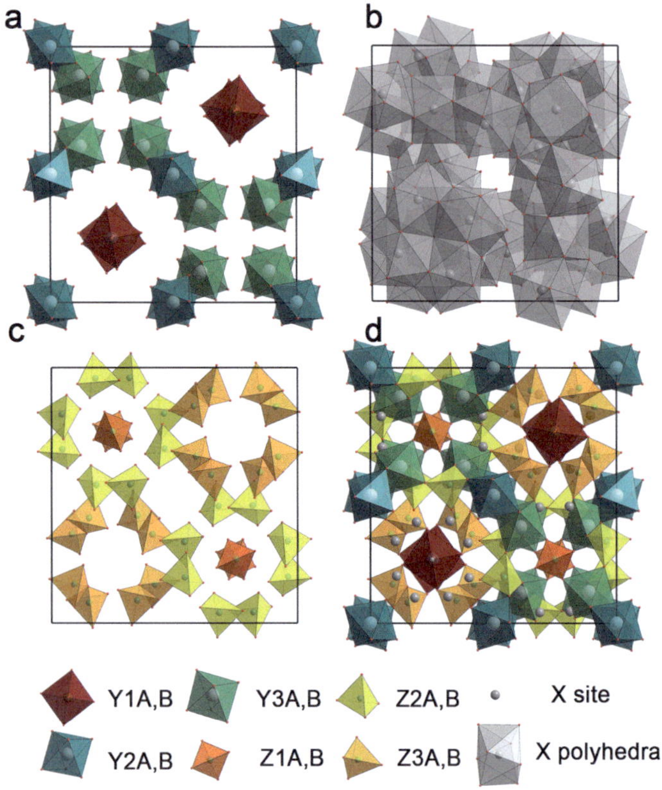

Fig. 3.18 The projection of the vesuvianite crystal structure along the *c* axis: dark green and light green octahedra denote the *Y*2A,B and *Y*3A,B sites and the square-pyramid *Y*1A,B is shown with red (**a**). *X* sites are shown with gray polyhedra and gray spheres (**b**); bright orange, orange and yellow tetrahedrons correspond to the Z1A,B; Z2A,B and Z3A,B sites (**c**), respectively. Figure 1.52d shows the general view of the structure

P4/nnc model

This structural model is characterized by the statistical disorder in the rod *Y*1-*X*4-*X*4-*Y*1 of the half-populated (4*e*) positions, which are situated on the fourfold axis (Fig. 3.19). Most often, this model describes crystal structures of VSMs from skarns or volcanic rocks, which are also characterized by the presence of partially populated additional *T*1 and *T*2 sites.

In order to assure a proper accuracy in determining the occupancy of channel positions, the refinement of the crystal structures of VSMs samples with a general symmetry *P4/nnc* was also performed in the space group *P4/n*. The deviation from the theoretical occupancy by 50% was within the error of ±2%. The overwhelming majority of samples with the *P4/nnc* symmetry crystallized under high-temperature conditions. The exceptions are the low-temperature crusts of "hydrovesuvianite" from the Wiluy River basin, Yakutia and Kovdor, Murmansk region as well as acicular

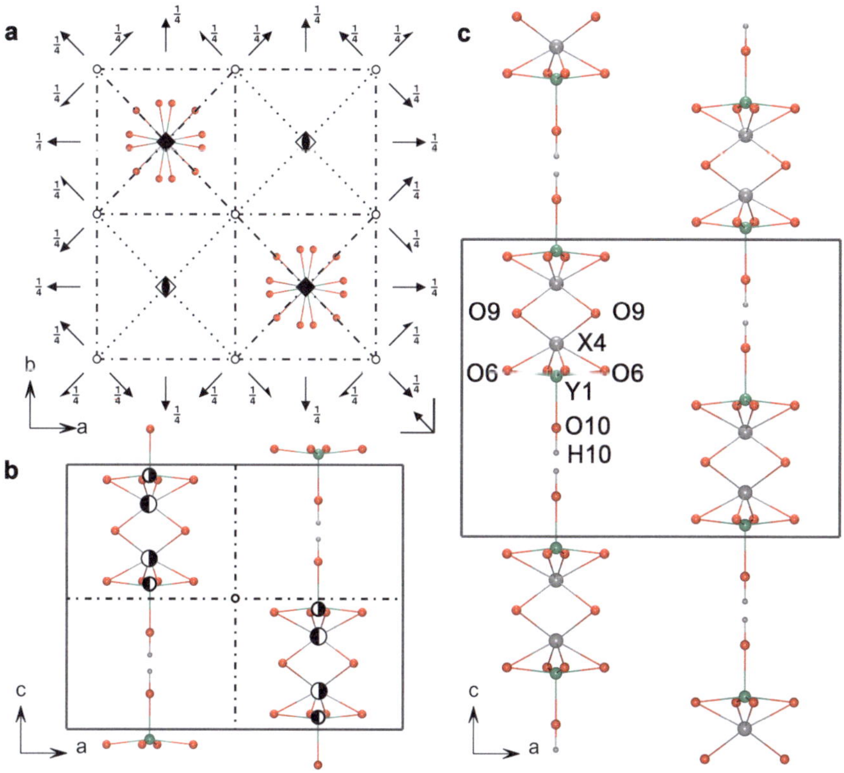

Fig. 3.19 The projections of cationic rods *Y*1-*X*4-*X*4-*Y*1 of positions along the *c* axis (**a**), onto the (001) plane with the indication of the elements of symmetry (**b**) and on (010) plane with the designations of positions (the unit cell is outlined) (**c**). The positions of hydrogen atoms are given in accordance with the structural model by Lager et al. (1999a)

crystals of F-rich vesuvianite from Tas–Khayakhtakh and Lupikko, Karelia (all the localities are in Russia).

For the samples described within the *P*4/*nnc* model, glide-plane violating reflections were often observed. For example, a total of 128 violating reflections were noted for the vesuvianite from the fenitized gabbroids of the Western Keivy, Kola Peninsula, Russia. Some of them (Fig. 3.20) belong to the *n* plane, perpendicular to the fourfold axis, the other part to the (100) plane. Weak intensity of these reflections, coupled with the absence of a detectable difference in the occupancy of the *Y*1A and *Y*1B sites (when refining in the space group *P*4/*n*), as well as a large number of positions with negative displacement factors while refining the structure in the space group *P*4*nc*, indicate the resulting *P*4/*nnc* symmetry.

According to the X-ray structural data refined in the *P*4/*nnc* space group, single-crystal XRD patterns of most VSMs do not contain violating reflections (Fig. 3.21). As a rule, high-symmetry samples show intense reflections of the 00*l* family with

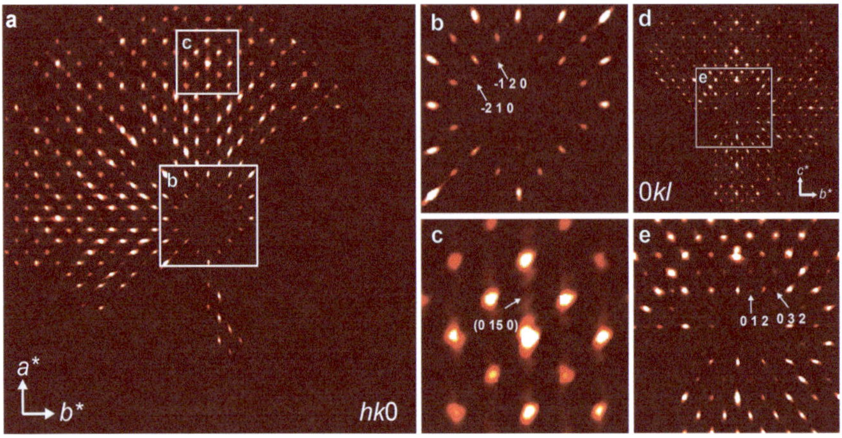

Fig. 3.20 Sections of the single-crystal XRD pattern in the reciprocal space (for the sample 976 MB): general view of the *hk*0 section (**a**); fragments of the pattern showing the presence of additional violating reflections for the *n* plane, perpendicular to the fourfold axis (**b**), and the *P*4/*nnc* symmetry (**c**), general view of the 0*kl* section (**d**) and a fragment of the pattern showing additional reflections that violate the *n* plane lying in (100) (**e**)

even *l*, as well as reflections *084*, *0-84*, *0-8-4* and *08-4*, the intensity of which exceeds the intensity of other reflections.

Acicular crystals of vesuvianite from skarn rocks of Tas–Khayakhtakh, Yakutia, Russia demonstrated under a polarizing microscope uneven extinction along the *c* axis. The study of the diffraction pattern of this sample showed the splitting of some reflections in the small-angle region (Fig. 3.22). An analysis of the *hk*0 projection of the reciprocal space section for this sample showed that most of the basic reflections (*0–20*, *−510*, *−400*) are duplicated, which may be due to the mechanical splitting of the vesuvianite crystals during their growth mainly along the [001] direction, as described by Galuskin (2005).

As an example of the *P*4/*nnc* model, we considered the structure of vesuvianite from Vesuvius (the sample 27844 FM), which has a volcanic genesis (Panikorovskii et al. 2017b). The structure of the sample under investigation was refined with an *R*-factor of 0.027 (Table A.12). Analysis of the occupancy of the *X*1 (4*c*), *X*2 (16 *k*), *X*3 (16 *k*) and *X*4 (4*e*) positions showed that they are fully occupied by Ca atoms (Table A.13). According to Panikorovskii et al. (2016b), Na atoms in the vesuvianite structure occupy the *X*4 and *X*1 positions. The average <*X*1–O>, <*X*2–O>, <*X*3–O>, and <*X*4–O> bond lengths are 2.424, 2.4724, 2.5586 and 2.460 Å, respectively (see Table A.14), which also agrees with the complete occupancy of these positions by Ca atoms. The value of the scattering factor of *Z* positions corresponds to their full occupancy by Si atoms.

It is known that in the vesuvianite structure, the octahedral *Y*2 (8*f*) site is predominantly populated by Al and Mg (Groat et al. 1992a; Galuskin 2005; Panikorovskii et al. 2016e), while the *Y*3 position (16 *k*) can be also occupied by Fe^{2+}, Fe^{3+}, Mn^{2+}, Zn and Cr (Phillips et al. 1987; Ohkawa et al. 1992; Armbruster et al. 2002). The refined

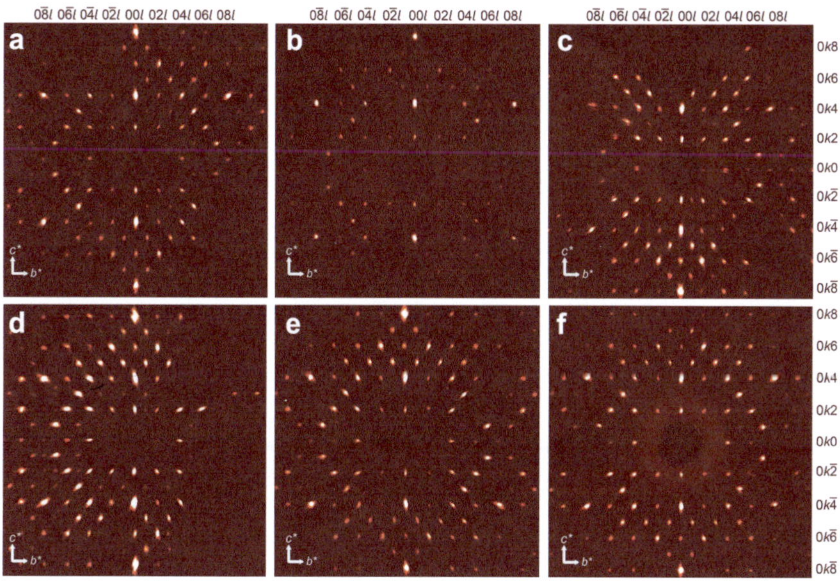

Fig. 3.21 Reconstructed (0*kl*) sections of the single-crystal XRD pattern in the reciprocal space for vesuvianite samples with the *P*4/*nnc* symmetry from Monzoni, Italy (**a**, sample 56-3057 SM), Tas–Khayakhtakh, Russia (**b**, E. V. Galuskin's sample), Kristiansand, Norway (**c**, sample 13975 FM), Oyta, Japan (**d**, sample 54541 FM), Piliola, Finland (**e**, E. V. Belogub's sample), and Jokum, Sweden (**f**, sample 13967 FM)

occupancy of the *Y*2 position is $Al_{1.00}$; the corresponding average bond length <*Y*2–O> is 1.896 Å. In the sample 27844 FM, the *Y*3 position has an admixture of iron, its refined occupancy is $Al_{0.90}Fe_{0.10}$, and the average <*Y*3–O> distance is 1.9662 Å. The tetragonal pyramid *Y*1 in vesuvianite from type locality is predominantly populated by Fe atoms. Its refined occupancy is $(Fe_{0.72}Mg_{0.28})_{1.00}$ or, taking into account the data of Mössbauer spectroscopy, $(Fe^{3+}{}_{0.50}Mg_{0.28}Fe^{2+}{}_{0.22})_{1.00}$. The distances between the half populated positions *Y*1–*X*4 and *X*4–*X*4 are 1.14 and 2.49 Å, respectively, which agrees well with the disordered structure model.

For vesuvianite from the type locality, a partial population of the additional *T*1 (8*h*) and *T*2 (2*a*) positions is characteristic, which, in addition to the X-ray diffraction analysis, is confirmed by the FTIR and NMR spectra (see Sects. 3.1.4 and 3.1.5). According to the data by Gnos and Armbruster (2006), partial occupancy of these positions is typical for samples with a temperature of formation in the range of 500–900 °C. For samples with the occupancy of the *T*1 position more than 50% (Groat et al. 1998; Bellatreccia et al. 2005a; Panikorovskii et al. 2017c), the O7 oxygen position splits into O7A and O7B, but for samples with a low population of the *T*1 position (as in our case) no splitting is observed (Galuskin et al. 2003b).

According to Groat et al. (1992a), in the structure of vesuvianite, fluorine replaces hydroxyl groups at the O10 and O11 sites. The obtained scattering factor of the O11

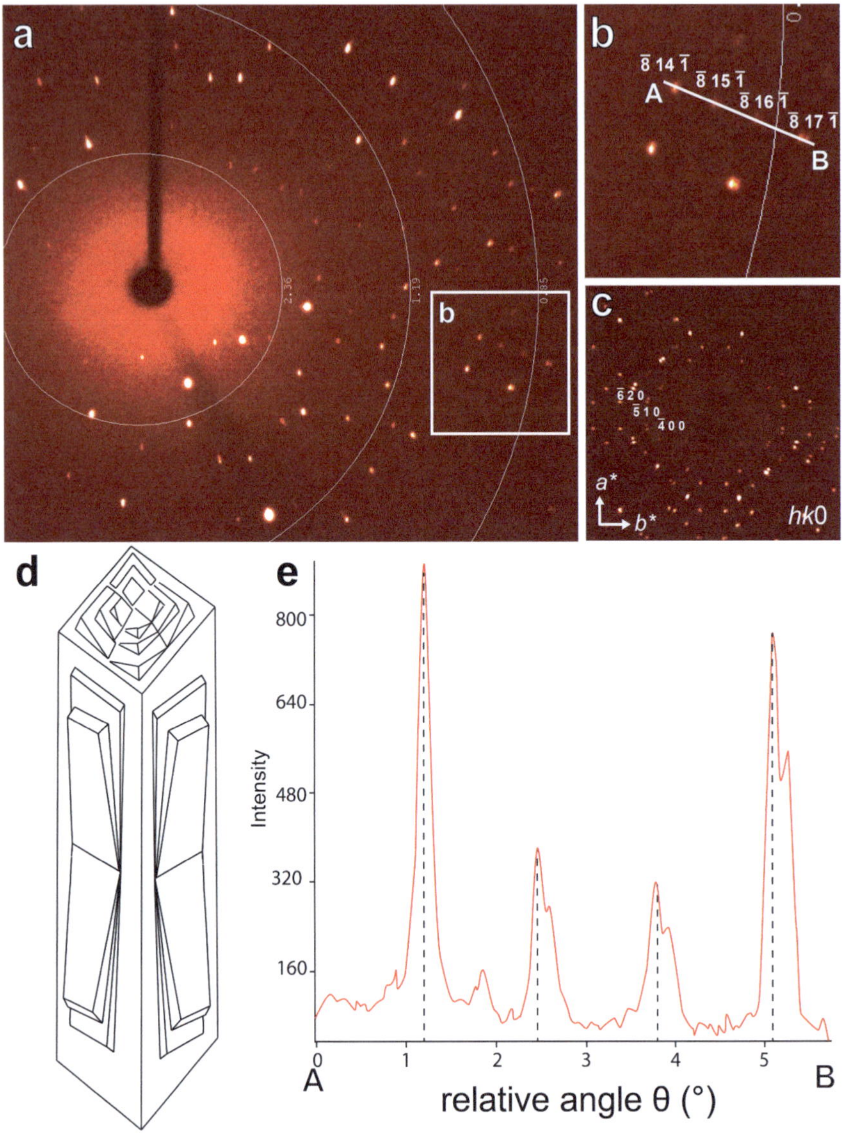

Fig. 3.22 Single-crystal X-ray diffraction pattern for a vesuvianite sample from Tas–Khayakhtakh, Russia (**a**, E. V. Galuskin's sample) showing splitting of some reflections in the small-angle θ region (**b**), doubling *0–20*, *−510*, *−400* reflections of the *hk*0 section of the reciprocal space (**c**), the model of the splitting of the vesuvianite crystal, according to Galuskin (2005) (**d**) and the profile of a series of *−8 k−1* reflections of a split crystal (**e**)

position for the sample 27844 FM corresponding to 8.25 ē confirms this conclusion; the refined occupancy of the O11 site is $O_{0.75}F_{0.25}$.

Taking into account the structural data, chemical composition, IR spectroscopy, NMR and Mössbauer spectroscopy data, the final crystal chemical formula of this sample can be written as follows (Panikorovskii et al. 2017b):

$$^{X1}(\mathrm{Ca})_{2.00}\ ^{X2}(\mathrm{Ca})_{8.00}\ ^{X3}(\mathrm{Ca})_{8.00}\ ^{X4}(\mathrm{Ca}_{0.97}\mathrm{Na}_{0.03})_{1.00}$$
$$^{Y1}\left(\mathrm{Fe}^{3+}{}_{0.50}\mathrm{Mg}_{0.28}\mathrm{Fe}^{2+}{}_{0.22}\right)_{\sum 1.00}\ ^{Y2}\left(\mathrm{Al}_{3.85}\mathrm{Fe}^{2+}{}_{0.15}\right)_{\Sigma 4.00}$$
$$^{Y3}\left(\mathrm{Al}_{5.26}\mathrm{Mg}_{1.83}\mathrm{Fe}^{2+}{}_{0.54}\mathrm{Fe}^{2+}{}_{0.26}\mathrm{Mn}_{0.11}\right)_{\Sigma 8.00}$$
$$^{Z1}(\mathrm{Si})_{2.00}\ ^{Z2}(\mathrm{Si})_{8.00}\ ^{Z3}(\mathrm{Si})_{8.00}(\mathrm{O})_{68.00}$$
$$^{T1+T2}(\mathrm{Al}_{0.44}\mathrm{B}_{0.25}\square_{4.31})_{\Sigma 5.00}\ ^{W}(\mathrm{OH}_{5.65}\mathrm{O}_{2.30}\mathrm{F}_{2.00}\mathrm{Cl}_{0.05})_{\Sigma 10.00}.$$

*P*4/*n* Model

An increase in the number of ordered domains in the VSMs structures is accompanied by the increasing order in the structure channels, that is, instead of the sequence *Y*1–*X*4–*X*4–*Y*1 containing sites with a population of ½, observed in the space group *P*4/*nnc*, the following configurations are realized: *Y*1–□–*X*4–□ or □–*X*4–□–*Y*1 (Armbruster and Gnos 2000b) (Fig. 3.23).

In the *P*4/*n* model, along with strong reflections, there are weak maxima in the diffraction pattern, for which $h + k \neq 2n$, as well as 00 *l* reflections, for which $l \neq 2n$ (Fig. 3.24). Their presence indicates the absence in the structure of glide planes *n* oriented parallel to (010), as well as planes *c* parallel to (110) that corresponds to lowered symmetry according to the *P*4/*nnc* → *P*4/*n* scheme.

Differences between the *P*4/*nnc* and *P*4/*n* models (with vacancies and full populations in the latter one) also appear on the graph of the electron density map (Fig. 3.25). For VSMs with *P*4/*n* symmetry, there is a partial ordering of the Mg^{2+} and Fe^{3+} cations in octahedral positions, which results in a significant difference between the average lengths of the *Y*3A–O and *Y*3B–O bonds, as well as in different scattering factors of the corresponding sites. This ordering in octahedral sites compensates the stresses in the structure resulting from the ordering of cations in channels (Panikorovskii et al. 2016d). With good-quality crystals, it is possible to localize the positions of protons within the *P*4/*n* model in the structure, while for the *P*4/*nnc* model, determination of the positions of hydrogen atoms using X-ray diffraction is almost impossible due to the statistical disorder.

As an example of the *P*4/*n* model, below we discuss the structure of a vesuvianite from the metasomatic rocks of the Akhmatovskaya pit (sample 28975 FM). The value of the final *R*-factor is 0.027 (Table A.15).

As noted above, the most important characteristic of low-symmetric species belonging to the vesuvianite supergroup is the population of the *X*4A, *X*4B, *Y*1A and *Y*1B positions. The total occupancy factor for the pairs of positions *X*4A + *X*4B, *Y*1A + *Y*1B, *Y*1A + *X*4A and *Y*1B + *X*4B is equal to 1. The refined population of the channel positions *X*4A and *X*4B are $Ca_{0.02}$ and $Ca_{0.98}$. The scattering factors of

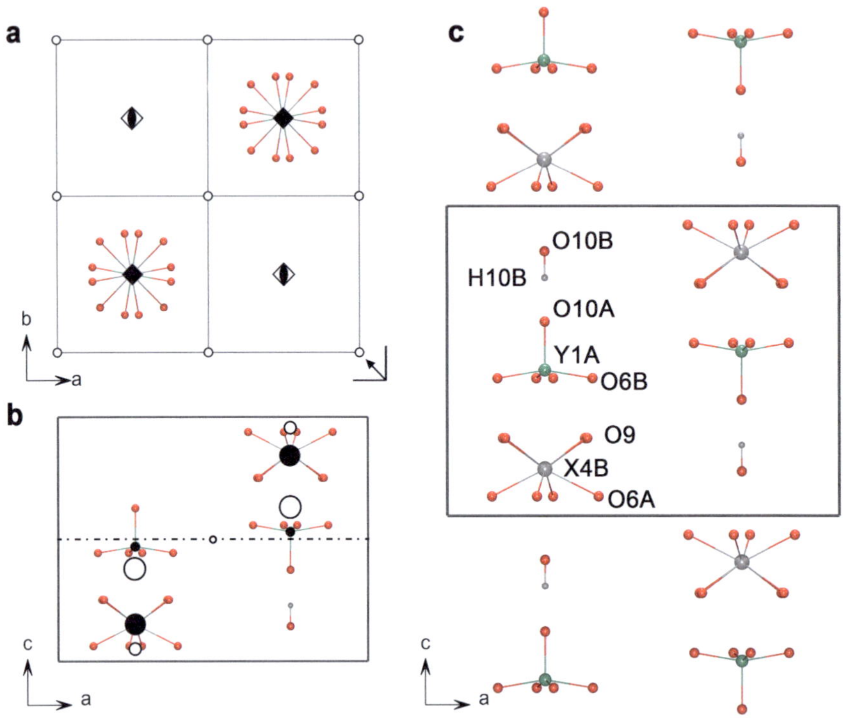

Fig. 3.23 Projection of cationic rods *Y*1-*X*4-*X*4-*Y*1 running along the *c* axis for the *P*4/*n* model of the vesuvianite structure on the (001) plane (**a**), projection on the (010) plane with the indication of the symmetry elements (**b**); the projection of cationic chains on the (010) planewith the designations of positions and the selection of the unit cell (**c**). The positions of the atoms are given in accordance with the structural model of vesuvianite from the Akhmatovskaya pit (sample 28975 FM)

the positions *X*1A,B (4*f*), *X*2A,B (8*g*), *X*3A,B (8*g*) and *X*4A,B (2*c*) correspond to their population exclusively with Ca atoms (Table A.16). Their average *X*–O bond lengths vary within 2.43–2.57 Å (Table A.17).

The octahedral *Y*2A,B positions (4*d* and 4*e*) are populated exclusively by Al atoms; the corresponding bond lengths <*Y*2A–O> and <*Y*2B–O> are 1.8946 and 1.9004 Å. In addition to the high degree of order of the channel positions (*X*4A,B, *Y*1A,B), the ordering of octahedral positions is observed with the refined occupancies $[(Al)_{0.87}(Fe)_{0.13}]_{1.00}$ and $[(Al,Mg)_{0.78}(Fe)_{0.22}]_{1.00}$ for *Y*3A (8*g*) and *Y*3B (8*g*), respectively. The incorporation of Mg in the *Y*3B position is confirmed by an increase in the average bond length <*Y*3B–O> for this site to a value of 2.005 Å compared to <*Y*3A–O>, which is equal to 1.936 Å. The refined occupancies of the *Y*1A and *Y*1B positions are $(Fe_{0.60}Al_{0.38})_{0.98}$ and $(Fe_{0.02})$, respectively.

The difference in the occupancies of the *Y*3A and *Y*3B octahedra, which are coordinated by the OH^- groups (Table 3.6), leads to a nonequivalent arrangement of the hydrogen atoms H11A and H11B; the angles for the strongest hydrogen bonds

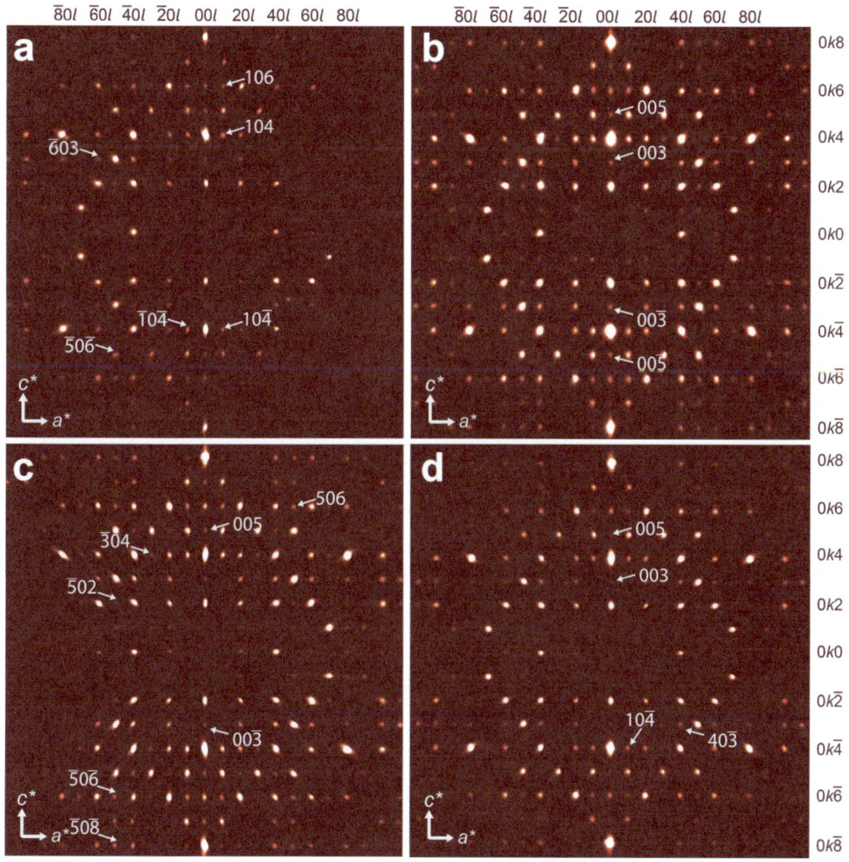

Fig. 3.24 Reconstructed ($h0l$) section of the single-crystal X-ray diffraction pattern in the reciprocal space at the level $k = 0$ for vesuvianite from Tirol, Italy (**a**, sample 51085 FM), Eremeevskaya pit, Urals, Russia (**b**, sample 18983 FM), Govorukhinsky quarry, Urals, Russia (**c**, sample 23677 FM) and Maderaner, Switzerland (**d**, sample 27726 FM). White arrows indicate violating diffraction maxima for the space group *P*4/*nnc*

O11A–H11A···O11B and O11B–H11B···O11A differ by 15°. A change in the geometry of the hydrogen bonds leads to the splitting of bands in the IR spectrum in the region of 3440–3470 cm^{-1} (see details in Sect. 3.1.5).

The refined and empirical formulae for the studied sample can be written as follows (Chukanov et al. 2018a):

$$Ca_{19.00}\ {}^{Y1}(Fe_{0.62}Al_{0.38})\ {}^{Y2}Al_{4.00}\ {}^{Y3}[(Al, Mg)_{6.68}(Fe, Ti, Mn)_{1.32}]Si_{18}O_{68}(OH_9O)_{10}$$

and $Ca_{19.00}\left(Al_{8.22}Mg_{1.83}Fe_{1.73}Ti_{1.54}Mn_{0.08}\right)_{\Sigma 13.40}Si_{17.88}O_{68}(OH, O)_{\Sigma 9.00}$.

***P*4*nc* model**

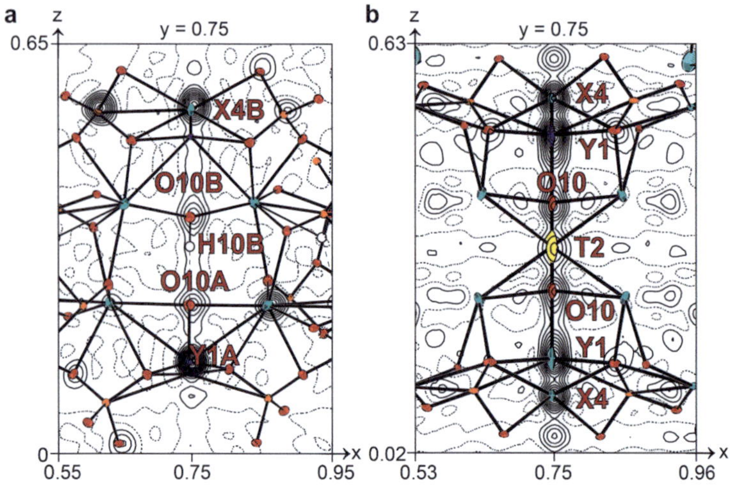

Fig. 3.25 Electron density maps (views along the *Y*1-*X*4-*X*4-*Y*1, chains located on the four-fold axis, with the coordinates 0.75, 0.75, *z*) for the *P*4/*n* vesuvianite from the Akhmatovskaya pit (**a**, sample 28975 FM) and *P*4/*nnc* vesuvianite from Vesuvius, Italy (**b**, sample 51062 FM). Thermal ellipsoids are drawn with a 50% probability level

Table 3.6 Hydrogen-bonding system for in the crystal structure of vesuvianite from the Akhmatovskaya pit, Urals, Russia (sample 28975 FM)

Bond	H···A	D···A	∠(DHA)
O11A–H11A···O7A	2.24(5)	2.770(3)	117(4)
O11A–H11A···O7B	2.37(5)	2.945(3)	121(4)
O11A–H11A···O11B	2.26(3)	3.088(3)	153(5)
O11B–H11B···O11A	2.33(4)	3.088(3)	138(4)

In most recent works, starting with the paper by Allen and Burnham (1992), the VSMs are conventionally divided into two groups: "high" (high-temperature, the *P*4/*nnc* space group) and "low" (low-temperature, the *P*4/*n* and *P*4*nc* space groups). This treatment is largely justified, because, as a rule, low-temperature vesuvianite contains both domains with the *P*4/*n* and *P*4*nc* symmetry (Table 1.5). In addition, most samples containing domains with the *P*4*nc* symmetry have the resulting *P*4/*n* symmetry. Both modifications are close in the structural complexity (Fig. 1.24) and were formed in the same temperature range. However, a minor part of the low-temperature VSMs has *P*4*nc* symmetry: only four structure models in this space group are described in the literature (Ohkawa et al. 1992; Armbruster and Gnos 2000a, b; Panikorovskii et al. 2016b).

The verification of the correctness of the structures refinement in the *P*4*nc* space group was carried out according to the method proposed by Armbruster and Gnos (2000b). The structures were refined in the *P*4/*nnc* space group and 50 reflections, for which the difference between the experimental and calculated intensities is the most significant were considered. In the cases when the structure with the *P*4*nc* symmetry was incorrectly refined in the space group *P*4/*nnc*, the F_o^2 parameter was always greater than F_c^2, which, as a rule, was observed for all reflections during the test. In addition, samples with the *P*4/*n* symmetry, when refined in the *P*4*nc* space group, demonstrated a significant number of sites with negative anisotropic displacement parameters.

The main difference between the *P*4*nc* and *P*4/*n* vesuvianite modifications is the relative orientation of the tetragonal pyramids. In the *P*4*nc* modification, all the tetragonal *Y*1 pyramids are oriented similarly, along the [001] or [00$\bar{1}$] direction (Fig. 3.26), whereas in the *P*4/*n* modification these pyramids are differently directed (Fig. 3.23).

The most important criterion for the structure refinement in the *P*4*nc* space group is the presence of diffraction maxima of the *hk*0 type, where $h + k \neq 2n$, which indicates the absence of a plane *n* perpendicular to the fourfold axis. The most intense reflections of this kind (Fig. 3.27) were observed for vesuvianite samples from Monzoni, Italy (sample 56-3057 SM, Fig. 1.31a, Yoko-Dovyren massif (E. V. Galuskin's sample) and the Klyuchevsky mine (sample 27952 FM Fig. 1.20c).

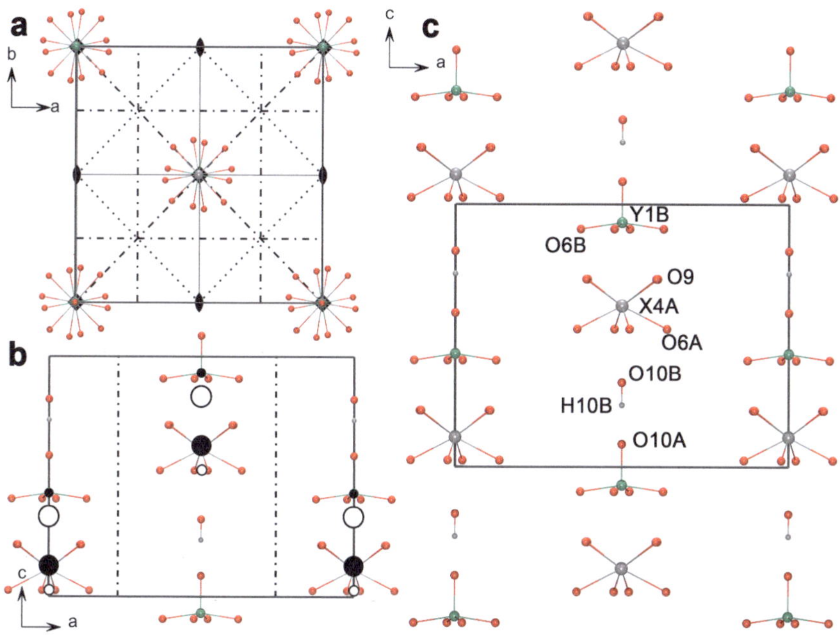

Fig. 3.26 Projection of the rods *Y*1-*X*4-*X*4-*Y*1 of cationic sites along on the (001) plane for the *P*4*nc* model of the vesuvianite structure (**a**), projection onto the (010) plane with the indication of symmetry elements (**b**); the projection of cationic chains on (010) plane with the indication of positions. The unit cell is outlined. The positions of atoms are given in accordance with the structural model of vesuvianite from the Karmankul'skii cordon, Urals, Russia (sample 178/11282 SM)

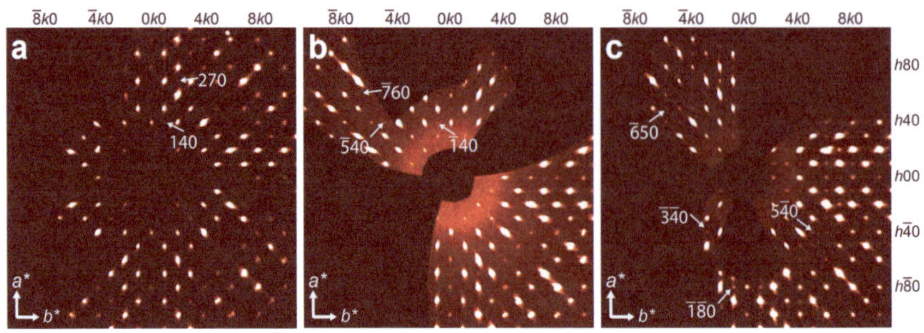

Fig. 3.27 Reconstructed (*hk*0) section of the single-crystal X-ray diffraction pattern in the reciprocal space at the level $l = 0$ for vesuvianite from Monzoni, Italy (**a**, sample 56-3057 SM), Yoko-Dovyren massif, Russia (**b**, E. V. Galuskin's sample) and Kluchevskiy mine, Middle Urals, Russia (**c**, sample 27952 FM). White arrows indicate diffraction maxima violating space group *P*4/*nnc*

As an example of the *P*4*nc* model, below we discuss the structure of vesuvianite from Iordanov, Poland (E. V. Galuskin's sample) which was refined to the R_1-factor of 0.027 (Table A.18). The bond lengths as well as the scattering factors of the *X*- and *Z*-groups of sites correspond to the population of these positions only by the Ca and Si atoms, respectively (Table A.19). The refined populations of the pairs of channel positions *X*4A and *X*4B are $Ca_{0.87}$ and $Ca_{0.13}$, and the calculated populations of the pairs of positions *Y*1A and *Y*1B are $(Fe_{0.08}Al_{0.05})_{0.13}$ and $(Fe_{0.75}Al_{0.12})_{0.87}$, respectively. The average bond lengths <*Y*1A–O> and <*Y*1B–O> are in agreement with their populations and amounted to 2.146 and 1.993 Å, respectively (Table A.20).

Among the crystal-chemical features of this structure, the selective entry of fluorine in the O10B position should be noted. This phenomenon is quite expected, because in the structure F replaces the position of the hydroxyl group (Groat et al. 1995b), which in the *P*4*nc* vesuvianite modification occupies an ordered position. It seems that F is located exclusively in the position of the hydroxyl group that forms a weaker hydrogen bond (with longer O…O distances), because it is energetically more favorable.

The refined formula for the studied sample can be written as follows:

$$Ca_{19.00}\,{}^{Y1}\left(Fe^{3+}_{0.83}Al_{0.17}\right)\,{}^{Y2}Al_{4.00}\,{}^{Y3}[(Al,Mg)_{7.28}(Fe,Mn)_{0.72}]Si_{18}O_{68}(OH_9O)_{10}$$

3.2 Wiluite and Boron Bearing Vesuvianite

In his work, Grigoriev (1991) notes that wiluite was brought by Eric Laxman from the 1790th expedition. The first wiluite description was made by Pallas (1793), as mentioned in the work by Lyachovich (1954), but the mineral was mistakenly defined as garnet. The correct description of the mineral was given only in 1797 (Klaproth 1797). The presence of boron in wiluite was determined by Jannasch (1884), and its anomalous biaxial optical properties were described by Prendel (1887). The first study of the vesuvianite crystal structure was carried out for a sample collected at Wiluy River (Warren and Modell 1931) and may be related to wiluite.

As noted in Sect. 1.1.3, wiluite can be found in akhtarandite rocks on the banks of the Wilyuy River, Yakutia, Russia. Wiluite was found in association with "akhtarandite" (another name "akhtaragdite"-"hydrogarnet"-bearing pseudomorphs after a wadalite-like mineral: (Galuskin and Galuskina 2003; Galuskin 2005). Figure 3.28a, b shows an achtarandite-bearing rock containing large aggregates of "akhtarandite" (1) and grossular crystals (2). The term "akhtarandite" is applied to aggregates of garnet containing $[H_4O_4]^{4-}$ groups associated with split crystals of vesuvianite (Fig. 3.28c) as well as clinochlore and clay minerals (Fig. 3.28d, e). Tetragonal-prismatic crystals of grossular were also found inside the tetrahedral spongy pseudo morphs of akhtarandite, the cause for the decrease in the overall symmetry of which is associated with the ordering of Mg^{2+} and Fe^{3+} in two independent octahedral positions (Panikorovskii et al. 2022).

The approval of wiluite as a separate mineral species caused a wide discussion of the properties of this mineral (Galuskin 2000; Galuskin and Galuskina 2000; Groat et al. 2000). Until then, all vesuvianite group minerals were described as vesuvianite.

Fig. 3.28 Photo of achtaragdite (1) bearing rock composed by grossular-diopside-chlorite rodingite-like agregates with admixture of minerals of the kaolinite-serpentine group (**a**); Grossular (2) crystals with epitaxial overgrowths of "achtaragdite" in rodingite-like rock (**b**); BSE image of minerals of the internal part of sponge achtaragdite pseudomorphs, (**c**–**e**): splitting $(H_4O_4)^{4-}$-substituted, B-bearing vesuvianite (3) crystals, clinochlore (4) agragates and octahedral grossular-katoite (2) crystals (**c**); short-prismatic crystals of hydrogrossular (**d**); long-prismatic hydrogrossular crystal elongated along the [110] (**e**)

Fig. 3.29 Intergrowth of 3 wiluite crystals with vicinal facets from Wiluy River (**a**, sample G. V. Barkhudarova's), wiluite from skarn rock of Titovskoe locality, Tas–Khayakhtakh, Yakutia Russia (**b**, sample 68750 FM), wiluite crystals in achtaragdite-like rock from Siki River basin, Siki-Yadun fault, Evenkiya, Siberia, Russia (**c**). 1—wiluite, 2—calcite, 3—"achtaragdite"

Since 2000, wiluite attracted special attention in mineralogical studies. Despite the fact that VSMs are, in fact, boron concentrators and samples with significant boron contents are quite common in nature (according to our statistics, ~10% of the samples studied in the collection contain more than 0.7 wt% of B_2O_3), wiluite with B_2O_3 contents more than 2.6 wt% is relatively rare.

Wiluite findings are quite rare; today, only nine localities of this mineral are known: Bill Valley (USA), Wiluy River basin (Fig. 3.29a) Titovskoe locality, Tas–Khayakhtakh, Yakutia (Fig. 3.29b) and Siki-Yadunskiy fault in the basin of the Nizhnyaya Tunguska river from trapps in Siberia (Fig. 3.29c), Solongo boron deposit in the Baikal Lake region (all in Russia), Arricia (Italy), and Templeton (Canada), Kozlov, Vysočina Region, (Czech) (Groat et al. 1994a, 1998, 2013; Bellatreccia et al. 2005a; Chukanov 2014; Panikorovskii et al. 2017c).

3.2.1 *Occurrence and Physical Properties*

Wiluite from the type locality in the Wiluy River basin, at the mouth of the Akhtaranda river was found as dark-green or greenish-brown short-prismatic crystals (up to 4 cm long) with numerous pits and furrows on different faces (Fig. 3.30a) in rodingite-like rocks.

Fig. 3.30 SEM (BSE) images of the {001} section of the creviced wiluite crystal with numerous etching pits (**a**, G. V. Barkhudarova's sample), overgrowths of thin epitaxial crust "hydrovesuvianite" on wiluite (**b**, E. V. Galuskins sample) and intergrowths of numerous grossular crystals with a wiluite crystal (**c**, G. V. Barkhudarova's sample). 1—wiluite, 2—"hydrovesuvianite", 3—grossular, 4—montmorillonite

The epitaxial thin (up to 50 μm) crusts of Si-deficient vesuvianite (Panikorovskii et al. 2016a), shown in Fig. 3.30b, are often observed as overgrowths on wiluite crystals. Almost all wiluite crystals contain numerous inclusions and overgrowths of various minerals: montmorillonite pseudomorphs filling prismatic gaps after gehlenite, as well as grossular (Fig. 3.30c), chlorite, fassaite, fluorapatite, goethite, pyrite, wollastonite, perovskite (inclusion in Ti-containing grossular), as well as an unknown mineral of the cancrinite group.

Oriented intergrowths of wiluite and grossular are known (Panikorovskii et al. 2014). Split crystals of Si-deficient vesuvianite shown in Fig. 1.7b were found in the paragenesis with wiluite (Galuskin et al. 2003b). Wiluite usually crystallizes in the *P*4/*nnc* space group. The refractive indices are: $\omega = 1.721$, $\varepsilon = 1.725$ for $\lambda = 589$ nm and 2 V angle attaining 30–35° independently of the growth sector.

3.2.2 Chemical Composition

The main feature of the chemical composition of wiluite (Fig. 3.31) is the presence of a significant number of B atoms (>2.5 apfu). Incorporation of boron into additional *T*1 and *T*2 sites in the vesuvianite structure via the $B^{3+} + Mg^{2+} \leftrightarrow 2H^{+} + Al^{3+}$ and $B^{3+} + 2O^{2-} \leftrightarrow H^{+} + O^{2-}$ substitutions, respectively (Groat et al. 1994a) results in the increase of the total positive charge of cations (without H) to 146^{+}–148^{+} in vesuvianite (Groat et al. 1992a) to 152^{+}–156^{+} in wiluite (Panikorovskii et al. 2017c). In addition, the incorporation of boron results in the inversion of the optical sign from (–) for vesuvianite to (+) for wiluite (Kurazhkovskaya et al. 2005). The optical sign changes near 2 B *apfu* or 2.5 B *apfu,* according to Gnos and Armbruster (2006) and Kurazhkovskaya et al. (2005), respectively.

In the chemical composition of wiluite the Ti content shows significant variation (0.26–0.83 apfu). The Al content also shows a range (6.43–8.20 apfu), indicating variability in its concentration across different samples depends of other *Y*-components (Mg, Fe and Ti). The Mg (3.35–4.70 apfu) content are strong correlate with boron (2.60–3.29 apfu). The Fe content also variable in the range (0.77–1.43 apfu). The chemical composition of the holotype wiluite (Fig. 3.30a) sample from Yakutia is given in Table 3.7.

3.2.3 Thermogravimetry and Differential Scanning Calorimetry

One of the distinctions of wiluite from the other VSMs is a low H_2O content, because oxygen dominates over hydroxyl groups at the O10 and O11 positions. The H_2O content in the mineral measured by TG/DSC (Fig. 3.32), is 0.89 wt% or 2.93 OH^- groups per unit cell and is consistent with the H_2O content of 0.61 wt% in wiluite

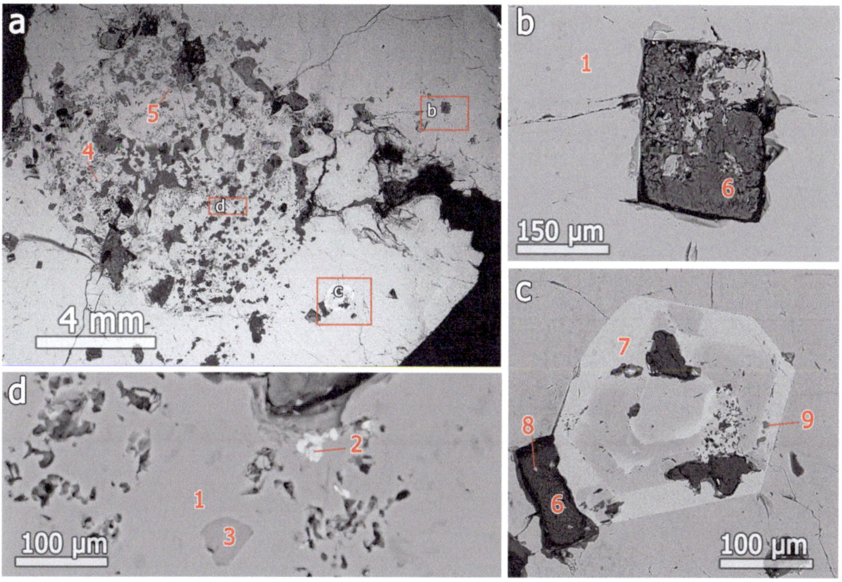

Fig. 3.31 SEM (BSE) images of the (001) section of a fissured wiluite crystal: **a** a general view of wiluite crystal with pyrite (5) and diopside (4) inclusions; **b** montmorillonite (6) filling prismatic gaps after gehlenite; **c** crystal of Ti-enriched grossular (7) containing inclusions a cancrinite-like phase (9) in association with apatite (8); **d** goethite (2) and calcite (3) inclusions in wiluite (1). Sample of G. V. Barkhudarova

according to the data by Groat et al. (1998). The endothermic effect at 1027 °C corresponds to the melting of the mineral with its subsequent recrystallization into åkermanite (~70%), quartz (~10%) and an amorphous phase (~20%) after calcination at 1200 °C.

3.2.4 Infrared Spectroscopy

The most important distinction of the wiluite infrared spectrum (Fig. 3.33) from the spectra of most other VSMs is the presence of additional bands that correspond to vibrations of the B–O bonds in the BO_3 and BO_4 groups. The BO_3 groups are bonded to the $Y1O6_4O10$ polyhedra via a common O10 oxygen atom, forming a $T2Y1O_7$ cluster, where $T2 = B$ and $Y1 = Fe^{3+}$, Fe^{2+}, Mn^{3+}, Cu^{2+}, Al, or Mg. Consequently, stretching vibrations with the participation of the BO_3 group involve four degrees of freedom, corresponding to the lengths of the $T2$–O12 (×2), $T2$–O10 and $Y1$–O10 bonds. As a result, four non-degenerate stretching modes, are realized (Chukanov et al. 2018a). Thus, the expected number of bands in the region of stretching vibrations groups BO_3 (1200–1570 cm^{-1}) is 4 (Fig. 3.11, see Sect. 3.1.5). In two of them, the O10 atom is involved to a minor degree and in some spectra corresponding bands are

Table 3.7 Chemical composition of wiluite studied in this work in comparison with the holotype sample (6) studied by Groat et al. (1998)

Component	1	2	3	4	5	6
SiO_2	36.72	36.52	37.07	37.57	35.96	36.11
TiO_2	1.67	1.50	2.22	0.89	0.92	0.80
Al_2O_3	14.17	12.31	11.93	11.94	13.02	12.03
FeO	2.31	1.86	1.88	2.43	3.42	3.21
MnO	0.07	0.13	0.12	0.06	0.06	
MgO	5.42	4.56	4.57	5.09	6.32	6.48
CaO	36.13	36.02	35.74	36.74	35.44	35.54
B_2O_3	3.11	3.20	3.89	3.07	3.12	3.06
Ce_2O_3	–	–	–	–	–	0.18
H_2O*	0.89**	0.89**	0.89**	1.21	1.21	0.61**
F	–	–	–	–	–	0.50
O≡F						0.21
Sum	100.49	96.99	98.31	99.00	99.47	98.31
apfu						
Si	18.02	17.98	18.39	18.13	17.99	17.99
Ti	0.62	0.56	0.83	0.32	0.35	0.30
Al	8.20	7.14	6.98	6.79	7.68	7.06
Fe	0.95	0.77	0.78	0.98	1.43	1.34
Mn	0.03	0.05	0.05	0.03	0.03	
Mg	3.97	3.35	3.38	3.66	4.71	4.81
Ca	19.00	19.00	19.00	19.00	19.00	18.97
Ce						0.03
B	2.64	2.72	3.33	2.56	2.70	2.63
H	2.91	2.92	2.94	3.89	4.03	2.03
F						0.79

1 to 3—sample 1, 4 to 5—sample 2
* The H_2O content was calculated bas on single-crystal XRD data
** H_2O content was measured by TG after heating up to 1100 °C
Dash—content below detection limit

not resolved. As a result, in some wiluite IR spectra only three bands are observed in the range of 1200–1500 cm^{-1} (Panikorovskii et al. 2017c).

Groat et al. (1994a) supposed the possibility of a partial linear coordination of boron atoms at the *T*2 site. However, our research (Panikorovskii et al. 2017c) completely confirmed the correctness of the assumption by Bellatraccia et al. (2005a) about the triangular and tetrahedral coordination of boron in the structure of wiluite. The wide shoulder at 1110 cm^{-1} was assigned to the asymmetric B–O stretching vibrations of the tetrahedral BO_4^{5-} groups whereas the remaining low-frequency

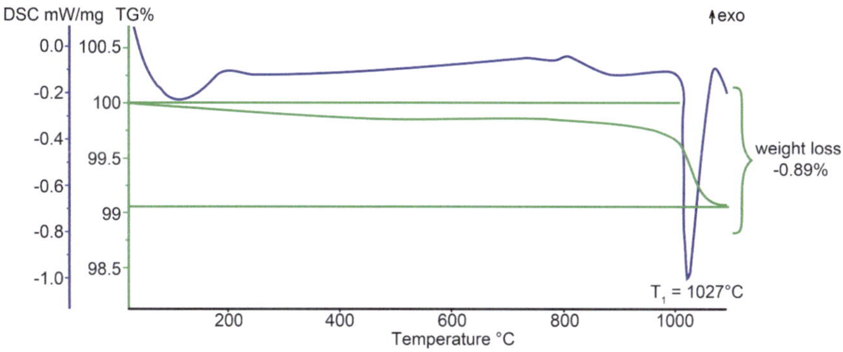

Fig. 3.32 Thermogravimetry curve (green line, wt%) and differential scanning calorimetry curve (blue line) of wiluite

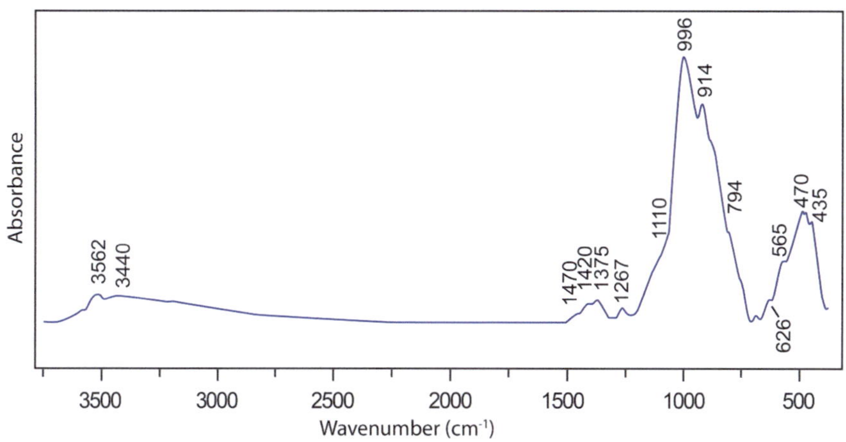

Fig. 3.33 Infrared spectrum of wiluite

bands of BO_4^{5-} are overlapped by the bands of the silicate groups (Kurazhkovskaya et al. 2005).

The region of the O–H stretching vibrations of the wiluite IR spectrum does not contain bands below 3300 cm^{-1}, corresponding to the vibrations of strong hydrogen bonds O10–H···O10 and in the range of 3300–3700 cm^{-1}, only weak bands are present because the position of the hydrogen atom is almost completely replaced by boron. According to the Libowutsky correlations (see above in Sect. 3.1.5), the well-resolved weak bands at 3338, 3481 and 3566 cm^{-1} observed in the IR spectrum of wiluite studied by Panikorovskii et al. (2017c) correspond to the O···O distances of 2.75, 2.89 and about 3 Å, respectively.

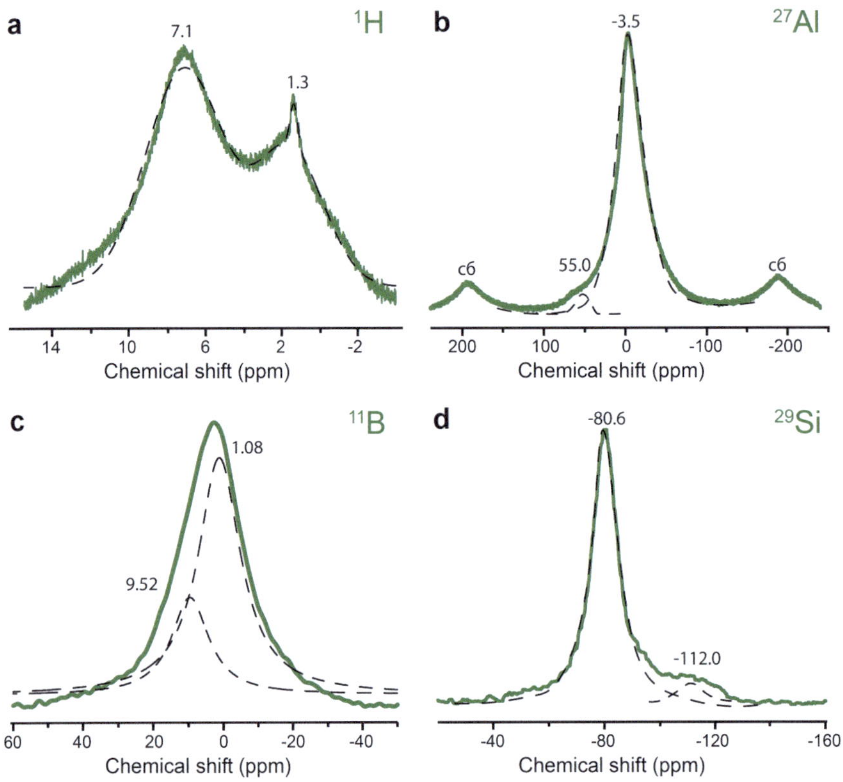

Fig. 3.34 MAS ^{1}H (**a**), ^{27}Al (**b**), ^{11}B (**c**) and ^{29}Si (**d**) NMR spectra of wiluite. The dashed lines show the approximating curves

3.2.5 Solid-State Nuclear Magnetic Resonance

The wiluite proton NMR spectrum (Fig. 3.34a) contains two peaks with the chemical shifts of 7.1 and 1.3 ppm. The O···O distances in the O–H···O hydrogen bonds calculated according to the formula δ_{iso} (ppm) = 79.05–0.255*d* (O–H···O) (nm), proposed by Yesinowski et al. (1988), gives the O···O distances of 2.82 and 3.05 Å. The most intense peak at 7.1 ppm correspond to the hydrogen bonds O11–H1*a*···O7 and O10–H2···O10 with the distances of 2.784 (4) and 2.74–2.82 Å (Ohkawa et al. 1992, 2009; Pavese et al. 1998a; Lager et al. 1999a). The second peak at 1.3 ppm corresponds to a weaker hydrogen bond O11–H1*b*···O11 with the O···O distances of 3.01–3.04 Å (Pavese et al. 1998a; Lager et al. 1999a). These data are in an excellent agreement with O···O distances estimated from the IR spectra.

The observed ^{27}Al NMR spectrum (Fig. 3.34b) is typical for a Fe-bearing vesuvianite. It contains one broad peak at –3.5 ppm, which is in the range of chemical shifts for octahedrally coordinated Al (^{VI}Al). It was assigned to Al at the *Y*2A,B and *Y*3A,B sites (Phillips et al. 1987; Olejniczak and Żabiński 1996). At the same time, broadening of the peak at 50–55 ppm can be assigned to the trace amount of Al in the 4-coordinated *T*1 site (<1 *apfu*). There are no bands which could correspond to the five-coordinated Al. The broad peaks at ± 190 ppm belong to spinning sidebands.

The broad peak observed in the ^{11}B NMR spectrum (Fig. 3.34c) consists of two components with the chemical shifts of 1.1 and 9.5 ppm. These components were assigned to boron with tetrahedral and triangular coordination at the positions *T*1 and *T*2, respectively. Similar chemical shifts with 3- and 4- coordinated boron were reported for minerals of the tourmaline supergroup (Lussier et al. 2009). The relative integrated intensities of the $^{[4]}B$ and $^{[3]}B$ peaks are 73.7% to 26.3%, respectively which corresponds to the $^{[3]}B$ and $^{[4]}B$ contents of 0.82 and 2.29 *apfu*, respetively. The $^{[4]}B$: $^{[3]}B$ ratio obtained from the NMR spectrum are in perfect agreement with the single-crystal X-ray structural data.

In the wiluite structure, there are three coordination tetrahedra, two of which (*Z*1 and *Z*3) share four vertexes with other polyhedra ($Si^{(4)}$ positions) and another one (*Z*2) is occupied by $Si^{(3)}$ atoms, i.e. has a pending O atom when the neighboring *T*1 position is vacant. The most intense peak in the ^{29}Si NMR spectrum (Fig. 3.34d) observed at –80.6 ppm is attributed to $^{[4]}Si$ at the positions *Z*1, *Z*3 and (partially) *Z*2 (Phillips et al. 1987). A weak band with a chemical shift of −112 ppm corresponds to $Si^{(3)}$ at the *Z*2 site (d'Espinose de la Caillerie 1994).

3.2.6 Powder Diffraction

The powder XRD pattern of wiluite, as well as its individuality by the characteristic reflections are shown in Fig. 3.35 in comparison with vesuvianite from the Bazhenovskoe deposit. The unit cell parameters of wiluite refined from the powder data are: $a = 15.7542(3)$, $c = 11.7210(3)$ Å and are in a good agreement with the data obtained earlier (Groat et al. 1998).

A significant difference between unit cell parameters of vesuvianite and wiluite ($a \approx 15.75$, $b \approx 11.72$ Å for wiluite and $a \approx 15.55$, $b \approx 11.82$ Å for vesuvianite) leads to significant differences in the positions of the main lines of the powder XRD pattern. As one can see in Fig. 3.7b, the 440, 600 and 620 peaks for wiluite from Yakutia are shifted from the analogous lines of the ordinary vesuvianite from rodingites of Bazhenovskoe Deposit, Urals towards smaller 2Θ angles. Data on intensities of the reflections, their positions and *hkl* indices for wiluite are given in Table A.21.

3.2.7 Mössbauer Spectroscopy

Experimental data on the wiluite Mössbauer spectroscopy are shown in Fig. 3.36. The spectrum can be described as a superposition of five symmetrical doublets (Table 3.8). The most intense *c* doublet with the parameters of isomeric shift IS = 0.36 mm/s, quadrupole splitting QS = 0.49 mm/s and area *S* = 39% corresponds to Fe^{3+} in the

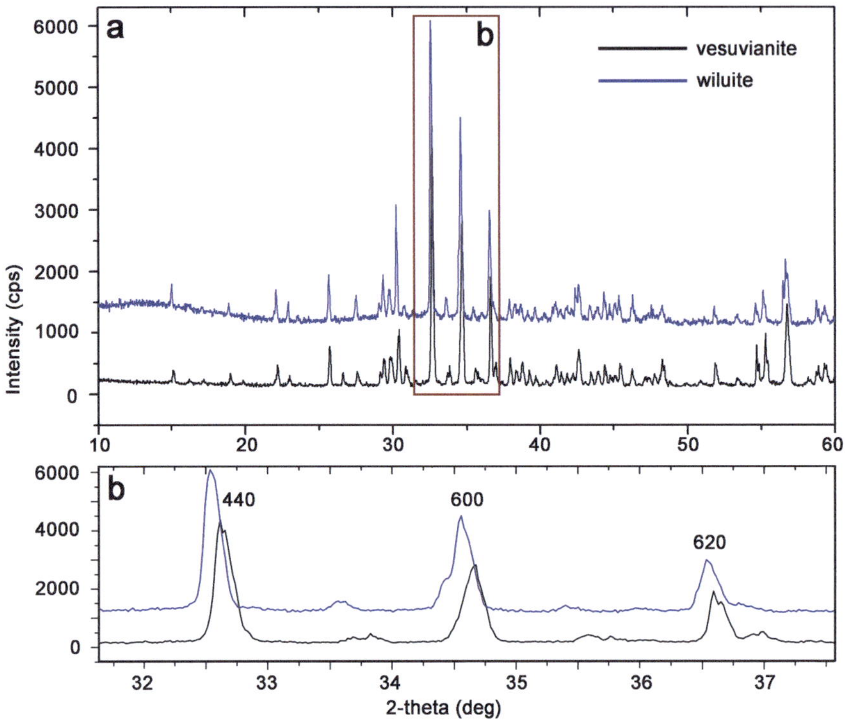

Fig. 3.35 Powder XRD pattern of wiluite ($a = 15.7542(3)$, $c = 11.7210(3)$ Å) and vesuvianite ($a = 15.5479(4)$, $c = 11.8176(3)$ Å)

octahedral coordination at the $Y3$ site (Kraczka and Żabiński 2003; Rusakov et al. 2006). The weak e doublet with IS = 0.65 mm/s and QS = 1.68 mm/s corresponds to Fe^{3+} atoms in the $Y1$ position. The intense d doublet with the parameters IS = 0.70 mm/s, QS = 0.62 mm/s and $S = 31\%$ corresponds to Fe^{2+} in the octahedral $Y3$ position. According to the literature data (Manning and Tricker 1975; Rusakov et al. 2006; Groat et al. 2013), the b doublet with IS = 1.12 mm/s, QS = 2.61 mm/s and $S = 12\%$ most probably corresponds to Fe^{2+} in the $Y1$ position. According to Burns (1994), Geiger et al. (2000), the g doublet with the parameters IS = 0.92 mm/s, QS = 1.89 mm/s and $S = 10\%$ in the Mössbauer spectra of the vesuvianite can be related to Fe^{2+} cations in the $Y2$ octahedron. Earlier this doublet was not observed in the Mössbauer spectrum of wiluite.

3.2.8 *Photoelectron Spectroscopy*

The recording of photoelectron spectra for the clarification of the valence state and the coordination of the main cations occurring in the wiluite structure was carried out

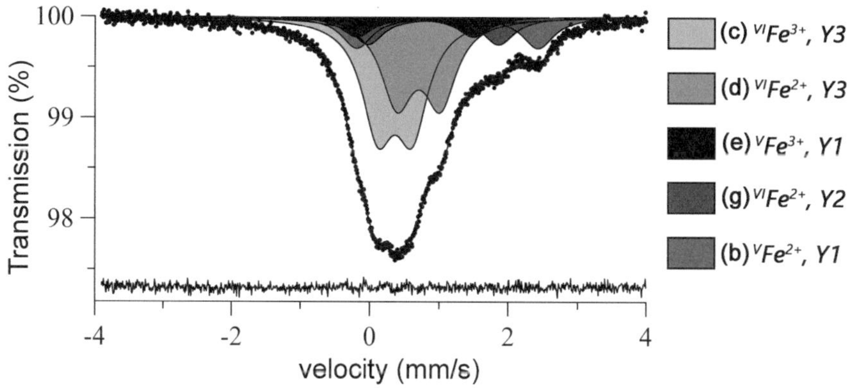

Fig. 3.36 Mössbauer spectrum of wiluite

Table 3.8 Hyperfine parameters of the wiluite Mössbauer spectrum and their interpretation*

Partial spectrum	Isomer shift, mm/s	Quadrupole splitting, mm/s	Line half width, mm/s	Peak area,%	Assignment
c	0.364 ± 0.003	0.492 ± 0.004	0.557 ± 0.007	39.4 ± 0.4	$^{VI}Fe^{3+}$ at *Y*3
e	0.653 ± 0.011	1.680 ± 0.020		7.6 ± 0.5	$^{V}Fe^{3+}$ at *Y*1
d	0.706 ± 0.003	0.621 ± 0.005		30.8 ± 0.4	$^{VI}Fe^{2+}$ at *Y*3
b	1.119 ± 0.005	2.616 ± 0.010		11.9 ± 0.3	$^{V}Fe^{2+}$ at *Y*1
g	0.918 ± 0.010	1.886 ± 0.018		10.4 ± 0.5	$^{VI}Fe^{2+}$ at *Y*2

* The symbols of the partial spectra are given in accordance with (Rusakov et al. 2006)

on the electronic levels of Si2p, Ca2p, Al2p, Mg1s, B1s and Fe2p (Fig. 3.37). The 2p spectrum of the electronic level of Si (Fig. 3.37a) is represented by a broad peak with a binding energy of 102.5 eV, which corresponds to Si^{4+} atoms in a tetrahedral environment (the *Z*1, *Z*2 and *Z*3 positions: (Elmi et al. 2011). In the 2p spectrum of the Ca level (Fig. 3.37b), the signal from the $Ca2p_{3/2}$ sublevel was divided into two components with binding energies of 347.98 and 346.96 eV. The most intense line is referred to Ca in the eight-fold coordination in the positions *X*1, *X*2, and *X*4 (Elmi et al. 2011), and a weaker line to Ca coordinated by nine oxygen atoms in the *X*3 position (Zakaznova-Herzog et al. 2008).

In the work by Barr et al. (1997), overlapping of the binding regions of Al2p with octahedral (73.5–74.8 eV) and tetrahedral (73.2–74.34 eV) coordination was noted. Therefore, the observed peak at a binding energy of 74.8 eV (Fig. 3.37c) can be attributed to Al at the tetrahedral *T*1 and octahedral *Y*2, *Y*3 positions.

The photoelectron Mg1s spectrum (Fig. 3.37d) is presented by two peaks with the binding energies of 1304.0 and 1300.2 eV. The most intense band corresponds to Mg in octahedral coordination (Elmi et al. 2010). The second peak can be assigned to Mg at the five-coordinated *Y*1 site, which is in agreement with the data of the X-ray structural analysis.

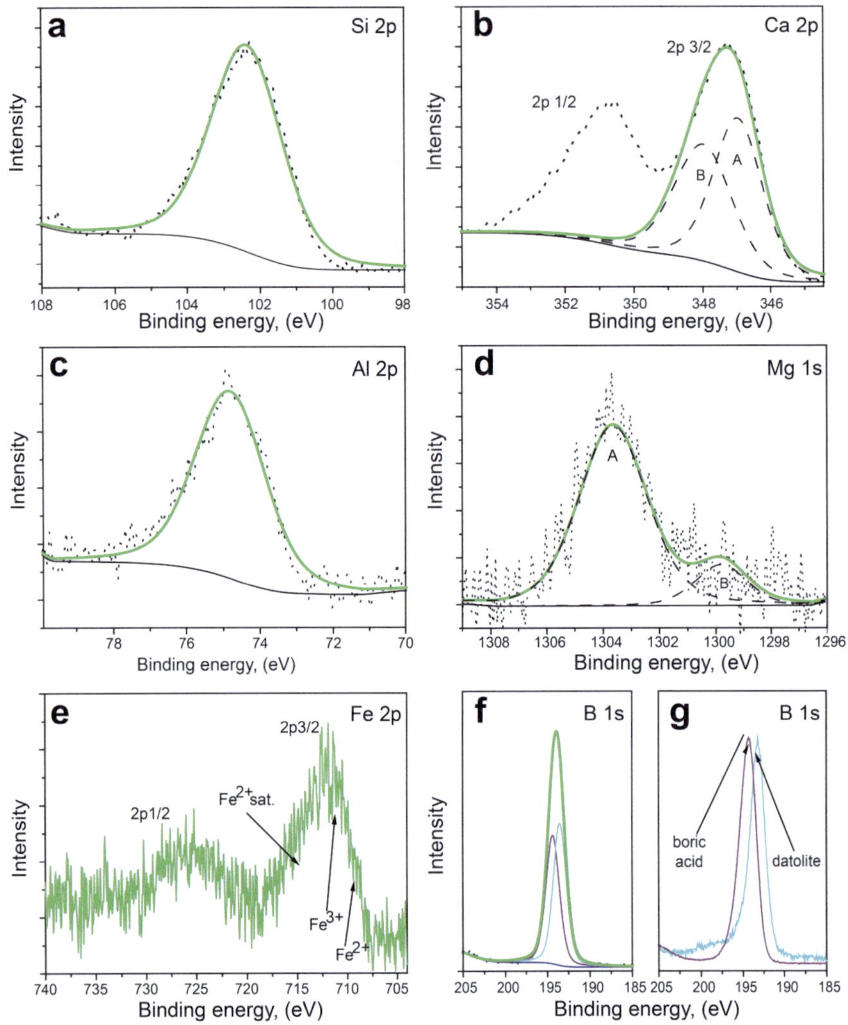

Fig. 3.37 Wiluite photoelectron spectrum: Si 2p level (**a**), Ca 2p level (**b**), Al 2p level (**c**), Mg 1 s level (**d**), Fe 2p level (**e**) B 1 s level (**f**), 1 s-level of standard boric acid and datolite (**g**). The dashed lines show the approximating curves

The Fe2p spectrum is very complex and consists of spin-orbitally splitted 2p3/2 and 2p1/2 peaks and their satellites. The $Fe2p_{3/2}$ signal is split into two components, which correspond to Fe^{2+} and Fe^{3+} (Fig. 3.37e). The overall profile is close to the spectrum described by Kim et al. (2014), where, due to overlaying of the components, it is impossible to unambiguously interpret the lines corresponding to $Fe^{2+/3+}$ with 5- and sixfold coordination For this reason, the valence state of the iron and its coordination have been studied in detail with the help of Mössbauer spectroscopy (see Sect. 2.2.10).

In order to obtain information on the boron coordination, B1s spectra of datolite and boric acid were measured as references. Four-coordinated (datolite) and three-coordinated (boric acid) boron have the binding energies of 193.2 and 194.3 eV, respectively (Fig. 3.37g). The B1s spectrum of wiluite (Fig. 3.37f) consists of two peaks with the binding energies of 193.5 and 194.3 eV, which can be attributed to tetrahedral and triangular boron coordination, respectively.

3.2.9 Single-Crystal X-Ray Structural Analysis

The refinement of the wiluite crystal structure was performed in three tetragonal space groups, *P4/nnc*, *P4/n* and *P4nc*. All wiluite samples (with a boron content > 2.5 *apfu*) correspond to the symmetry *P4/nnc*, but a significant proportion of samples with boron content in the range of 1–2 *apfu* belongs to the *P4/n* or *P4nc* modifications (Table A.11), which contradicts one of the defended positions of the Ph.D. thesis by Borovikova (2005). One such example illustrating the symmetry violation is the boron-containing vesuvianite (space group *P4/n*) forming epitaxy on wiluite (Fig. 3.30b): the projection of the single-crystal XRD pattern in the reciprocal space is shown in Fig. 3.38. Low symmetry for boron-containing low-temperature VSMs is noted in the monograph by Galuskin (2005).

The wiluite crystal structure (Fig. 3.39), in contrast to other VSMs, contains additional tetrahedral and triangular $T1$ and $T2$ sites, which are more than half populated by boron (with more than 2.5 B *apfu*) (Groat et al. 1998). The study of the VSMs collection using X-ray structural analysis revealed 19 samples structurally close to wiluite, however, after comparison with chemical analysis data, only 12 samples were identified as wiluite sensu *strictu* since additional $T1$ positions may be partially filled with the Al^{3+} cations. In the remaining 7 samples the X-ray scattering of the $T1$ and $T2$ positions corresponded to wiluite, but the boron content was less than 1.5 *apfu* and the splitting of the O7 oxygen position characteristic of wiluite was not observed.

The refinement in the *P4/nnc* model was chosen because only four violating reflections with the intensities in the range $2\sigma_I < I < 29\sigma_I$ were observed. The structure was refined to $R_1 = 0.022$ for 1643 reflections with the intensities $|F_o| \geq 4\sigma_F$. The model proposed by Groat et al. (1996a) was chosen as initial one. Crystallographic data and parameters of the refinement are given in Table A.22, atomic coordinates and site occupancies in Table A.23, and interatomic distances in Table A.24. Since the atomic coordinates and occupancies of most positions in wiluite is similar to those in the *P4/nnc* vesuvianite model, described in Sect. 3.1.9, only populations of the $Y1$, $T1$, $T2$, O7A,B, O12 positions are considered in detail below.

In the wiluite model (Groat et al. 1996a), the five-coordinated $Y1$ position is occupied exclusively by Fe atoms and demonstrates negative anisotropic displacement parameters, which may be due to the admixture of lighter elements. In other structural models published elsewhere (Groat et al. 1994a, 1996a, 1998; Bellatreccia et al. 2005a) the $Y1$ site is also dominated by Fe. However, in our model, the scattering of

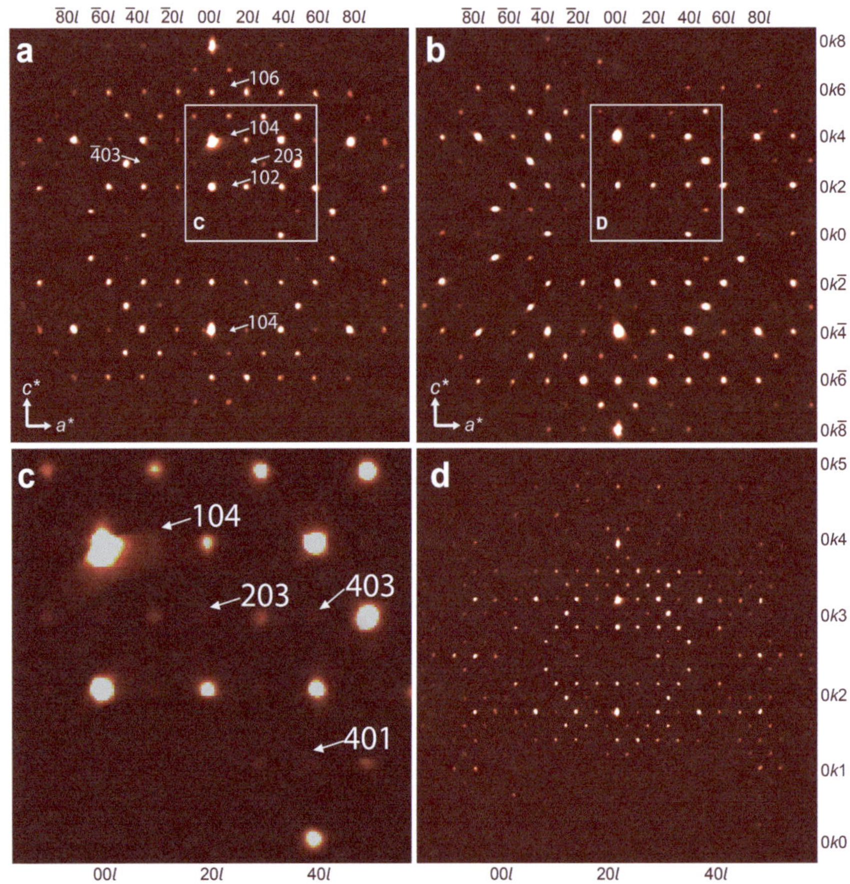

Fig. 3.38 Reconstructed (010) sections of the single-crystal XRD patterns in the reciprocal space for a sample of an auto-epitaxial B-containing vesuvianite (with 1.5 B *apfu*) (**a**) and a parent wiluite crystal (**b**) and their enlarged fragments (C, D). The white arrows show the diffraction peaks violating the space group *P*4/*nnc*

the *Y*1 position corresponds to 18.16 e^-, and its total population is $(Mg_{0.56}Fe_{0.44})_{1.00}$ (Mg was chosen as the lighter element, since the NMR data indicate the absence of ^{V}Al in the studied sample). The predominance of magnesium over iron in the five-coordinated position in structural models by Groat et al. (1994a, 1996a, 1998), Bellatreccia et al. (2005a) is confirmed by the Mössbauer spectroscopy data.

In the VSMs titanium occurs only in the octahedral positions *Y*2 and *Y*3 (Aksenov et al. 2016), therefore the possibility of incorporation of Ti into the *Y*1 position, as proposed by Groat et al. (1994a), was excluded. The average length of the *Y*1–O bond is 2.080 Å, which is in a good agreement with its refined population by cations other than Ti^{4+}. Moreover, Ti^{4+} in the five-fold square pyramid coordination forms a short apical "titanyl" bond Ti=O which is 1.61–1.71 Å long (Chukanov et al. 2023b)

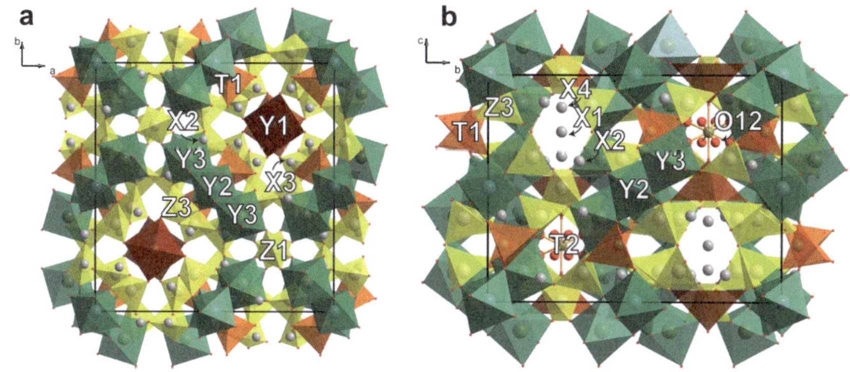

Fig. 3.39 The projection of the wiluite crystal structure along the *c* (**a**) and *a* (**b**) axes. The *Z*- and *Y*-centered polyhedra are highlighted in yellow, and red, respectively, the *X*-positions are shown with gray balls, and additional *T*1 tetrahedra are highlighted in orange

whereas in the most Ti–rich VSMs this bond length is close to 2 Å (Aksenov et al. 2016).

Incorporation of boron and aluminum in the additional *T*1 and *T*2 positions leads to the appearance of an additional O12 position with a partial occupancy, as well as splitting of the O7 position into O7A and O7B (Fig. 3.40). Similar splitting of the O7 and O11 positions associated with the introduction of B atoms into the *T*1 position is noted in the works by Groat et al. (1994a), Bellatreccia et al. (2005a). Both split O7A and O7B positions were refined anisotropically; the distance between them is 0.49 Å.

The scattering factor for the *T*1 position corresponds to 4.9 e^- and could correspond to the situation when boron atoms almost completely (with the population $^{T1}B_{0.98}$) occupy this position. However, for pure borate tetrahedra the <B–O> bond length should be in the range of 1.46–1.50 Å (Filatov and Bubova 2008), which is significantly less than the observed <*T*1–O> length of 1.557 Å. Taking into account the total boron content in the studied sample and the presence of Al^{3+} in tetrahedral coordination (according to the ^{27}Al solid-state NMR data), the final occupancy of the *T*1 position was refined as $(B_{0.51}Al_{0.18}\square_{0.31})_{1.00}$. This study confirms the hypothesis about the partial occurrence of Al in the *T*1 position of wiluite, advanced by Groat et al. (1994a, b).

Splitting of the O7 position depends on the degree of occupancy the *T*1 position (Fig. 3.41a). For example, when *T*1 is filled by 7%, the <*T*1–O7> mean bond length is equal to 2.01 Å (Galuskin et al. 2007a). In the case of 62% (B + Al) occupancy, the <*T*1–O7B> bond length is equal to 1.64 Å (Groat et al. 1996a). When the O7 position is split, there are two possible local configurations (Fig. 3.41b). The O7B position is populated only if the *T*1 position is populated. The O7A position is populated when the *T*1 site vacant. It should be noted that the splitting in the O7 position is in no way connected with the occupancy of the *Z*2 position (the lengths of the *Z*2–O7A and *Z*2–O7B bonds are 1.672 and 1.654 Å, respectively).

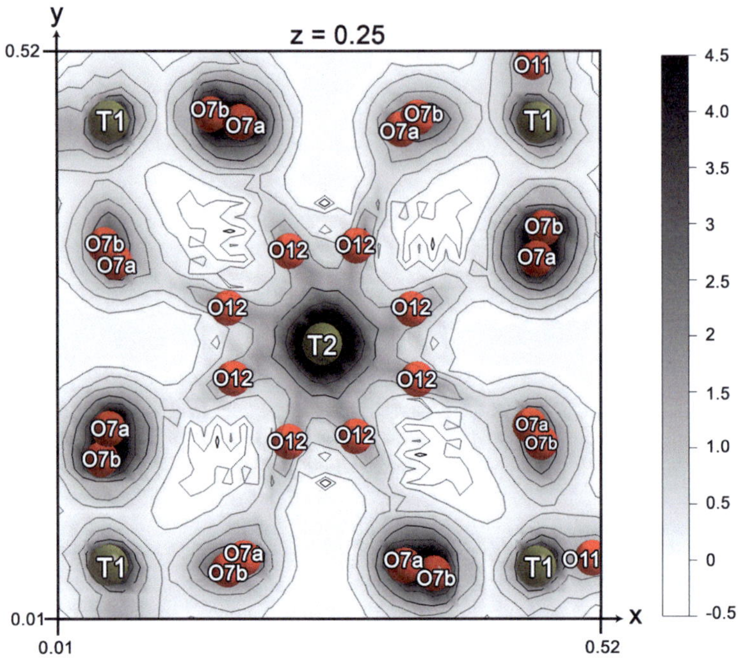

Fig. 3.40 The (001) section of the electron density distribution map for wiluite around the fourfold axis, at the 0.25*z* level. Contour intervals are drawn with a step of 0.5 e Å^{-3}

The *T*2 position is coordinated by the low-occupied O12 (×8) and O10 (×2) sites (Fig. 3.42). In each local situation with this configuration, only two O12 sites and one O10 site can be populated. Any other population is unlikely due to short O12–O12 distances in the 1.296–1.703 Å range. The average <*T*2–O> distance is equal to 1.374 Å, which is in a good agreement with the bond lengths for boron in triangular coordination. The refined populations of the positions O10 and O12 are 0.64 and 0.19, respectively, i.e. relatively close to the calculated theoretical values of 0.5 and 0.25.

The final crystal-chemical formula of the studied wiluite sample is:

$$^{X1}(\mathrm{Ca})_{2.00}\,^{X2}(\mathrm{Ca})_{8.00}\,^{X3}(\mathrm{Ca})_{8.00}\,^{X4}(\mathrm{Ca})_{1.00}$$
$$^{Y1}\left(\mathrm{Mg}_{0.56}\mathrm{Fe}^{2+}{}_{0.27}\mathrm{Fe}^{3+}{}_{0.17}\right)_{\Sigma 1.00}\,^{Y2}\left(\mathrm{Al}_{3.90}\mathrm{Fe}^{2+}{}_{0.10}\right)_{\Sigma 4.00}$$
$$^{Y3}(\mathrm{Al}_{3.82}\mathrm{Mg}_{3.14}\mathrm{Ti}_{0.63}\mathrm{Fe}^{3+}{}_{0.21}\mathrm{Fe}^{2+}{}_{0.16}\mathrm{Mn}_{0.04})_{\Sigma 8.00}$$
$$^{Z1}(\mathrm{Si})_{2.00}\,^{Z2}(\mathrm{Si})_{8.00}\,^{Z3}(\mathrm{Si})_{8.00}(\mathrm{O})_{68.00}$$
$$^{T1+T2}(\mathrm{B}_{3.04}\mathrm{Al}_{0.72}\square_{1.24})_{\Sigma 5.00}\,^{\mathrm{W}}(\mathrm{O}_{8.32}\mathrm{OH}_{0.96})_{\sigma 9.28}12\mathrm{O}_{1.52}$$

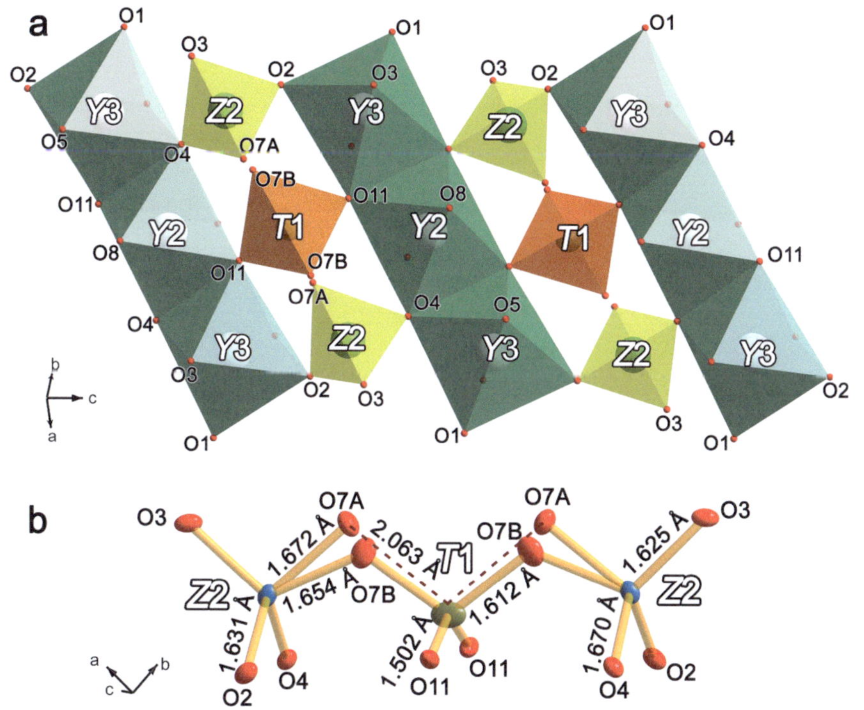

Fig. 3.41 A fragment of the wiluite structure formed by the *Y*3–*Y*2–*Y*3 trimers of octahedra connected via *Z*2 tetrahedra and an additional $T1O_4$ tetrahedron ($T1 = $ B, Al, □) by shared vertices (**a**), and the local configuration of the *Z*2–*T*1–*Z*2 group of tetrahedra sharing common vertices with split O7A and O7B positions (**b**). Thermal ellipsoids are drawn with a 50% probability

3.3 Cyprine and Cu-Bearing Vesuvianite

In 1821, the famous Swedish chemist J. Berzelius described a copper-bearing blue vesuvianite-related mineral from the province of Sauland, Norway under the name "cyprine" (Berzelius 1821). Later it was shown that samples of "cyprine" from Norway contain only 0.73 wt% of CuO (Lindström 1888). The specimen of blue Cu-bearing vesuvianite studied in this work is shown in Fig. 1.25. Detailed crystal chemistry studies of Cu-containing VSMs (Shannon 1922; Fitzgerald et al. 1986b; Ohkawa et al. 1992; Dyrek et al. 1992; Panikorovskii et al. 2016e) showed that copper occupies the *Y*1 position in the tetragonal pyramid.

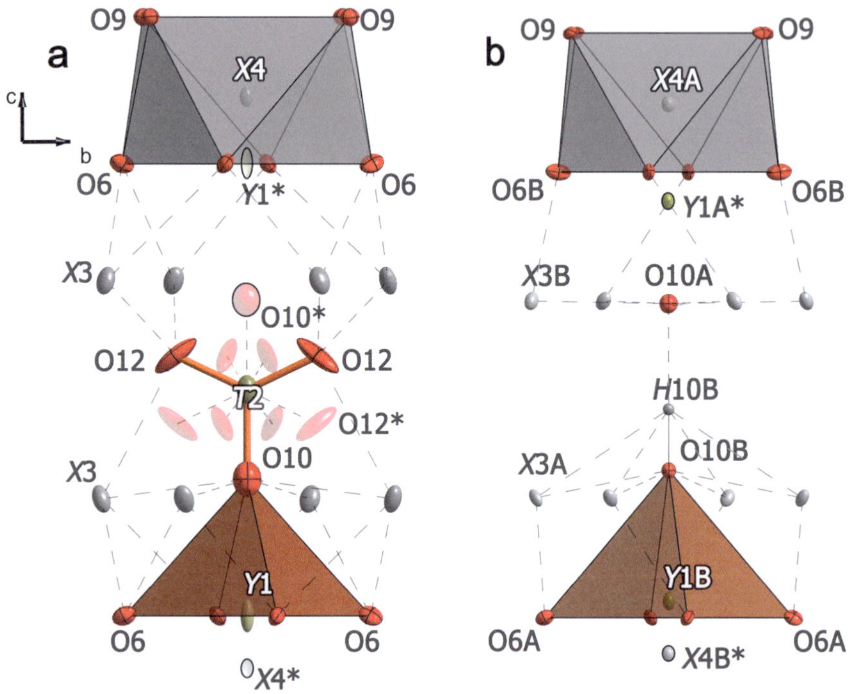

Fig. 3.42 The local configuration of cationic rods along the *a* axis in the structure of wiluite (**a**) and in the structure of alumovesuvianite, according to Panikorovskii et al. (2017a) (**b**). Positions marked with an asterisk are vacant in this configuration. Thermal ellipsoids are drawn with a 50% probability level

3.3.1 Occurrence and Physical Properties

A Cu-dominant VSM was discovered in specimens from the Wessels mine, Kalahari manganese field, South Africa and approved by the IMA CNMNC as a new mineral species with the historical name cyprine (Panikorovskii et al. 2017e). Cyprine occurs in association with apatite, calcite, and henritermierite. The idealized formula of the mineral is $Ca_{19}Cu^{2+}(Al_{11}Mg)(SiO_4)_{10}(Si_2O_7)_4O(OH)_9$. Cyprine (Fig. 3.43) forms irregular aggregates of deep red prismatic crystals (up to 1 cm in length and 3 mm in thickness) with the {100}, {110} and {331} faces.

The refractive indices of cyprine are slightly higher than those of most other minerals of the vesuvianite supergroup: ω = 1.744(2), ε = 1.732(2). The mineral shows strong pleochroism: from O = dark purple, E = pale red to O = dark red, E = pale yellow (O >> E). In addition to the Wessels field in South Africa, ciprine was found at the Franklin mine field, New Jersey, USA.

Fig. 3.43 Chaotic aggregates of dark red prismatic crystals of cyprine in association with pink calcite aggregates (holotype cyprine, A, I. V. Pekov's sample)

3.3.2 Chemical Composition

Cyprine exhibits distinct chemical zonality with significant variations of the Mg, Al, and Mn (Fig. 3.44a–f) components. The manganese-rich edge zones of its crystals are characterized by the ordering of divalent and trivalent manganese in the of the *Y*2A,B and *Y*3A,B octahedral positions. The relative increase in the manganese content occurs according to the substitution schemes $Mg^{2+} + Al^{3+} \leftrightarrow Fe^{3+} + Mn^{2+}$ and $Al^{3+} \leftrightarrow Mn^{3}$, which leads to an increase in the unit-cell parameters of the marginal zone enriched in manganese (a = 15.5699 (2) Å, c = 11.8042 (2) Å) relative to the more magnesian axial zone (a = 15.5173 (4) Å, c = 11.8230 (5) Å). The composition of individual zones of cyprine crystals shown in Fig. 3.44f is given in Table 3.9.

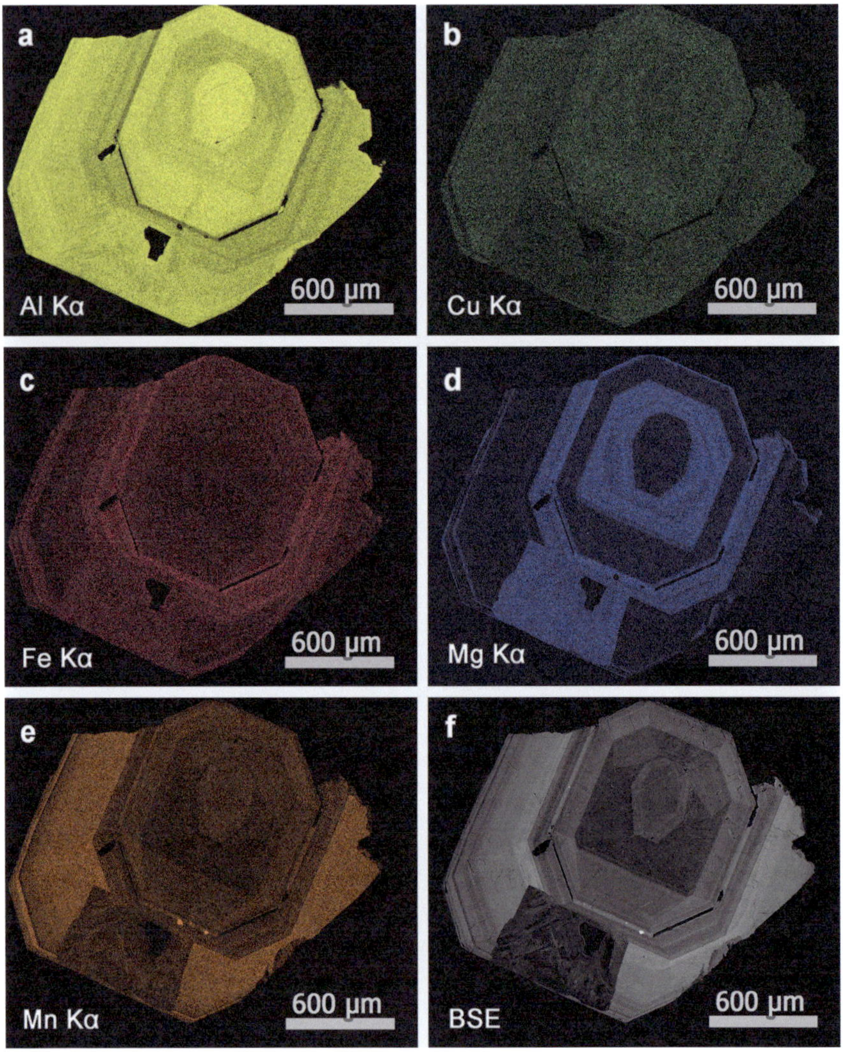

Fig. 3.44 Cross-section images (001) of one of the cyprine crystals in the characteristic rays Al *K*α (**a**), Cu *K*α (**b**), Fe *K*α (**c**), Mg *K*α (**d**), Mn *K*α (**e**) and back-scattered electrons (**f**)

3.3.3 Thermogravimetry and Differential Scanning Calorimetry

The H_2O content in cyprine was determined using thermogravimetry combined with differential scanning calorimetry (Fig. 3.45). A total weight loss of 3.30 wt% of H_2O corresponds to 10.7 OH^- groups per formula unit, which is somewhat higher than the theoretical content $(OH)_9$. A possible excess of hydroxyl-ion may be due to the

Table 3.9 Chemical composition of cyprine (major components)

Sample	1	2	3	4	5	6
SiO_2	36.88	37.22	36.99	36.54	36.56	36.44
Al_2O_3	14.82	14.61	15.83	13.64	13.73	13.73
Fe_2O_3	0.44	0.72	0.83	2.84	2.54	2.13
Mn_2O_3	2.62	2.82	1.56	5.85	6.07	6.22
MgO	3.23	3.03	3.21	0.94	1.25	1.13
CaO	36.71	36.81	36.62	36.28	36.18	36.09
CuO	2.76	2.36	1.89	1.87	1.77	1.96
Cr_2O_3	0.16	0.33	0.23	0.02	0.04	0.05
H_2O*	2.79	2.80	2.79	2.76	2.76	2.75
Sum	97.62	97.9	97.16	97.98	98.14	97.75
apfu						
Si	17.82	17.93	17.91	17.86	17.92	17.9
Al	8.44	8.3	9.03	7.86	7.93	7.95
Fe^{3+}	0.16	0.26	0.3	1.04	0.94	0.79
Mn	0.96	1.03	0.57	2.18	2.26	2.33
Mg	2.33	2.18	2.32	0.68	0.91	0.83
Ca	19.00	19.00	19.00	19.00	19.00	19.00
Cu	1.01	0.86	0.69	0.69	0.66	0.73
Cr	0.06	0.13	0.09	0.01	0.02	0.02
H	9.00	9.00	9.00	9.00	9.00	9.00

* The H_2O content was calculated bas on single-crystal XRD data

presence of additional OH groups at the O10A position in the channels of the crystal structure or a garnet-type substitution (Panikorovskii et al. 2017e).

3.3.4 Infrared Spectroscopy

The infrared spectrum of holotype cyprine from the Wessels mine with empirical formula $Ca_{19}(Cu_{0.91}Mg_{0.09})_{\Sigma1.00}(Al_{8.38}Mg_{1.64}Mn^{3+}_{1.87}Fe^{3+}_{0.29}Cr_{0.10})_{\Sigma12.28}$ $Si_{17.86}O_{67.86}[(OH)_{9.28}O_{0.72}]$ shown at (Fig. 3.46). The spectrum very similar to the spectrum of Mn- and Cu-containing VSM from the N'Chaning III mine (South Africa), but differs significantly from the spectrum of Norwegian "cyprine" in the region of bending and stretching vibrations of silicate groups, stretching vibrations of the M···O bonds, and stretching vibrations of the O–H bonds.

An additional band is observed in the spectrum of cyprine at 3375 cm^{-1}, which is absent in the review by Groat et al. (1995b), probably, corresponds to O–H stretching vivrations of the Cu···O10A–H···O10B fragment (Panikorovskii et al. 2017b). On the

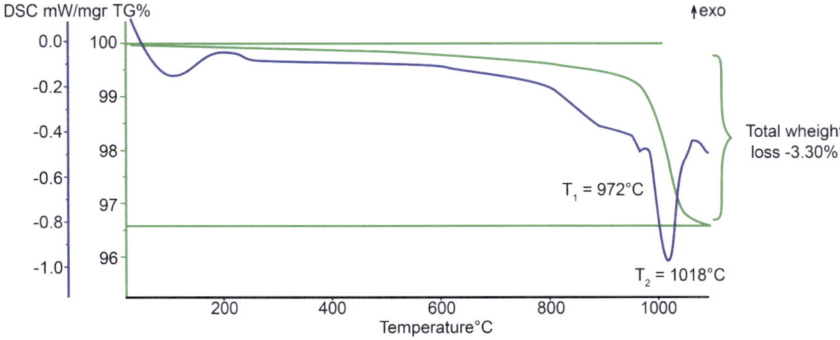

Fig. 3.45 Thermogravimetry (green line, wt%) and differential scanning calorimetry (blue line) curves of cyprine

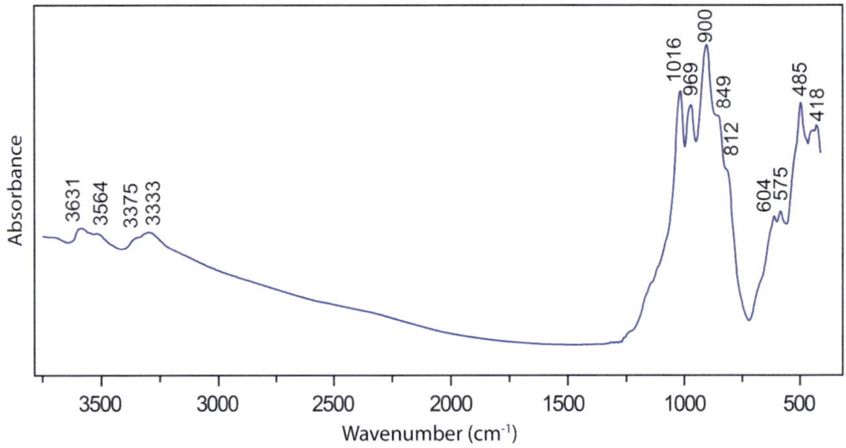

Fig. 3.46 Infrared spectrum of cyprine

contrary, in the spectra of Norwegian Cu-bearing vesuvianite there are no absorption lines in the most high-frequency region of 3600–3700 cm^{-1}, corresponding to weak O10–H…O7 bonds in low-symmetry VSMs (Chukanov et al. 2018a). In the region of stretching and bending vibrations of the apical Si–O bonds and Si–O–Si bridges, there are two bands of approximately equal intensity located at 607 and 580 cm^{-1} which are characteristic of the VSMs with the *P*4/*n* symmetry (Borovikova and Kurazhkovskaya 2006).

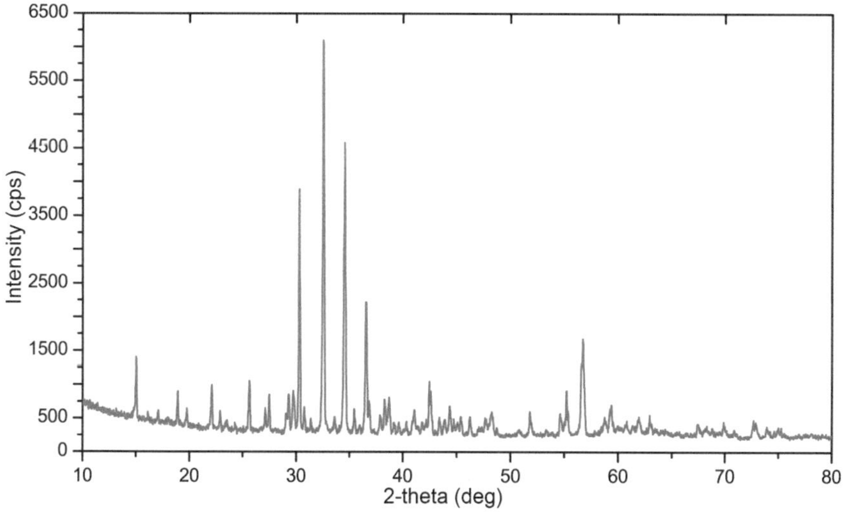

Fig. 3.47 Powder X-ray diffraction pattern of cyprine

3.3.5 Powder X-Ray Diffraction

The powder X-ray diffraction pattern of cyprine (Fig. 3.47) is similar to that of most VSMs. The values of the unit cell parameters for cyprine refined from the powder X-ray diffraction data for a zoned crystal (a = 15.5652(5) Å, c = 11.7921(4) Å) lie between the values for its Mn-enriched (a = 15.5699, c = 11.8042 Å) and Mg-enriched (a = 15.5173, c = 11.8230 Å) zones determined from their single crystal XRD data. On the graph of the a and c parameters, these values do not substantially differ from those of the majority of VSMs, since the substitution of cations in the $Y1$ position does not have a significant effect on the unit cell parameters. However, the parameters may be essentially different if a significant portion in the $Y2$ and $Y3$ positions of Al is substituted by a larger Mn^{3+} cation. Data on intensities and positions of lines and their *hkl* indices are given in Table A.25.

3.3.6 Optical Absorption Spectroscopy

The optical absorption spectra of cyprine (Fig. 3.48) enriched in Mg contain lines at 23500, 21800, 18000, 15400 and 12000 cm^{-1}, and an additional peak at 24000 cm^{-1} is present in the spectra of its high manganese zone. All bands, with the exception of the peak at 23500 cm^{-1}, have E||O polarization. These lines, except the peak at 15400 cm^{-1}, correspond to d–d transitions characteristic of Mn^{3+} with the coordination numbers of 5 and 6 (Platonov et al. 1995). Heating the samples to 600 °C in a reducing atmosphere leads to a decrease in the intensities of the peaks at 18000,

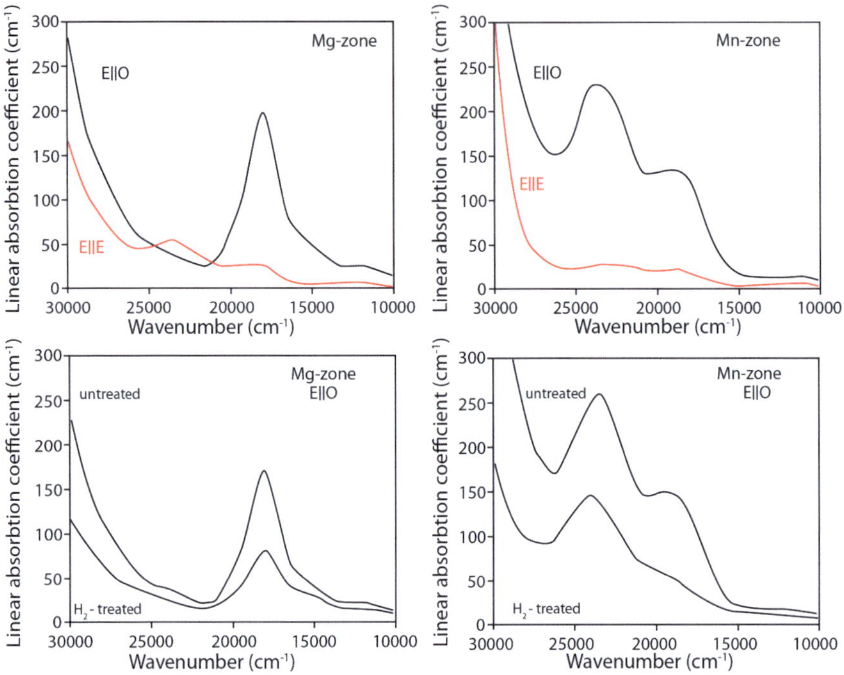

Fig. 3.48 Optical absorption spectra of cyprine obtained for zones enriched in Mg and Mn in initial samples (in the upper figures) and their comparison with the spectra of these samples after heating up to 600 °C in a hydrogen atmosphere (in the lower figures)

12000 and 21800 cm^{-1}, corresponding to the reduction of Mn^{3+} in the *Y*2 and *Y*3 positions. The peak at 15400 cm^{-1} corresponds to Cu^{2+} in the five-coordinated *Y*1 position (Dyrek et al. 1992).

3.3.7 Single-Crystal X-Ray Structural Analysis

Cyprine crystallizes in the *P*4/*nnc* and *P*4/*n* space groups. A high-symmetric variety of this mineral was found in Franklin Mine, New Jersey (USA) and Kleppan, Sauland (Norway) deposits (Fitzgerald et al. 1986b; Hålenius et al. 2016), while the low-symmetry modification is characteristic of the holotype cyprine from the Wessels mine, Kalahari (South Africa) (Panikorovskii et al. 2017e).

For the X-ray diffraction studies, fragments of cyprine from magnesian (1) and manganese (2) zones were selected (Fig. 3.49). For the magnesian rich cyprine, only three 0*kl*-type reflections with $k + l \neq 2n$ and six *hk*0-type reflections with $h + k \neq 2n$ were observed. For the manganese zone of a cyprine crystal, 23 0*kl*-type reflections with $k + l \neq 2n$ and one *hk*0-type reflection with $h + k \neq 2n$ were observed. The analysis of the glide plane violating reflections showed that for both samples the *P*4/

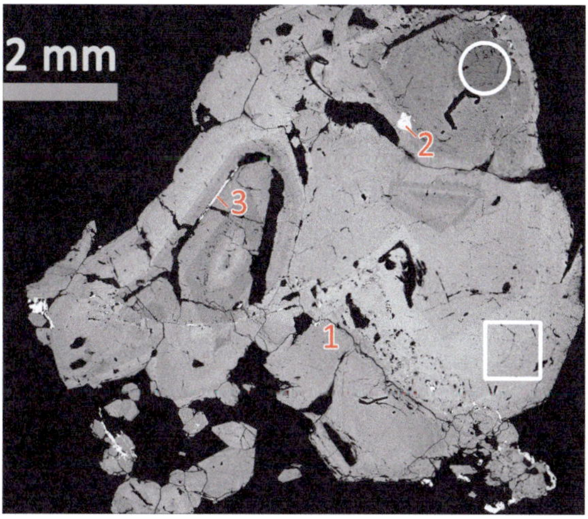

Fig. 3.49 Image of a cyprine aggregate (1) with inclusions of henritermierite (2) and hydroxylapatite (3) in back-scattered electrons. The white circle indicates the area of Mg-containing cyprine, while the square shows Mn-rich cyprine, both studied by single-crystal X-ray diffraction

n space group is most likely. A small number of violating reflections is associated with a minor difference between the population of the pairs of the channel positions *Y*1A, *X*4B and *Y*1B, *X*4A. The structures were refined to $R_1 = 0.034$ and 0.035 for 2848 and 2071 independent reflections with the intensities $|F_o| \geq 4\sigma$ for the Mg- and Mn-enriched zones, respectively (Table A.26).

In the crystal structure of cyprine, the positions *Z*1A,B, *Z*2A,B and *Z*3A,B are populated by silicon atoms (Tables A.27 and A.28). The refined occupancy factors of the Z1A,B positions for both zones are almost equal to each other: 0.94 for Z1A and 0.97 for Z1B. This indicates the presence of a small number of hydrogarnet $[H_4O_4]^{4-}$ defects substituting SiO_4^{4-} groups (Galuskin 2005) which is characteristic of low-temperature VSMs (Armbruster and Gnos 2000c; Galuskin et al. 2003b).

The positions *X*1, *X*2A,B, *X*3A,B and *X*4A,B in the structures of both zones of the mineral are populated exclusively by Ca atoms. The channel positions (*X*4A and *X*4B) were refined with the occupancy factors of 0.475 and 0.525 for zone-1, and 0.663 and 0.347 for zone-2, respectively.

The main difference between the structures of magnesian and manganese varieties of cyprine is the population of the *Y*-group and *X*4A,B positions. In the structure of cyprine, Cu^{2+} atoms (Fig. 3.50) occupy the *Y*1A,B positions, as evidenced by their scattering factor which corresponds to ~29 e^-, and the Jan–Teller distortion of the Cu–O bonds *Y*1A–O6(4×) = 2.043(3) Å and *Y*1A–O10 = 2.258(6) Å for zone-1, and *Y*1B–O6(4×) = 2.010(4) Å and *Y*1B–O10 = 2.274(7) Å for zone-2 (Tables A.29 and A.30).

The crystal chemical formulas for magnesian (1) and manganese-rich (2) zones of cyprine are:

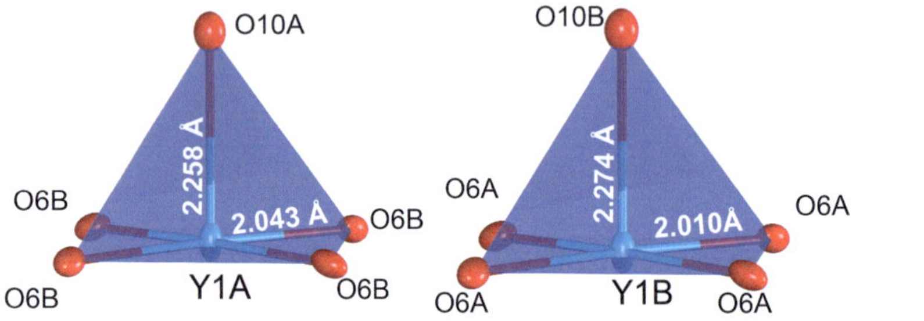

Fig. 3.50 Coordination and bond lengths for the *Y*1A,B positions in magnesian cyprine (zone 1) from the Wessels mine, South Africa

$^{X1}(Ca)_{\Sigma 2.00}$ $^{X2A}(Ca)_{\Sigma 4.00}$ $^{X2B}(Ca)_{\Sigma 4.00}$ $^{X3A}(Ca)_{\Sigma 4.00}$ $^{X3B}(Ca)_{\Sigma 4.00}$
$^{X4A}(Ca)_{\Sigma 0.475}$ $^{X4B}(Ca)_{\Sigma 0.525}$
$^{Y1A}(Cu_{0.50}Mg_{0.025})_{\Sigma 0.525}$ $^{Y1B}(Cu_{0.45}Mg_{0.025})_{\Sigma 0.475}$ $^{Y2A}(Al)_{\Sigma 2.00}$ $^{Y2B}(Al)_{\Sigma 2.00}$
$^{Y3A}(Al_{2.80}Mn^{3+}{}_{0.56}Mg_{0.50}Fe^{3+}{}_{0.10}Cr_{0.04})_{\Sigma 4.00}$
$^{Y3A}(Al_{2.70}Mn^{3+}{}_{0.63}Mg_{0.50}Fe^{3+}{}_{0.12}Cr_{0.05})_{\Sigma 4.00}$
$^{Z1A}(Si_{0.94}[H_4O_4]_{0.06})_{\Sigma 1.00}$ $^{Z1B}(Si_{0.97}[H_4O_4]_{0.03})_{\Sigma 1.00}$
$^{Z2A}(Si)_{\Sigma 4.00}$ $^{Z2B}(Si)_{\Sigma 4.00}$ $^{Z3A}(Si)_{\Sigma 4.00}$
$^{Z3B}(Si)_{\Sigma 4.00}(O)_{\Sigma 68.00}(OH_{9.00}O)_{\Sigma 10.00}$for zone-1;
$^{X1}(Ca)_{\Sigma 2.00}$ $^{X2A}(Ca)_{\Sigma 4.00}$ $^{X2B}(Ca)_{\Sigma 4.00}$ $^{X3A}(Ca)_{\Sigma 4.00}$ $^{X3B}(Ca)_{\Sigma 4.00}$
$^{X4A}(Ca)_{\Sigma 0.663}$ $^{X4B}(Ca)_{\Sigma 0.337}$ $^{Y1A}(Cu_{0.30}Mg_{0.037})_{\Sigma 0.337}$
$^{Y1B}(Cu_{0.60}Mg_{0.063})_{\Sigma 0.663}$ $^{Y2A}(Al_{1.48}Mn^{3+}{}_{0.40}Fe^{3+}{}_{0.12})_{\Sigma 2.00}$ $^{Y2B}(Al_{1.74}Mn^{3+}{}_{0.20}Fe^{3+}{}_{0.06})_{\Sigma 2.00}$
$^{Y3A}(Al_{1.90}Mn^{3+}{}_{0.75}Fe^{3+}{}_{0.72}Mn^{2+}{}_{0.40}Mg_{0.22}Cr_{0.01})_{\Sigma 4.00}$
$^{Y3B}(Al_{2.82}Mn^{3+}{}_{0.45}Fe^{3+}{}_{0.34}Mn^{2+}{}_{0.20}Mg_{0.18}Cr_{0.01})_{\Sigma 4.00}$ $^{Z1A}(Si_{0.94}[H_4O_4]_{0.06})_{\Sigma 1.00}$
$^{Z1B}(Si_{0.97}[H_4O_4]_{0.03})_{\Sigma 1.00}$ $^{Z2A}(Si)_{\Sigma 4.00}$ $^{Z2B}(Si)_{\Sigma 4.00}$ $^{Z3A}(Si)_{\Sigma 4.00}$
$^{Z3B}(Si)_{\Sigma 4.00 \Sigma 68.00}(OH_{9.00}O)_{\Sigma 10.00}$for zone-2.

3.4 Alumovesuvianite

The content of Al_2O_3 in vesuvianite varies within the limits 5.54–21.62 wt% (Fitzgerald et al. 1986a; Groat et al. 1992a). In the VSMs structures, Al atoms usually occupy octahedral *Y*2 and *Y*3 positions (Groat et al. 1992a); the presence of five-coordinated aluminum (at the *Y*1 position) was noted only in samples of VSMs of low-temperature genesis with low iron content from Asbestos (Canada), Luning (USA), Georgetown, (USA) and Pease Lunghin (Switzerland) (Fitzgerald et al. 1986a; Phillips et al. 1987; Olejniczak and Żabiński 1996).

Fig. 3.51 Aggregate of alumovesuvianite crystals

3.4.1 Occurrence and Physical Properties

Alumovesuvianite was found in in the rodingite zone of an abandoned mine situated in the Jeffrey Mine field, Canada, in association with diopside and prehnite (Panikorovskii et al. 2017a). The idealized formula of the mineral can be written as $Ca_{19}Al(Al_{10}Mg_2)(SiO_4)_{10}(Si_2O_7)_4O(OH)_9$. Alumovesuvianite (Fig. 3.51) forms chaotic aggregates of crystals on granular aggregate of diopside. The size of individual crystals is up to 4 × 4 × 6 mm. The crystals are usually colorless or with a pinkish tinge, associated with the presence of trace amounts of trivalent manganese.

Crystals of alumovesuvianite are short-prismatic or tabular, sometimes with numerous dipyramids. The main crystal forms are {100}, {110}, {210}, {111}, {101}, {001}, {112}, and {221}. Alumovesuvianite has rather low refractive indices: $\omega = 1.725(2)$, $\varepsilon = 1.722(2)$. Pleochroism is not observed.

3.4.2 Chemical Composition

In the chemical composition of alumovesuvianite, as well as magnesiovesuvianite iron is practically absent (Table 3.10). In this mineral, Mg is the only bivalent charge-balancing component at the octahedral sites and Mn^{3+} cations occurring in minor

Table 3.10 Chemical composition of alumovesuvianite from Asbestos, Canada (1–4) and Georgtown, USA (5, sample 27952 FM), wt%

Sample	1	2	3	4	5
SiO_2	37.08	37.45	37.10	37.34	37.42
Al_2O_3	18.82	18.90	19.15	18.85	18.34
Fe_2O_3	0.22	0.14	0.20	0.21	0.21
Mn_2O_3	0.67	0.50	0.45	0.39	
MgO	2.48	2.21	1.69	2.30	2.68
CaO	36.57	36.92	36.75	36.85	37.21
Cr_2O_3	–	–	–	–	1.14
H_2O*	2.78	2.81	2.80	2.81	2.83
Sum	98.62	98.93	98.14	98.75	99.83
apfu					
Si	17.98	17.99	17.90	17.97	17.83
Al	10.76	10.70	10.89	10.69	10.30
Fe	0.08	0.05	0.07	0.08	0.07
Mn	0.25	0.18	0.17	0.14	
Mg	1.79	1.58	1.21	1.65	1.90
Ca	19.00	19.00	19.00	19.00	19.00
Cr					0.43
H	9.00	9.00	9.00	9.00	9.00

* The H_2O content was calculated based on single-crystal XRD data
Dash—content below detection limit

amounts is the main chromophore. The Cr_2O_3 impurity can also act as a chromophore (Fig. 1.27c), its maximum content was 1.14 wt% in sample 51073 MF from Georgetown (USA).

3.4.3 *Thermogravimetry and Differential Scanning Calorimetry*

The water content in the mineral was measured using DSC/TG analysis (Fig. 3.52). The total weight loss was 2.61 wt%, which is consistent with the theoretical H_2O content of 2.82 wt%. The main step of the mass loss corresponds to the endothermic peak at 1035 °C, which is associated with the disappearance of OH groups.

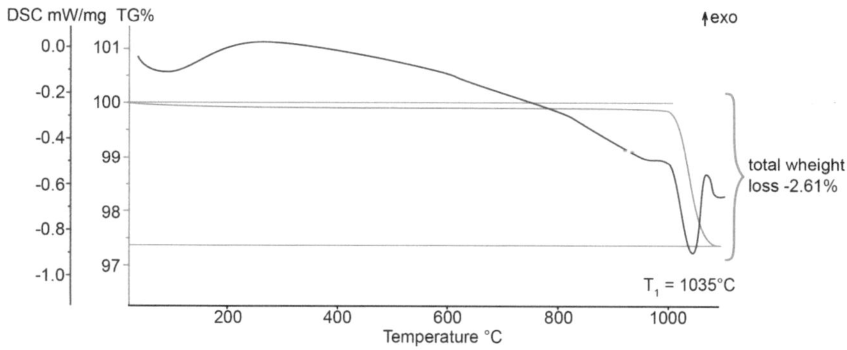

Fig. 3.52 TG (green line, wt%) and DSC (blue line) curves of alumovesuvianite

3.4.4 *Infrared Spectroscopy*

One of the most informative methods for identifying magnesio- and alumovesuvianite is infrared spectroscopy, since data on the chemical composition are insufficient for the correct diagnostics these two minerals. In this context, one should pay attention to the significant difference in the spectra of these minerals in the region of stretching vibrations of the O–H bonds. In this region, the spectra of alumovesuvianite (Fig. 3.53) show the presence of the peaks at 3671, 3632, 3527, 3407, 3212 and 3051 cm^{-1} corresponding to the bands A, B, E, G, J, and M in accordance with the work by Groat et al. (1995b), whereas in the spectrum of magnesiovesuvianite there are only three bands at 3630, 3450 and 3200 cm^{-1}. It should also be noted that in the region of Si–O stretching vibrations of isolated SiO_4 tetrahedra of the alumovesuvianite, the band at 908 cm^{-1} is split into two bands with the maxima at 919 and 897 cm^{-1}. It is also important to note the presence of the band at 713 cm^{-1}, which is rare for the VSMs, and corresponds to the bending vibrations involving OH^- groups (OH libration mode).

3.4.5 *Solid-State Nuclear Magnetic Resonance*

Similar to the spectra of other low-temperature low-iron VSMs, the solid-state ^{27}Al NMR spectrum of alumovesuvianite contains two bands (Fig. 3.54). The peak with a chemical shift of –7.96 ppm corresponds to Al with octahedral coordination and the peak at 39.2 ppm is assigned to Al with five-fold coordination. The relative integrated intensities of the ^{VI}Al and ^{V}Al bands correspond to 91.8 to 8.2%, which corresponds to the total population of the *Y*1A,B positions mainly by aluminum. The bands at ±120 ppm are the spinning sidebands.

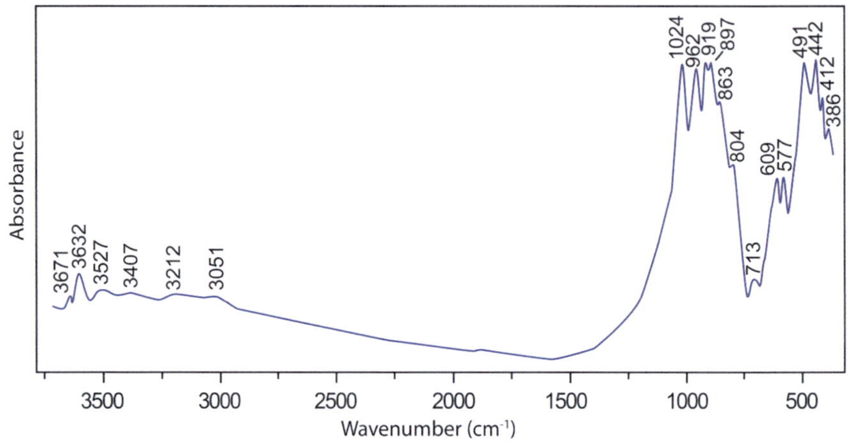

Fig. 3.53 Infrared spectrum of alumovesuvianite

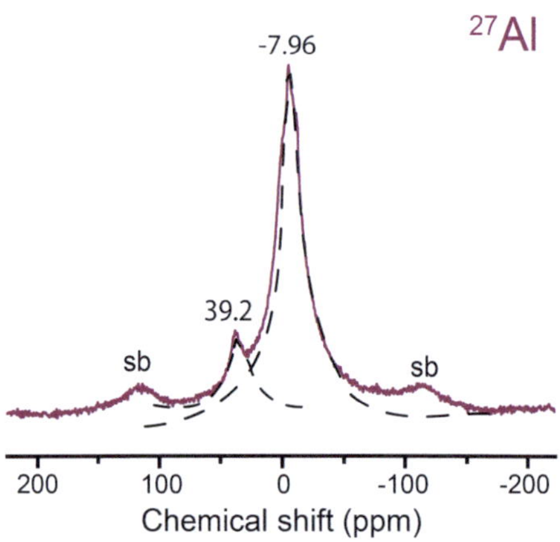

Fig. 3.54 The ^{27}Al solid-state NMR spectrum of alumovesuvianite. The dashed lines show the approximating curves

3.4.6 Powder X-Ray Diffraction

The powder X-ray diffraction pattern of alumovesuvianite is shown in Fig. 3.55. The refined parameters of the unit cell are $a = 15.5603(5)$ Å and $c = 11.8467(4)$ Å, which is in a good agreement with the single-crystal X-ray diffraction data. Data on the intensities and positions of the lines and their *hkl* indices are given in Table A.31.

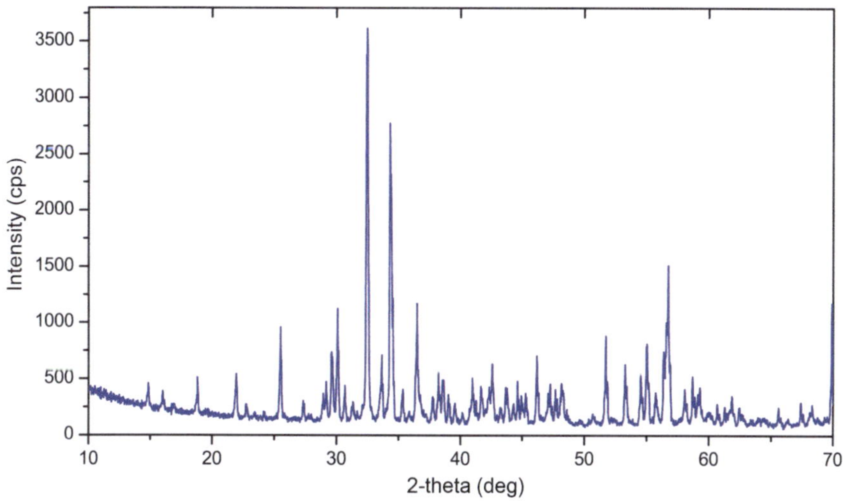

Fig. 3.55 Powder X-ray diffraction pattern of alumovesuvianite

3.4.7 Single-Crystal X-Ray Structural Analysis

For the structurally characterized sample of alumovesuvianite, 624 violating reflections with intensities $I > 3\sigma(I)$ for the *P4/nnc* space group were observed. Most of them (616) corresponded to 0*kl* and *hkl* reflections with $k + l \neq 2n$, where $h = k$ and $l \neq 2n$ (Armbruster and Gnos 2000a), which led to the choice of the space group *P4/n*. The presence of eight *hk*0-type reflections with intensities in the range $3\sigma(I) < I < 8\sigma(I)$, for which $h + k \neq 2n$, is also noted. The presence of these reflections indicates the presence of a small number of domains with the *P4nc* symmetry. The structure of alumovesuvianite was refined to $R_1 = 0.036$ for 3098 independent reflections with $|F_o| \geq 4\sigma$ (Table A.32).

The average <Z–O> bond lengths, as well as the scattering factors of the *Z*1A,B; *Z*2A,B and *Z*3A,B sites correspond to the population of the *Z*-positions exclusively by silicon atoms (Tables A.33 and A.34). The positions *X*1, *X*2A,B, *X*3A,B and *X*4A,B with coordination numbers 7–9 are populated by calcium atoms. The refined population of the channel positions *X*4A and *X*4B is 0.07 and 0.93, respectively.

The occupancy of the channel positions *Y*1A,B in the tetragonal–pyramidal coordination with aluminum is the species-defining feature of alumovesuvianite. Previous structural studies (Fitzgerald et al. 1986a) of aluminum-enriched iron-free vesuvianite from Asbestos (Québec, Canada) showed that Al^{3+} cations can occupy the *Y*1A,B positions. In the structure of the studied sample, Al dominates in the *Y*1A,B positions in total (Fig. 3.56). The average bond lengths <*Y*1A–O> and <*Y*1B–O> are 1.916 (with the refined occupancy of $Al_{0.93}$) and 2.150 (with the refined occupancy of $Fe_{0.07}$) Å, respectively. The <*Y*1A–O> bond length of 1.916 Å is significantly greater than the theoretical <Al–O> bond length of 1.842 Å for Al^{3+} with five-fold

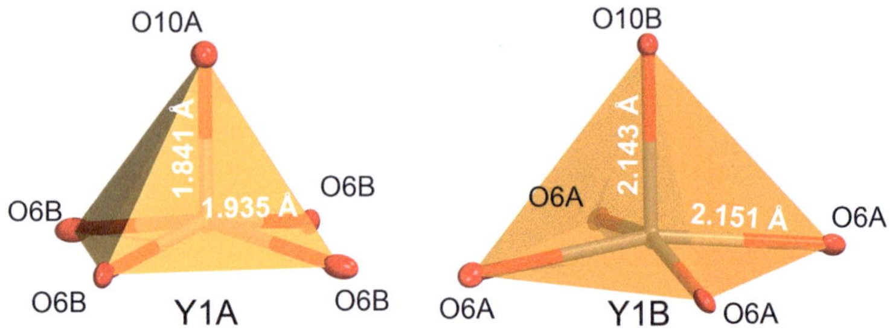

Fig. 3.56 Coordination and bond lengths for *Y*1A,B positions in alumovesuvianite from Jeffrey Mine, Canada

coordination (Hawthorne and Gagné 2024) which may be due to the presence of Mn^{3+} admixture at the *Y*1A site. The total scattering factor for the *Y*1A,B positions corresponds to 16.31 e^{-}, which, taking into account chemical and solid-state NMR data, corresponds to the total occupancy $(Al_{0.73}Mn_{0.20}{}^{3+}Fe_{0.07}{}^{3+})_{\Sigma 1.00}$ for this pair of positions. The octahedral positions are populated by aluminum atoms, as evidenced by the corresponding <*Y*2A–O> and <*Y*2A–O> bond lengths of 1.894 and 1.898 Å, respectively. The bond lengths <*Y*3A–O> and <*Y*3B–O> was 1.925 and 1.974 Å, and their significant difference is explained by the non-equivalent population of the *Y*3A and *Y*3B positions: $(Al_{0.90}Mg_{0.10})_{\Sigma 1.00}$ and $(Al_{0.60}Mg_{0.40})_{\Sigma 1.00}$, respectively.

For alumovesuvianite, positions of hydrogen atoms were determined (Table 1.7). The geometry of hydrogen bonds is similar to that of low-temperature VSMs without F (Pavese et al. 1998a; Lager et al. 1999a).

Thus, alumovesuvianite is an almost completely ordered representative of *P*4/*n* VSMs: the coupled positions *Y*1A and *X*4B are 93% populated whereas the occupancy factor of the *Y*1B and *X*4A positions is 7%. The different occupancy of the positions of cationic rods causes local stresses in the structure and affects the distribution mechanisms of magnesium and aluminum in octahedral positions, which is discussed in detail in the next chapter. The partial occupancy of the *Y*1A,B positions by aluminum in the structure of the magnesiovesuvianite indicates the possible existence of a continuous isomorphic series between these two minerals.

Taking into account the data of solid-state NMR, the crystal chemical formula of alumovesuvianite, can be written as follows:

$$^{X1}(Ca)_{2.00}\,^{X2A}(Ca)_{4.00}\,^{X2B}(Ca)_{4.00}\,^{X3A}(Ca)_{4.00}\,^{X3B}(Ca)_{4.00}\,^{X4A}(Ca)_{0.07}\,^{X4B}(Ca)_{0.93}\,^{X1A}(Al_{0.73}Mn_{0.20})_{\Sigma 0.93}\,^{Y1B}(Fe)_{0.07}\,^{Y2A}(Al)_{2.00}\,^{Y2B}(Al)_{2.00}\,^{Y3A}(Al_{3.60}Mg_{0.40})_{\Sigma 4.00}\,^{Y3B}(Al_{2.40}Mg_{1.60})_{\Sigma 4.00}\,^{Z1A}(Si)_{1.00}\,^{Z1B}(Si)_{1.00}\,^{Z2A}(Si)_{4.00}\,^{Z2B}(Si)_{4.00}\,^{Z3A}(Si)_{4.00}\,^{Z3B}(Si)_{4.00}O_{68.00}(OH_{9.00}O)_{\Sigma 10.00}.$$

3.5 Magnesiovesuvianite

Magnesium plays a significant role in VSMs and is always present in their chemical composition. This fact helps to distinguish VSMs from the minerals of the grossular-andradite series using routine EDS analyses. Studies of VSMs using EXAFS and single crystal X-ray diffraction analysis (Valley et al. 1985; Ohkawa et al. 1992) showed that Mg dominates in the *Y*1 position of some VSMs from Chichibu (Japan). The role of magnesium in petrology and vesuvianite chemistry is discussed by Hoisch (1985), who noted for the first time that among 13 cations per formula unit, occupying *Y* positions with octahedral and tetragonal-pyramidal coordination, there are two divalent cations, primarily magnesium. Peculiarities of the synthesis of Mg-enriched vesuvianite-type compounds are discussed by Hochella et al. (1982).

3.5.1 Occurrence and Physical Properties

Magnesiovesuvianite (Panikorovskii et al. 2017d) was discovered during the study of samples from the systematic collection of the A. E. Fersman Mineralogical Museum RAS in Moscow. According to the label, the holotype mineral sample (catalog number 59783, available from 1957) was collected at the Tuydo ravine, not far from Lojane in Macedonia.

The mineral was found in association with grossular, calcite, diopside, clinochlore and an unknown manganese mineral, has a hydrothermal origin and forms pink fibrous and random columnar aggregates (Figs. 3.57 and 3.58) of crystals 5–40 μm thick and up to 7 mm long, with fiber optic effect. Individual magnesiovesuvianite crystals (Fig. 3.33d, g) have a [001] elongation and are faceted with the following simple forms: {100}, {110}, {111}, {101}, and {001}. Some crystals are characterized by a more complex habit (Fig. 3.33e).

3.5.2 Chemical Composition

Iron is practically absent in magnesiovesuvianite (Table 3.11), as a result of which Mg and subordinate Al occupy the five-coordinated *Y*1 position. Among other impurities, a minor amount of Mn should be noted. As in the case of alumovesuvianite, Mn^{3+} may be the main chromophore in magnesiovesuvianite determining its pink color. According to the IR spectroscopy data, boron is absent in the holotype sample of magnesiovesuvianite.

Fig. 3.57 Photo of magnesiovesuvianite sample 59783 FM

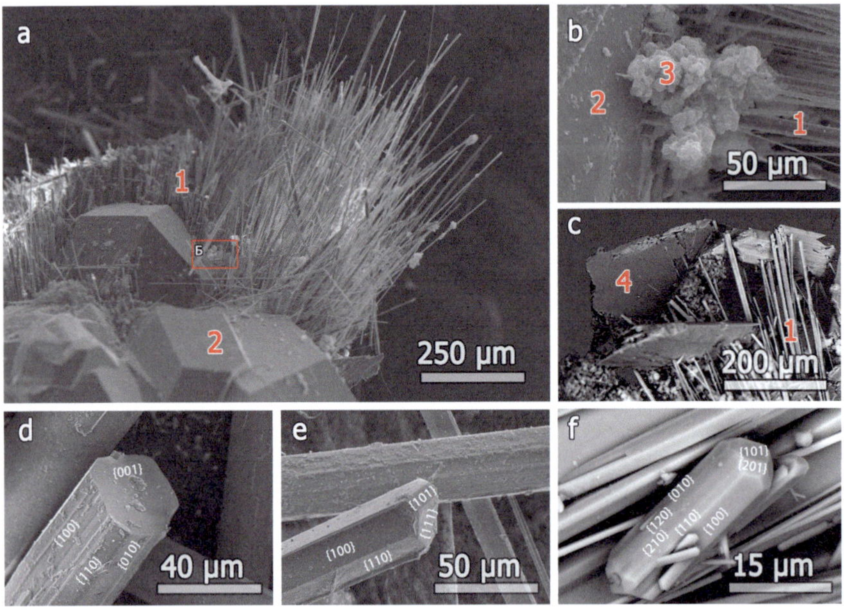

Fig. 3.58 BSE image of a fibrous aggregate of magnesiovesuvianite (sample 59783 FM) on grossular (**a**), aggregates of an unknown manganese mineral (**b**), diopside overgrowth on magnesiovesuvianite (**c**) and magnesiovesuvianite crystals (**d–f**). 1—magnesiovesuvianite, 2—grossular, 3—unknown manganese mineral, 4—diopside

Table 3.11 Chemical composition of magnesiovesuvianite from Tuydo (wt%)

Sample	1	2	3	4	5	6
SiO_2	36.27	36.50	36.72	36.64	36.45	36.85
Al_2O_3	19.80	19.95	20.01	19.82	20.20	19.55
FeO	–	0.05	0.03	–	0.09	0.05
MnO	0.11	0.15	0.14	0.10	0.15	0.12
MgO	1.64	1.70	1.65	1.88	1.63	1.75
CaO	35.85	36.28	36.52	36.22	36.10	36.52
Na_2O	0.01	0.08	0.05	0.14	0.07	0.02
H_2O*	2.73	2.78	2.79	2.78	2.76	2.78
Sum	96.41	97.49	97.91	97.58	97.45	97.64
apfu						
Si	17.93	17.77	17.79	17.81	17.84	17.88
Al	11.54	11.45	11.42	11.36	11.65	11.18
Fe	0	0.02	0.01	0	0.04	0.02
Mn	0.05	0.06	0.06	0.04	0.06	0.05
Mg	1.21	1.23	1.19	1.36	1.19	1.27
Ca	18.99	18.92	18.95	18.87	18.93	18.98
Na	0.01	0.08	0.05	0.13	0.07	0.02
H	9.00	9.00	9.00	9.00	9.00	9.00

* The H_2O content was calculated bas on single-crystal XRD data
Dash—content below detection limit

3.5.3 Thermogravimetry and Differential Scanning Calorimetry

The water content in the mineral (3.10 wt%), determined using TGA (Fig. 3.59), is in accordance with the theoretical content of 2.83 wt% The thermal data correspond to a low-symmetry VSM (Żabiński et al. 1996). The TG curve contains one significant mass loss step, which is accompanied by the endothermic peak at 1024 °C.

3.5.4 Infrared Spectroscopy

In the IR spectrum of the holotype magnesiovesuvianite sample (Fig. 3.60) three peaks (at 3630, 3450, and 3200 cm^{-1}) and two shoulders (at 3508 and 3050 cm^{-1}) are observed in the region of stretching vibrations of O–H bonds.

According to Borovikova and Kurazhkovskaya (2006), the presence of the IR band at 3630 cm^{-1} is characteristic of low-temperature varieties of VSMs (space groups *P*4/*n* and *P*4*nc*). In the region of O–Si–O bending vibrations of silicate groups, there

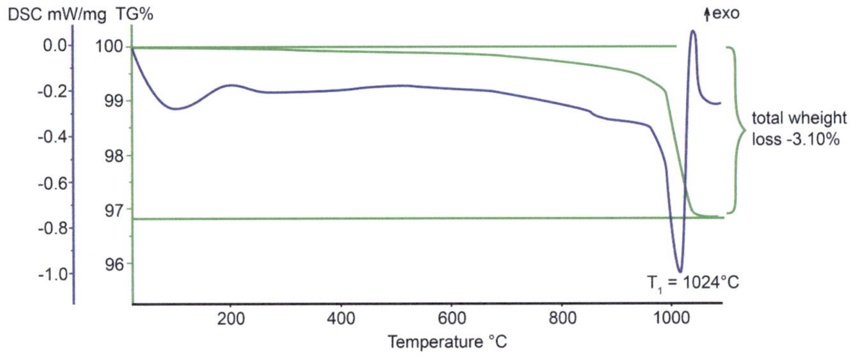

Fig. 3.59 TG (green line, wt%) and DSC (blue line) curves of magnesiovesuvianite

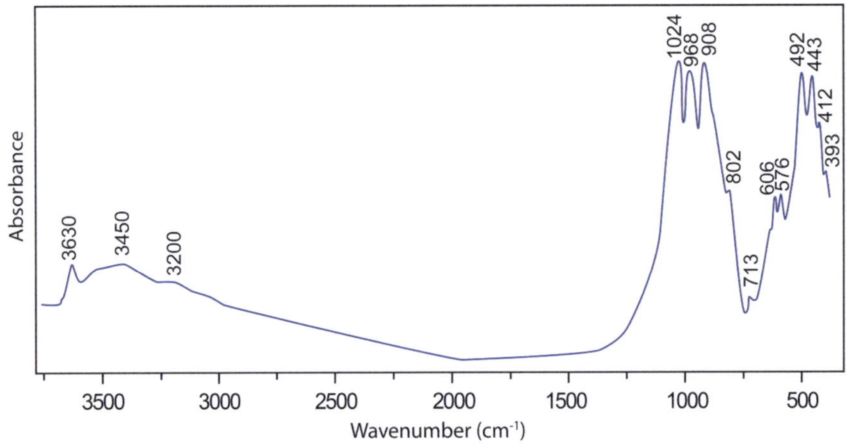

Fig. 3.60 Infrared spectrum of magnesiovesuvianite

are two bands of approximately equal intensities (at 606 and 576 cm^{-1}), characteristic of VSMs with the *P*4/*n* symmetry (Chukanov et al. 2018a).

3.5.5 Solid-State Nuclear Magnetic Resonance

Unlike the proton NMR spectrum of "hydrovesuvianite" (Panikorovskii et al. 2016a), which contains three lines (at 6.74, 3.38 and –0.95 ppm), the analogous spectrum of magnesiovesuvianite (Fig. 3.61a) contains only two peaks with the chemical shifts of 6.52 and –0.95 ppm. The absence of a peak at 3.38 ppm indicates a minor role (or absence) of the hydrogarnet-type substitution in magnesiovesuvianite. Calculations of the O···O distances in hydrogen bonds using the formula by Yesinowski et al.

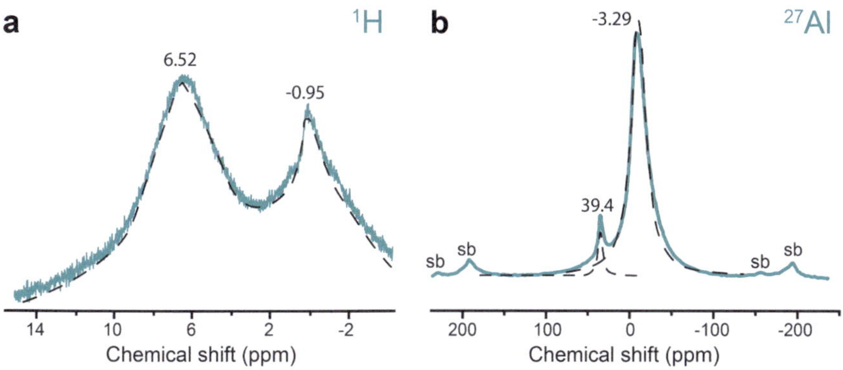

Fig. 3.61 ^{1}H MAS NMR spectrum (**a**), ^{27}Al spectrum (**b**) of magnesiovesuvianite. The dashed lines show the approximating curves

(1988) gave the values of 2.84 and 3.13 Å. According to the correlation by Libowitzky (1999), these values correspond to the strongest IR absorption peaks at 3450 and 3630 cm^{-1}, respectively. The broad peak at of 6.52 ppm corresponds to the O(11)–H(1*a*)···O(7) and O(10)–H(2)···O(10) hydrogen bonds with the O···O distances of 2.784 (4) and 2.74–2.82 Å, respectively. As in the spectra of other VSMs, the peak at –0.95 ppm corresponds to O(11)–H(1b)···O(11) bonds with the O···O distances of 3.01–3.04 Å (Pavese et al. 1998a; Lager et al. 1999a).

The ^{27}Al NMR spectrum of magnesiovesuvianite (Fig. 3.61b) contains two peaks with the chemical shifts of –3.29 and 39.41 ppm. The most intense peak at –3.29 ppm corresponds to ^{VI}Al in the octahedral *Y*2(A,B) and *Y*3(A,B) positions. The weaker peak at 39.41 ppm corresponds to ^{V}Al in the five-coordinated *Y*1(A,B) site. The integrated normalized intensities of these peaks are 96.5 to 4.5%, respectively. Two pairs of weak peaks in the region of ±(150–200) ppm are assigned as spinning sidebands. It should be noted that similar ^{27}Al NMR spectra with the presence of the ^{V}Al and ^{VI}Al lines were reported by Phillips et al. (1987), Olejniczak and Żabiński (1996) for samples of vesuvianite from Asbestos (Canada) and Piz Lunghin (Switzerland).

3.5.6 Powder X-Ray Diffraction

The powder XRD pattern of magnesiovesuvianite is characteristic for VSMs (Fig. 3.62). The refined unit cell parameters are $a = 15.5026(3)$ Å and $c = 11.7856(5)$ Å, which is in a good agreement with the unit cell parameters obtained using single-crystal XRD data. Data on the intensity, position of the lines and their *hkl* indexes are given in Table A.35.

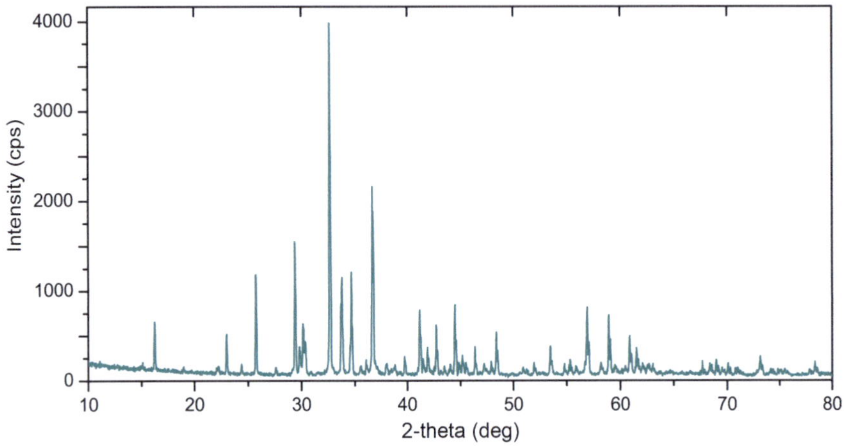

Fig. 3.62 Powder X-ray diffraction pattern of magnesiovesuvianite

3.5.7 Single-Crystal X-Ray Structural Analysis

For the studied sample of magnesiovesuvianite, 208 reflections of the 0*kl* and *hhl* types were observed, for which $k + l \neq 2n$ and $l \neq 2n$, and 14 reflections of *hk*0 type, for which $h + k \neq 2n$. Thus, the observed systematic violations are in agreement with the *P*4/*n* symmetry (Armbruster and Gnos 2000a). Attempts to refine the structure in the *P*4*nc* space group led to a large number of sites with negative displacement parameters.

The magnesiovesuvianite structure was refined to $R_1 = 0.027$ for 3266 independent reflections with intensities $|F_o| \geq 4\sigma_F$ as a merohedral twin (with twinning on the two-fold axis along [110]) with a component ratio of 0.485/0.515 (Table A.36).

All tetrahedral positions are occupied by Si atoms (Table A.37), but the refined populations for the *Z*1A,B positions, which are $Si_{0.95}$ and $Si_{0.88}$, confirm the presence of a small number of garnet defects in accordfnce to the substitution scheme $SiO_4^{4-} \leftrightarrow [H_4O_4]^{4-}$. The mean bond lengths <Z1A–O> and <Z1B–O> are 1.647 and 1.666 Å, respectively (Table A.38), and the scattering factors correspond to 13.3 e^- for *Z*1A and 12.3 e^- for *Z*1B.

The refined populations of the positions of the *X*-group cations correspond to their complete occupancy by Ca atoms. The occupancy factors of the channel positions *X*4A and *X*4B were refined as 0.73 and 0.27, respectively.

In magnesiovesuvianite, the tetragonal–pyramidal coordinated *Y*1A,B positions are predominantly populated by Mg (Fig. 3.63). The mean <*Y*1A–O> and <*Y*1B–O> distances are equal to 2.140 and 2.002 Å, respectively, which is in agreement with the refined occupancies of the *Y*1A,B positions: $Mg_{0.19}Al_{0.08}$ and $Mg_{0.41}Al_{0.32}$. The total scattering of the *Y*1A,B positions corresponds to 11.55 e^- (the estimated value from composition was 12.40 e^-). Taking into account the data of solid-state NMR, the final total population of the Y1A, B positions is $(Mg_{0.60}Al_{0.40})_{\Sigma 1.00}$. The octahedral

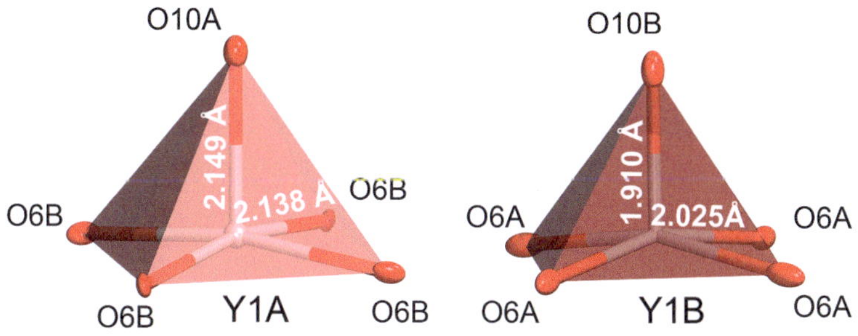

Fig. 3.63 Coordination and bond lengths for the *Y*1A,B positions in magnesiovesuvianite from Tuydo, Lojane, Macedonia

positions of *Y*2A and *Y*2B are fully populated with Al atoms; the corresponding mean bond lengths <*Y*2A–O> and <*Y*2A–O> are 1.904 and 1.898 Å. The mean bond lengths <*Y*3A–O> = 1.953 and <*Y*3A–O> = 1.932 Å are in a good agreement with the estimated population of these positions: $(Al_{0.86}Mg_{0.125}Mn_{0.015})_{\Sigma 1.00}$ for *Y*3A and $(Al_{0.905}Mg_{0.09}Fe_{0.005})_{\Sigma 1.00}$ for *Y*3B.

The refined crystal-chemical formula of magnesiovesuvianite derived taking into account solid-state NMR data is:

$$^{X1}(Ca)_{\Sigma 2.00}\,^{X2A}(Ca)_{\Sigma 4.00}\,^{X2B}(Ca)_{\Sigma 4.00}\,^{X3A}(Ca)_{\Sigma 4.00}\,^{X3B}(Ca)_{\Sigma 4.00}\,^{X4A}(Ca)_{\Sigma 0.73}$$
$$^{X4B}(Ca)_{\Sigma 0.27}\,^{Y1A}(Mg_{0.19}Al_{0.08})_{\Sigma 0.27}$$
$$^{Y1B}(Mg_{0.41}Al_{0.32})_{\Sigma 0.73}\,^{Y2A}(Al)_{\Sigma 2.00}\,^{Y2B}(Al)_{\Sigma 2.00}\,^{Y3A}(Al_{3.62}Mg_{0.36}Fe_{0.02})_{\Sigma 4.00}$$
$$^{Y3B}(Al_{3.44}Mg_{0.50}Mn_{0.06})_{\Sigma 4.00}$$
$$^{Z1A}(Si_{0.95}[H_4O_4]_{0.05})_{\Sigma 1.00}\,^{Z1B}(Si_{0.88}[H_4O_4]_{0.12})_{\Sigma 1.00}\,^{Z2A}(Si)_{\Sigma 4.00}$$
$$^{Z2B}(Si)_{\Sigma 4.00}\,^{Z3A}(Si)_{\Sigma 4.00}\,^{Z3B}(Si)_{\Sigma 4.00}(O)_{\Sigma 68.00}(OH_{8.46}O_{1.54})_{10.00}.$$

3.6 Fluorvesuvianite and F-Bearing Vesuvianite

The presence of fluorine in vesuvianite is known since the nineteenth century (Jannasch and Weingarten 1896; Weibull 1896). In particular, a significant fluorine content (up to 3.2 wt%) is typical for VSMs samples from Richardson (USA) and Kleppan (Norway). The crystal-chemical mechanism for the substitution of hydroxyl groups by fluorine and its effect on the infrared spectra of VSMs is studied in detail in the works by Groat (1992b, 1995b).

Fig. 3.64 The photograph of an aggregate of acicular crystals of fluorvesuvianite (A, V. N. Yakovenchuk's sample) from the Lupikko deposit. 1—fluorvesuvianite, 2—calcite

3.6.1 Occurrence and Physical Properties

Fluorvesuvian was discovered by S. N. Britvin and co-authors in the vicinity of the Lupikko ore field in Karelia (Britvin et al. 2003). Radial aggregates of fluorvesuvianite (Fig. 3.64) were found in chloritized diopside skarn in association with calcite, sphalerite, and clinochlore. Acicular crystals of the mineral have a length of up to 1.5 cm and 20–30 μm in thickness. Unlike other VSMs, fluorvesuvianite has particularly low refractive indices: $\omega = 1.702$ (1) and $\varepsilon = 1.699$ (1) for $\lambda = 589$ nm. The unit cell parameters refined from powder XRD data are $a = 15.516(2)$ Å, $c = 11.772(3)$ Å.

Genesis of fluorvesuvianite is not associated with the formation of rock-forming vesuvianite, which is widespread in Lupikko. The mineral crystallized at the late stage of the hydrothermal-metasomatic alteration of vesuvianite-containing skarns at a temperature of about 400 °C which corresponds to the typical conditions of the formation of *P*4*nc* and *P*4/*n* VSMs modifications (Allen and Burnham 1992).

Low-temperature vesuvianite with a high content of fluorine and *P*4/*nnc* symmetry was discovered by Galuskin et al. (2003a) in voids of modified magnesian skarn from the Dokuchan River basin (North Yakutia) in association with datolite. The authors of this work proposed two possible causes of the formation of low-temperature disordered vesuvianite: low-temperature F-metasomatism itself (Britvin et al. 2003) or epitaxial growth of low-temperature fluorvesuvianite on the already existing subtrate of high-temperature VSM (*P4/nnc*) (i.e. structural disorder could be inherited from the substrate).

3.6.2 Chemical Composition

Currently, only one locality of fluorvesuvianite (in Lupikko skarns in Karelia, Russia) is known. Its chemical composition in comparison with that of the F-containing vesuvianite from various deposits is given in Table 3.12. In fluorvesuvianite, the main charge-balancing bivalent cations with the coordination numbers 5 or 6 are Mg^{2+} and Fe^{2+}. The empirical formula of holotype fluorvesuvianite calculated on the basis of 50 cations per formula unit is (Britvin et al. 2003): $Ca_{19.03}(Al_{10.38}Mg_{1.39}Fe^{2+}{}_{1.15}Mn^{2+}{}_{0.04})_{12.96}Si_{18.01}O_{68.00}(F_{7.16}OH_{1.64}O_{0.80})_{9.60}$.

3.6.3 Infrared Spectroscopy

F-rich VSMs samples, most of which originate from skarns, usually contain significant amounts of boron (up to 2–3 wt%), and this is confirmed by IR spectrum of fluorvesuvianite from Lupikko (Fig. 3.65). Additional weak bands at 1575 and 1420 cm^{-1} may correspond to asymmetric stretching vibrations in distorted BO_3 triangles, but their assignment to asymmetric stretching vibrations of distorted carbonate anions cannot be excluded.

The A, B and C bands of O11–H stretching vibrations in the IR spectra of F-rich VSMs are weak or are not observed. These bands correspond to the weakest hydrogen bonds. This phenomenon has an obvious explanation: breaking of strong hydrogen bands as a result of the substitution of OH froups with F would be energeyically unfavourable. In the fluorovesuvianite structure, fluorine ions replace hydroxyl groups in the O11 position, which is in agreement with previously published works (Groat et al. 1992b; Ohkawa 1994).

In the region of stretching vibrations of the O–H bonds in the IR spectrum of the holotype fluorvesuvianite sample, bands at 3625, 3555, 3400, and 3170 cm^{-1} are observed. The band at 3625 cm^{-1} in the spectrum of fluorvesuvianite may correspond to unresolved bands *B* (3635 cm^{-1}) and *C* (3596 cm^{-1}), the band 3555 cm^{-1} to the superposition of *D* (3567 cm^{-1}) and *E* (3524 cm^{-1}) bands, the and at 3400 cm^{-1} to the superposition of *G* (3430 cm^{-1}) and *H* (3383 cm^{-1}), and the band at 3170 cm^{-1} to the superposition of *J* (3210 cm^{-1}) and *K* (3156 cm^{-1}) bands. However, the classification of O–H stretching bands suggested for vesuvianite (Groat et al. 1995b; Bellatreccia et al. 2005b) may be inapplicable to fluorvesuvianite. Relatively high intensities

Table 3.12 Chemical composition of F-bearing vesuvianite compared with holotype fluorvesuvianite (6), wt%

Sample	1	2	3	4	5	6
SiO_2	35.65	35.64	35.93	36.02	35.82	36.60
TiO_2	–	0.72	0.31	1.24	1.30	
Al_2O_3	19.70	19.20	16.52	17.04	16.86	17.90
FeO	2.91	1.00	4.65	4.79	4.85	2.80
MnO	–	2.27	1.92	1.37	1.43	0.10
MgO	1.40	0.57	0.40	0.34	0.44	1.90
CaO	34.81	34.53	35.01	33.50	33.72	36.10
Na_2O	–	–	–	0.54	0.49	–
CuO	–	0.82	–	–	–	–
H_2O*	1.52	1.55	1.57	1.59	1.62	0.50**
F	3.87	3.68	3.71	3.55	3.47	4.60
2F≡O	1.63	1.55	1.56	1.49	1.46	1.94
Sum	98.23	98.43	98.46	98.49	98.54	98.56
apfu						
Si	18.16	18.30	18.20	18.52	18.35	17.98
Ti		0.28	0.12	0.48	0.50	
Al	11.83	11.62	9.86	10.33	10.18	10.36
Fe^{2+}	1.24	0.43	1.97	2.06	2.08	1.15
Mn		0.25	0.17	0.15	0.19	0.04
Mg	1.06	1.74	1.45	1.05	1.09	1.39
Ca	19.00	19.00	19.00	18.46	18.51	19.00
Na				0.54	0.49	
Cu		0.32				
H	5.13	5.32	5.29	5.45	5.53	1.64
F	4.39	4.21	4.18	4.06	3.96	7.14

1—Lupikko, 2—Ostebo, Norway (sample 67940 FM), 3 –Kristiansand (sample 13975 FM), 4—Keivy (sample 50578 FM), 5—Keivy (sample 707 BM), 6—Lupikko, holotype (Britvin et al. 2003)
* The H_2O content was calculated based on single-crystal XRD data
** The H_2O content was measured by TG after heating up to 1100 °C
Dash—content below detection limit

of the bands of O–H stretching vibrations of fluorvesuvianite can be explained by polarization of O–H bonds due to hydrogen bonding of the type of O–H···F.

In the IR spectrum of rock-forming F-containing vesuvianite, there is no band at 3400 cm^{-1}, and the band 3170 cm^{-1} is offset in a higher wavenumbers region, which may be due to a different local coordination for hydrogen atoms coordinating the O11 oxygen position. The spectral lines in the region of 1020–400 cm^{-1} of the spectrum of the holotype fluorvesuvianite (Britvin et al. 2003).

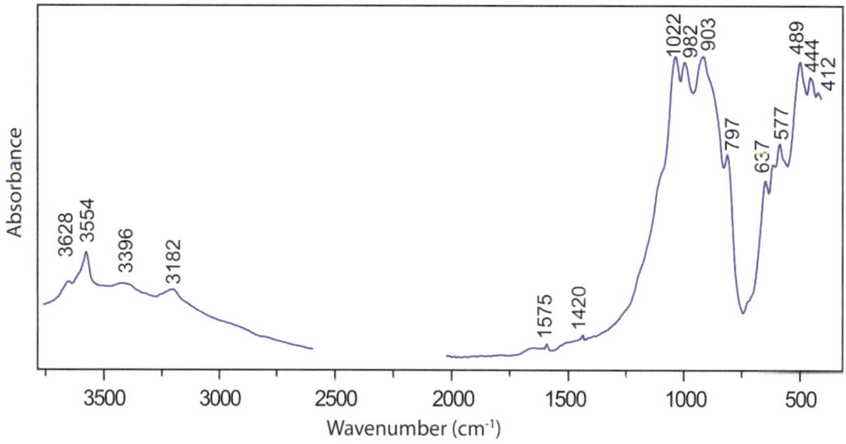

Fig. 3.65 The infrared spectrum of fluorvesuvianite

3.6.4 Single-Crystal X-Ray Diffraction Analysis

An X-ray diffraction study of all VSMs crystals shown in Fig. 3.66 showed their compliance with the *P*4/*nnc* group, despite differences in morphology, formation time, as well as Fe, Al, Mg and B contents. Thus, here and below we report data for an acicular crystal which is most similar in composition to holotype fluorvesuvianite.

The crystal structure of the F-containing vesuvianite (with 4.39 F *apfu*) was refined with $R_1 = 0.029$ (Table A.39). The relative elongation of the mean <*Z*1–O> distance (1.664 Å) and the incomplete population of the Z1 position ($Si_{0.87}\square_{0.13}$) (Table A.40)

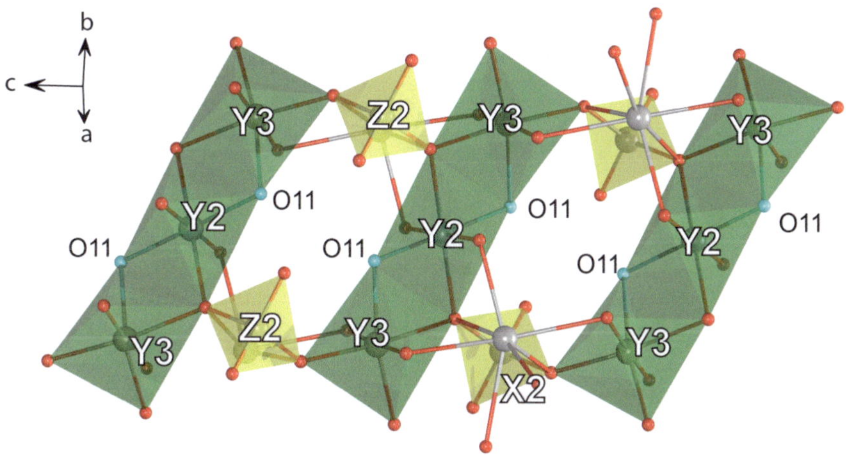

Fig. 3.66 Fragment of the structure of F-containing vesuvianite from Lupikko. The fluorine-bearing O11 position is shown with bluish-green balls

are in agreement with the previously obtained data and the low-temperature genesis of the sample (Britvin et al. 2003).

Similar to the Si-deficient vesuvianite varieties (Galuskin et al. 2003b; Panikorovskii et al. 2016a), the mean <*Z*1–O> bond lengths in fluorvesuvianite reached 1.664 Å and incomplete population of the *Z*1 site of 0.803(8) should be also noted. In this connection, the authors proposed the following substitution model: $(SiO_4)^{4-} \leftrightarrow 4F^-$. However, for a vesuvianite, substitution is more likely according to the $H_4O_4 \rightarrow SiO_4$ scheme, which explains the lack of silicon in the structure of fluorine-enriched vesuvianite from skarn of Dokuchan River (Galuskin et al. 2003a).

In the studied sample, the *Z*2 and *Z*3 positions are completely occupied by Si atoms; the corresponding bond mean Si–O lengths are 1.650 and 1.631 Å. The additional *T*1 and *T*2 positions are vacant, whereas in the rock-forming prismatic F-containing vesuvianite they are partially occupied by Al and B atoms. The scattering of cations at the *X* sites corresponds to that for calcium atoms. The *Y*1 position is predominantly populated by Fe^{2+} ions. The refined population of this position is $(Fe_{0.44}Mg_{0.06})_{0.50}$. Iron valence was determined based on the recommendations presented in Sect. 1.5. The occupancies of the octahedral positions *Y*2 and *Y*3 are $Al_{1.00}$ and $Al_{0.96}Fe_{0.04}$, respectively, which is in a good agreement with the average <*Y*2–O> and <*Y*3–O> bond lengths, which are equal to 1.901 and 1.940 Å, respectively (Table A.41).

According to Groat et al. (1992b), Ohkawa et al. (1992), Britvin et al. (2003), in the VSMs structures, fluorine ions substitute OH^- groups at the O10 and O11 positions, but according to the IR spectroscopy data, fluorine preferally occupies the O11 site (see comment in the preceding section) The local environment of the O11 site in the structure of the F-containing vesuvianite from Lupikko is shown in Fig. 3.66. The calculated population of the O11 position is $O_{0.60}F_{0.40}$.

The general crystal-chemical formula of the stusied sample of F-bearing vesuvianite from Lupikko is:

$$^{X1}(Ca)_{2.00}\,^{X2}(Ca)_{8.00}\,^{X3}(Ca)_{8.00}\,^{X4}(Ca)_{1.00}\,^{Y1}\left(Fe^{2+}{}_{0.88}Mg_{0.12}\right)_{\Sigma 1.00}$$
$$^{Y2}(Al)_{\Sigma 4.00}\,^{Y3}\left(Al_{7.00}Fe^{2+}{}_{0.32}Mg_{0.68}\right)_{\Sigma 8.00}$$
$$^{Z1}(Si_{1.74}[H_4]_{0.26})_{2.00}\,^{Z2}(Si)_{8.00}\,^{Z3}(Si)_{8.00}(O)_{69.00}\,^{W}(OH_{5.80}F_{3.20})_{\Sigma 9.00}.$$

3.7 Manganivesuvianite and Mn-Bearing Vesuvianite

The presence of up to 1.5 wt% Mn_2O_3 in pink translucent vesuvianite crystals from Jordanow, Poland (Fig. 3.67) was noted by Lasaulx (1883). The highest manganese contents up to 14.3 wt% were reported for the VSMs from the the N'Chwaning mine, Kalahari manganese field in South Africa (Armbruster and Gnos 2000c). Significant manganese contents (up to 1.5 wt% Mn_2O_3) are also recorded for Bi-containing vesuvianite samples from Långban, Sweden (Groat and Evans 2012; Hålenius et al.

Fig. 3.67 Aggregates of prismatic Mn-bearing vesuvianite crystals from Jordanow, Poland

2013). To date, three localities of manganivesuvianite are known: N'Chwaning and Wessels mines in South Africa and Pajsberg in Sweden.

3.7.1 Occurrence and Physical Properties

Manganivesuvianite is the first mineral of the vesuvianite supergroup distinguished by the predominant cation at the five-coordinated *Y*1 site (Armbruster et al. 2002). The mineral was found in the zone of hydrothermal alteration (at 250–400 °C) of manganese ores (Cairncross et al. 1997). It fills cracks in the ore body and

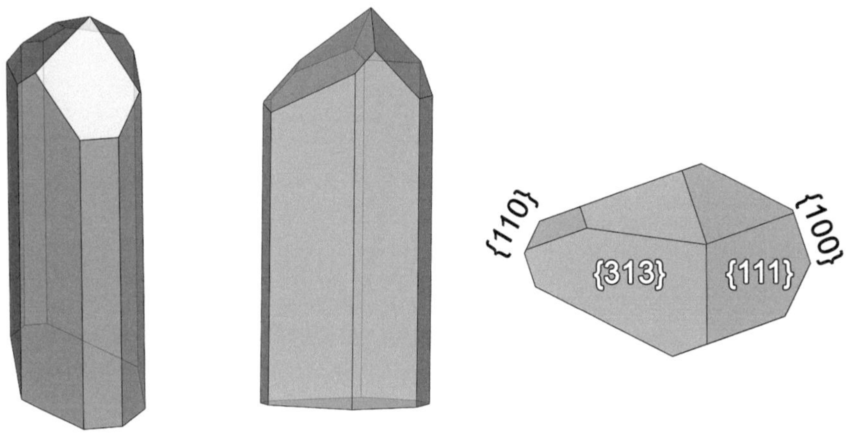

Fig. 3.68 The habit of manganivesuvianite crystals from N'Chwaning III mine

associates with apatite, garnet-supergroup minerals (holstamite-henritermierite or grossular), xonolite, serandite, tweddelite, and calcite. Manganivesuvianite forms long-prismatic (up to 1.5 cm) dark red crystals with a clear chemical zonality from the center to the edge zones and distinct growth striation on the faces of the prisms {100}, {110} and {120} (Fig. 3.68). The refractive indices are: $\omega = 1.731(1)$ and $\varepsilon = 1.719(1)$ for $\lambda = 589$ nm. The physical properties of manganivesuvianite are given in Tables 1.1 and Table 3.2. The unit cell parameters refined from the powder XRD data are: $a = 15.575(2)$ Å and $c = 11.824(2)$ Å (space group *P*4/*n*).

3.7.2 Chemical Composition

The chemical composition of holotype manganivesuvianite is given in Table 3.13. Our analyses of Mn-rich VSMs from Kalahari Manganese Fields show that all of them contain copper and there is a continuous solid-solution series between manganivesuvianite and cyprine (see the IR spectrum and empirical formula of an intermediate member of this isomorphous series in the next section).

It should be noted that the affinity of copper to this position is higher than that of the Fe^{3+} and Mn^{3+} ions; therefore, at Cu contents > 0.5 *apfu*, the mineral should be identified as cyprine, and not manganvesuvianite or vesuvianite. Taking into account the data of X-ray structural analysis, chemical composition, ESR, absorption spectra and photoluminescence for VSMs (Platonov et al. 1995; Armbruster and Gnos 2000c; Hålenius 2000; Armbruster et al. 2002; Panikorovskii et al. 2017a, e; Czaja et al. 2018), we can construct the following series of cations according to their degree of affinity (from the highest to the lowest) to the five-vertex coordination: $Cu^{2+} \rightarrow Fe^{3+} \rightarrow Mn^{3+} \rightarrow Fe^{2+} \rightarrow Mg^{2+} \rightarrow Cr^{3+} \rightarrow Al^{3+}$.

Table 3.13 Chemical composition of Mn-bearing VSM compared with holotype manganvesuvianite 4—(Armbruster et al. 2002), мас. %

Sample	1	2	3	4
SiO_2	35.48	36.35	36.79	36.15
Al_2O_3	14.02	15.98	15.13	14.73
Fe_2O_3	2.29	1.88	4.60	1.12
Mn_2O_3	8.08	4.04	2.11	6.79
CuO	1.87	1.58		0.02
MgO	0.40	1.63	2.49	2.35
CaO	35.18	35.74	36.13	35.73
SrO				0.11
Na_2O		0.07		0.03
H_2O*	2.70	2.73	2.75	2.67
F				0.12
O≡F				0.05
Sum	100.02	100.00	100.00	99.77
apfu				
Si	17.88	17.97	18.06	17.91
Al	8.33	9.31	8.75	8.60
Fe^{3+}	0.87	0.70	1.70	0.42
Mn	3.10	1.52	0.79	2.56
Mg	0.30	1.20	1.82	1.74
Ca	19.00	18.93	19.00	18.97
Na		0.07		0.03
Cu	0.71	0.59		0.01
Sr				0.03
H	9.00	9.00	9.00	8.29
F				0.19

1,2—N'chwaning III mine, 3—Monte Rosa, 4—N'chwaning II mine (Armbruster et al. 2002)
* The H_2O content was calculated on F + Cl + OH = 9

3.7.3 Infrared Spectroscopy

Unlike IR spectra of most low-temperature VSMs, the IR spectra of manganivesuvianite from N'chwaning III, South Africa (Fig. 3.69) contain two bands of approximately equal intensity (at 600–610 and 570–580 cm^{-1}) in the region of bending vibrations of silicate groups. In the range of Si–O and Si–O–Si stretching vibrations, the spectrum is characteristic of the VSMs with *P*4/*n* symmetry (Chukanov et al. 2018a).

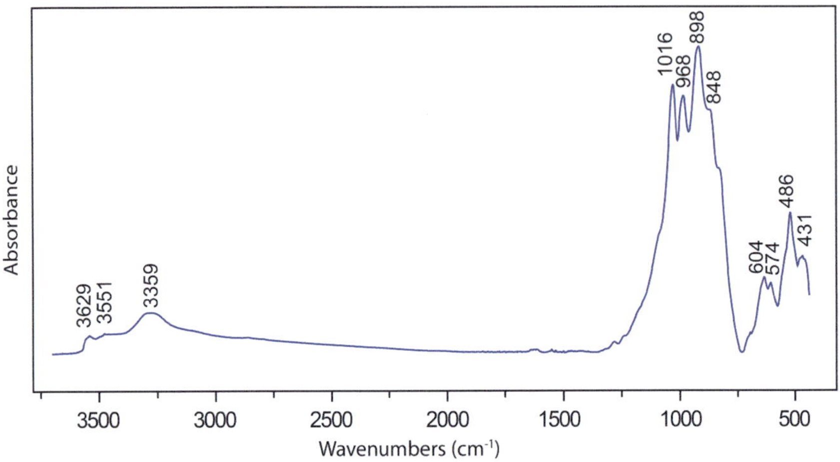

Fig. 3.69 Infrared spectrum of Cu-bearing manganivesuvianite with the empirical formula $Ca_{19}Cu_{0.40}Mg_{0.19}Mn_{3.25}Fe_{0.43}Al_{8.73}Si_{17.21}O_{69}(OH)_9$ from the N'Chaning III mine (South Africa)

In the IR spectrum of the manganivesuvianite, in the region of stretching vibrations of O–H bonds (Fig. 3.69), in addition to the bands at 3629(B) and 3551(F), there is a broad peak at 3359 cm^{-1} probably, corresponding to the superposition of O–H stretching bands including that of the Mn^{3+}–O–H···O configuration (Panikorovskii et al. 2017e).

3.7.4 Single-Crystal X-Ray Diffraction Analysis

For the holotype manganivesuvianite sample from the N'chwanning II mine, South Africa, *P*4/*n* symmetry was determined (Armbruster et al. 2002), however, the presence of domains with *P*4*nc* symmetry was noted. The atom coordinates and occupancies for the *X* and *Z* positions in the Mn-containing VSM are similar to those given in Sect. 3.1.10, therefore, below we discuss only the population of the *Y* positions.

When refining the crystal structure of the Mn-containing vesuvianite from the N'chwanning III, we also found the presence of *P*4*nc* symmetry domains, but refinement in the *P*4*nc* model resulted in a significant number of positions with negative displacement parameters. In the single-crystal XRD pattern, there are 86 violating reflections for the *P*4*nnc* space group and only 18 *hk*0-type reflections with where $h + k \neq 2n$ were observed, which indicates the correctness of the choice of the space group.

The structure of the manganivesuvianite was refined to $R_1 = 0.035$ (Table A.42). In most samples of VSMs, the octahedral *Y*2A,B positions are usually populated exclusively by Al atoms, but in manganivesuvianite these positions are partially filled by Mn as in our case. The refined occupancies of the *Y*2A and *Y*2B positions are

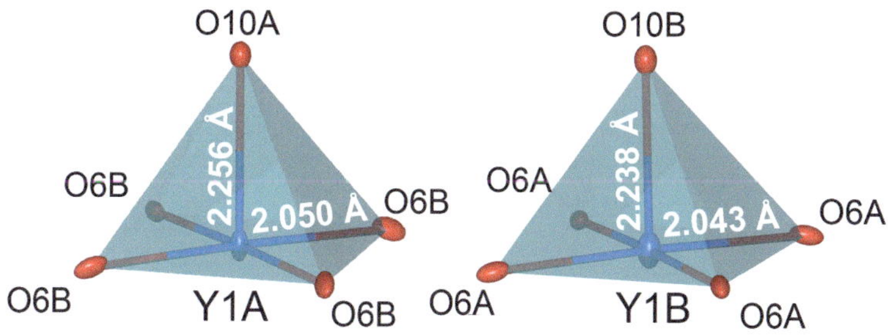

Fig. 3.70 Coordination and bond lengths for the *Y*1A,B positions in manganivesuvianite from N'chwanning II, South Africa

$Al_{0.90}Mn_{0.10}$ and $Al_{0.95}Mn_{0.05}$, respectively (Table A.43). The average bond lengths <*Y*2A–O> and <*Y*2B–O> are almost equal (1.908 and 1.900 Å), the occurrence of Mn^{3+} in the first of them is compensated by the occurrence of Mg^{2+} (refined as Al) in the second. The refined occupancies of the octahedral *Y*3A and *Y*3B positions are $Al_{0.81}Mn_{0.19}$ and $Al_{0.84}Mn_{0.16}$, and the mean length of the corresponding <*Y*3A–O> and <*Y*3B–O> bonds are equal to 1.982 and 1.960 Å, respectively (Table A.44). Thus, manganese occurs at the octahedral *Y*2 and *Y*3 positions in the manganivesuvianite structure.

The most interesting aspect of the studied structure is the predominant occupancy of the *Y*1A and *Y*1B positions by Mn^{3+} atoms; the total scattering of these positions is 25e^-. The mean lengths of the <*Y*1A–O> and <*Y*1B–O> bonds are 2.005 and 2.094 Å, respectively (Fig. 3.70). The pronounced elongation of the *Y*1A–O10A and *Y*1B–O10B bonds along the *c* axis, as well as the relative compression along *Y*1A–O6B and *Y*1B–O6A bonds, caused by Jahn–Teller effect, is characteristic for Mn^{3+}-centered polyhedra.

The crystal-chemical formula of the manganivesuvianite from N'chwanning II is:

$$^{X1}(Ca)_{2.00}\,^{X2A,B}(Ca)_{8.00}\,^{X3A,B}(Ca)_{8.00}\,^{X4A,B}(Ca)_{1.00}$$
$$^{Y1A,B}(Mn^{3+}{}_{0.67}Mn^{2+}{}_{0.33})_{\Sigma1.00}\,^{Y2A,B}(Al_{3.70}Mn^{2+}{}_{0.30})_{\Sigma4.00}$$
$$^{Y3A,B}(Al_{6.60}Mn^{2+}{}_{1.40})_{\Sigma8.00}\,^{Z1A,B}(Si)_{2.00}\,^{Z2A,B}(Si)_{8.00}$$
$$^{Z3A,B}(Si)_{8.00}(O)_{68.00}\,^{W}(OH_{9.03}O_{0.97})_{\Sigma10.00}$$

3.8 Hongheite

Despite the fact that this new vesuvianite supergroup mineral was approved in 2017 (IMA 2017-027), its status is being actively discussed by the vesuvianite nomenclature working group and the IMA CNMNC. The description of the mineral and the interpretation of experimental data are taken from the paper by Xu et al. (2019).

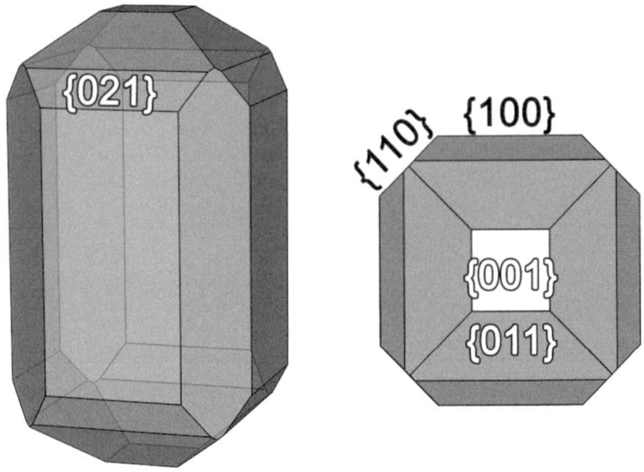

Fig. 3.71 Forms of hongheite crystals from the Malage skarn deposit, China

3.8.1 Occurrence and Physical Properties

Hongheite, ideally $Ca_{19}Fe^{2+}Al_4(Fe^{3+},Mg,Al)_8(\square,B)_4BSi_{18}O_{69}(O,OH)_9$, was found in the endoskarn zone of the Malage mine, located on the northern border of the Gejiu tin ore field, Hong Prefecture, Yunnan Province, China (Xu et al. 2019). The unit cell parameters of this mineral are a = 15.667(1) Å and c = 11.725(2) Å. Hongheite forms radial aggregates (with a diameter of 4 to 25 mm) composed of acicular crystals as well as well-formed crystals up to 2 cm in length in voids of the rock. The crystal forms (Fig. 3.71) are typical for the B-rich VSMs. The associated minerals are hydroxylapatite, danburite, fluorite, axinite-(Fe), datolite, cuspidine, as well as rare calcium and tin borates, including nordenskiöldine.

The mineral is optically positive, with $\omega = 1.720$ (2), $\varepsilon = 1.725$ (2) in white light; pleochroism is weak.

3.8.2 Chemical Composition

The chemical composition of hongheite (according to Xu et al. (2019)) is given in Table 3.14. The mineral is characterized by a significant content of iron (more than 10 wt% of FeO + Fe_2O_3) and a low content of H_2O due a rather high content of boron (2.82 wt% of B_2O_3).

Table 3.14 Chemical composition of hongheite, after Xu et al. (2019)

Component	Component	Minimal and maximal means	*apfu*
SiO_2	35.85	35.24–36.48	18.00
TiO_2	0.01	0.00–0.02	0.00
Al_2O_3	11.00	10.54–11.71	6.51
FeO*	2.14		0.90
Fe_2O_3*	7.92		2.99
CaO	33.57	32.26–34.94	18.06
MnO	0.42	0.37–0.46	0.18
MgO	3.48	3.06–3.83	2.61
Na_2O	0.01	0.00–0.03	0.01
B_2O_3**	2.82		2.44
Cr_2O_3	0.01	0.00–0.07	0.01
F	0.40	0.24–0.63	0.64
Cl	0.14	0.09–0.24	0.12
H_2O***	0.75		2.51
–F≡O	–0.17		
–Cl≡O	–0.03		
Sum	98.32		

* According to Mössbauer data
** According to ICP MS data
*** According to TG/DSC data

3.8.3 *Thermogravimetry and Differential Scanning Calorimetry*

The H_2O content of 1.45 wt% in hongheite, determined using DSC data (Xu et al. 2019), exceeds the theoretical value of 0.75 wt%. The TG curve contains one significant mass loss step, which is accompanied by the endothermic peak with a minimum at 1028 °C corresponding to the dehydroxylation and decomposition of the mineral.

3.8.4 *Infrared Spectroscopy*

The IR spectrum of hongheite is shown in Fig. 3.72. According to Xu et al. (2019), the band at 479 cm^{-1} corresponds to Fe^{3+}–O stretching vibrations in the *Y*3 octahedra. The peaks in the wave region of 530–1002 cm^{-1} are assigned to different modes of the SiO_4^{4-} and $Si_2O_7^{6-}$ groups. The band at 1271 cm^{-1} is related to the vibrations of B–O bonds in the BO_3 group. The lines at 1385, 1472 and 1558 cm^{-1} were assigned to the stretching vibrations inside the carbonate groups, but taking into

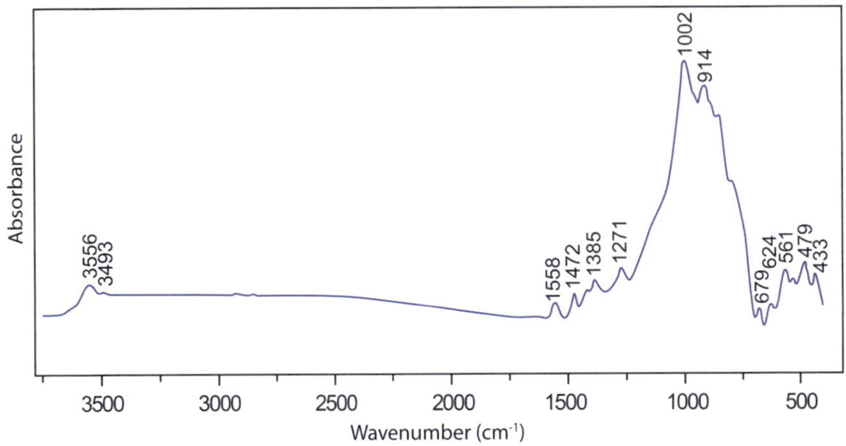

Fig. 3.72 Infrared spectrum of hongheite, after Xu et al. (2019)

account a strong distortion of the BO_3 triangle (with short B–O bonds) in B-bearing VSMs, the assignment of these bands (typical for B-bearing VSMs) to B–O stretching vibrations would be more convincing. In the region of stretching vibrations of O–H bonds, bands at 3556 and 3493 cm^{-1} are observed. The latter band is close to the F band in IR spectra of other VSMs whereas the wavenumber of 3556 cm^{-1} is intermediate between the wavenumbers of the D and E bands and may correspond to the Fe^{3+}–O–H···O configuration.

3.8.5 *Mössbauer Spectroscopy*

Experimental Mössbauer spectroscopy data are shown at Fig. 3.73. The obtained spectrum is described as a superposition of two symmetric doublets corresponding to Fe^{2+} at the *Y*1 site and Fe^{3+} at the *Y*3 site (Table 3.15).

3.8.6 *Single-Crystal X-Ray Diffraction Analysis*

The crystal structure of hongheite has been refined in the space group *P*4/*nnc* with $R_1 = 0.063$. The distinctive features of this mineral among the other VSM are the dominance of vacancy over Ca in the *X*4 position, the dominance of Fe^{3+} in the *Y*3 position and the dominant role of vacancy in the *T*1 position whereas the *T*2 position is completely populated by B atoms. Data on the occupancy of the key sites are given in (Table A.45).

3.9 Manaevite-(Ce) and *REE*-Bearing Vesuvianite

To date, several VSMs localities with a significant content of rare earth elements (up to 21 wt% of REE_2O_3) are known: San Benito and Seward (USA), Keivy and Dahu Unur (Russia) (Orlov and Martyanov 1960; Crook and Oswald 1979; Himmelberg and Miller 1980; Batieva et al. 1987; Fitzgerald et al. 1987). All of them are associated with altered alkaline rocks.

3.9.1 Occurrence and Physical Properties

Manaevite-(Ce) was found in the Phlogopite quarry of the Kovdor alkaline ultrabasic massif, Kola Peninsula, Russia. The mineral occurs in lenticular bodies which were formed in the final stages of the formation of skarnoid (Kaverin et al. 1988; Seredkin 2001; Ivanyuk et al. 2002). The associated minerals are calcite, natrolite, pectolite, vesuvian, pargasite, phlogopite, diopside, andradite and glagolevite. Manaevite-(Ce) forms dark orange outer zones (up to 2 mm thick) of short-prismatic dark brown *REE*-containing vesuvianite crystals up to 5 mm across in coarse-grained calcite aggregate (Fig. 3.74). The mineral is optically uniaxial, negative, its refractive indices significantly exceed those for all other VSMs: $\omega = 1.760$ (5) and $\varepsilon = 1.750$ (5). Some areas are slightly biaxial.

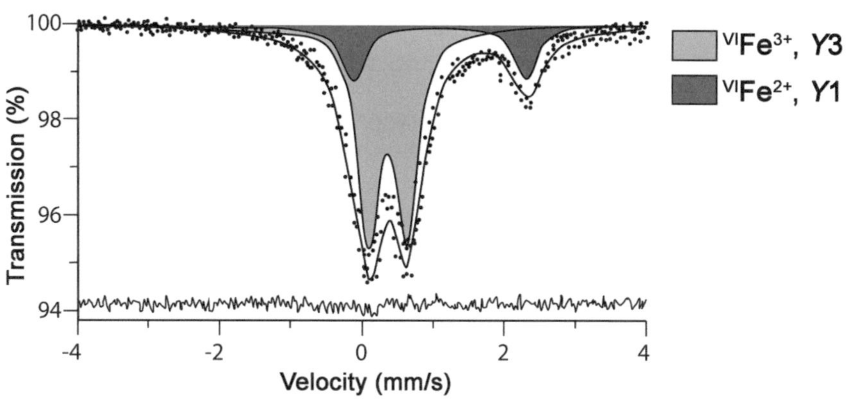

Fig. 3.73 Mössbauer spectrum of hongheite

Table 3.15 Hyperfine parameters of the Mössbauer spectrum of hongheite

Isomer shift, mm/s	Quadrupole shift, mm/s	Line width, mm/s	Peak area, %	Assignment
1.12 ± 0.02	2.41 ± 0.04	0.39 ± 0.07	23 ± 5	Fe^{2+} in *Y*1
0.39 ± 0.02	0.54 ± 0.02	0.36 ± 0.02	77 ± 5	Fe^{3+} in *Y*3

3.9.2 Chemical Composition

REE-bearing vesuvianite and manaevite-(Ce) have a distinct chemical zonality (Fig. 3.75). The marginal zones are enriched by Ti, *REE* and Mg and depleted in Al as compared with the central parts of the crystals (Table 3.16). A significant (up to 3 wt%) ThO_2 content in manaevite-(Ce) did not cause complete metamictization of the mineral, which indicates its high resistance to self-radiation.

3.9.3 Thermogravimetry and Differential Scanning Calorimetry

Manaevite-(Ce) demonstrates anomalous for the VSMs two-step weight loss (Fig. 3.76). Earlier this phenomenon was also observed for epitaxial crusts of Si-deficient B-containing vesuvianite described by Panikorovskii et al. (2016a). The first stage of weight loss (1.87 wt%) in the temperature range of 500–700 °C may correspond to the loss of the $(H_4O_4)^{4-}$ groups (Rivas-Mercury et al. 2008). The second stage of weght loss (4.24 wt%) accompanied by an endothermic peak with a minimum at 1018 °C corresponds to the loss of OH groups at the O10 and O11 positions (Karwowski et al. 1996; Żabiński et al. 1996). Endothermal features in the DSC curve in the temperature range of 20–500 °C may be due to the loss of water molecules occurring at the partly vacant *X*3 position or in metamict blocks.

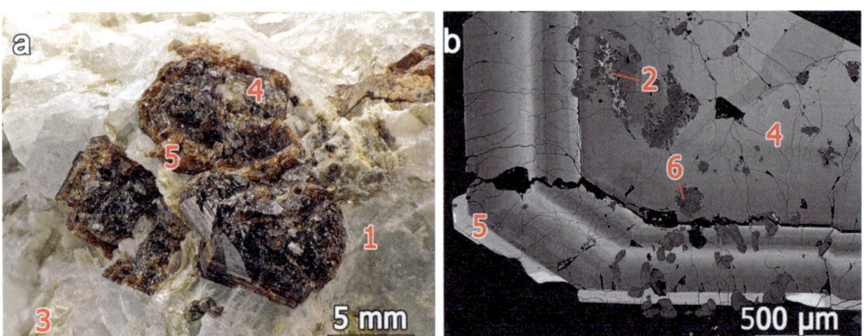

Fig. 3.74 Manaevite-(Ce) photograph (**a**) and BSE image of a crystal with *REE*-bearing vesuvianite and manaevite-(Ce) zones (**b**). 1—calcite, 2—andradite inclusions, 3—glagolevite, 4—vesuvianite, 5—manaevite-(Ce), 6—hydroxylapatite

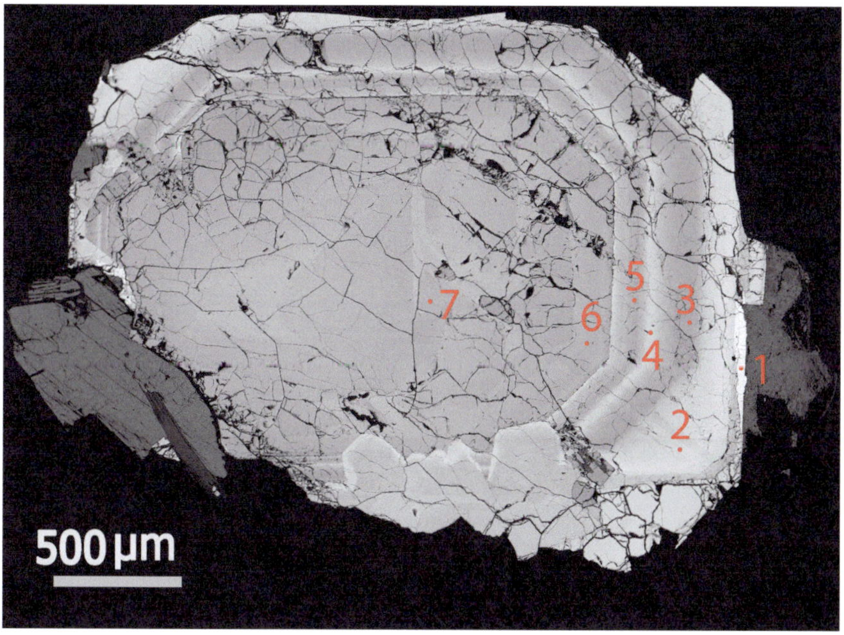

Fig. 3.75 The BSE image of the zonal *REE*-bearing VSM from the Kovdor massif. The red circles indicate points of analyses given in Table 1.25

3.9.4 CHN Analysis

According to the CHN analysis data, manaevite-Ce contains 1.57(5) wt% of CO_2 and 5.8(1) wt% of H_2O. The measured CO_2 content was attributed to calcite inclusions.

3.9.5 Infrared Spectroscopy

Manaevite-(Ce) has an unusual for the VSMs IR spectrum (Fig. 3.77). In addition to stretching vibrations of O–H bonds (the group of unresolved bands in the range of 3100–3700 cm^{-1}) it contains a peak at 1630 cm^{-1} corresponding to H–O–H bending vibrations of water molecules. A complex character of the spectrum and a poor resolution of the absorption bands in the regions of stretching and bending vibrations of Si–O bonds may be due to a partial metamictization of the mineral and partial substitution of Si by H_4O_4 and Al.

Table 3.16 Chemical composition of *REE*-bearing VSM and manaevite-(Ce) from Kovdor (wt%)

Sample	1	2	3	4	5	6	7
SiO_2	26.96	30.62	34.12	28.95	32.10	34.23	33.31
TiO_2	5.60	0.75	1.06	1.86	0.93	0.49	0.30
Al_2O_3	4.19	5.82	9.64	7.25	9.02	10.46	9.32
Fe_2O_3	5.30**	0.00	0.00	0.00	0.00	0.00	0.00
FeO	0.78**	8.58	7.20	7.01	6.95	6.82	6.59
MnO	0.05	0.03	0.04	0.02	0.05	0.08	0.08
MgO	4.42	4.69	4.14	4.30	4.11	4.25	4.47
CaO	18.39	23.54	30.21	24.59	29.05	31.42	29.07
Na_2O	0.29	0.64	0.40	0.50	0.16	0.30	0.41
SO_3	0.13	0.11	0.05	0.08	0.12	0.12	0.19
SrO	0.40	0.47	0.52	0.41	0.44	0.54	0.37
La_2O_3	2.97	3.65	1.62	3.82	2.41	1.97	2.35
Ce_2O_3	10.04	7.06	2.47	5.28	3.15	2.72	3.51
Pr_2O_3	1.52	0.91	0.32	0.76	0.23	0.28	0.32
Nd_2O_3	3.23	1.28	0.45	1.27	0.63	0.30	0.40
Sm_2O_3	0.35	0.20	0.00	0.00	0.00	0.00	0.00
Gd_2O_3	0.24	0.12	0.06	0.13	0.00	0.00	0.00
PbO	0.72	0.00	0.00	0.00	0.00	0.00	0.00
ThO_2	2.86	1.95	0.52	1.58	0.79	0.27	0.41
H_2O*	6.11**	3.91	4.34	4.63	4.50	4.74	4.22
F	0.31	0.27	0.28	0.24	0.29	0.27	0.28
Cl	0.08	0.12	0.15	0.12	0.12	0.10	0.11
–O≡F	0.13	0.11	0.12	0.10	0.12	0.11	0.12
–O≡Cl	0.02	0.03	0.04	0.03	0.03	0.02	0.03
Total	94.79	95.61	98.30	93.51	97.56	100.47	96.83
apfu							
Si	16.43	16.70	16.71	15.99	16.46	16.46	16.73
Ti	2.57	0.31	0.39	0.77	0.36	0.18	0.11
Al	3.01	3.74	5.57	4.72	5.45	5.93	5.52
Fe^{3+}	2.43	0.00	0.00	0.00	0.00	0.00	0.00
Fe^{2+}	0.40	3.91	2.95	3.24	2.98	2.74	2.77
Mn	0.03	0.01	0.02	0.01	0.02	0.03	0.04
Mg	4.01	3.82	3.03	3.54	3.14	3.04	3.35
Ca	12.01	13.76	15.86	14.56	15.96	16.19	15.65
Na	0.34	0.68	0.38	0.53	0.16	0.28	0.40
S^{6+}	0.06	0.04	0.02	0.03	0.05	0.04	0.07
Sr	0.14	0.15	0.15	0.13	0.13	0.15	0.11

(continued)

Table 3.16 (continued)

Sample	1	2	3	4	5	6	7
La	0.67	0.73	0.29	0.78	0.46	0.35	0.43
Ce	2.24	1.41	0.44	1.07	0.59	0.48	0.65
Pr	0.34	0.18	0.06	0.15	0.04	0.05	0.06
Nd	0.70	0.25	0.08	0.25	0.11	0.05	0.07
Sm	0.07	0.04	0.00	0.00	0.00	0.00	0.00
Gd	0.05	0.02	0.01	0.02	0.00	0.00	0.00
Pb	0.12	0.00	0.00	0.00	0.00	0.00	0.00
Th	0.40	0.24	0.06	0.20	0.09	0.03	0.05
H	24.83	14.20	14.16	17.04	15.16	15.16	14.08
F	0.60	0.47	0.43	0.42	0.47	0.41	0.44
Cl	0.08	0.11	0.12	0.11	0.10	0.08	0.10

The calculation of the coefficients in the formula was carried out on 46 cations (with the exception of H) using X-ray structural data for sample 1
* H_2O calculated based on single-crystal XRD data
** Ratio Fe^{2+}/Fe^{3+} determined from Mössbauer data

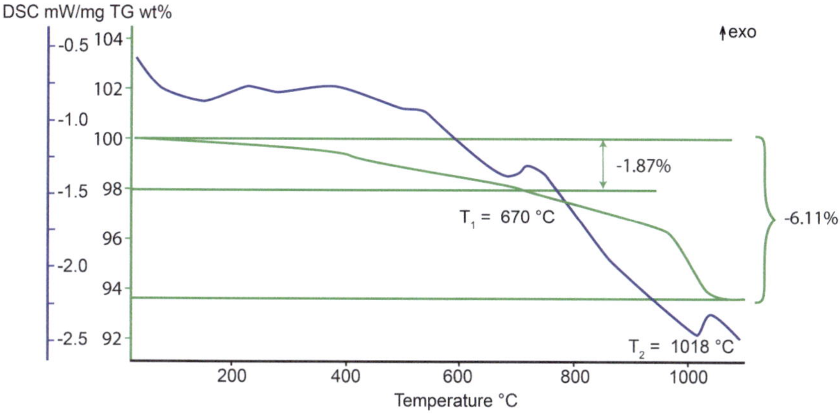

Fig. 3.76 TG (green line) and DSC (blue line) curves of manaevite-(Ce)

3.9.6 Solid-State Nuclear Magnetic Resonance

The solid-state ^{27}Al NMR spectrum of manaevite-(Ce) (Fig. 3.78a) contains two peaks with different half-widths. The broad peak at 1.66 ppm is attributed to ^{VI}Al in the *Y*2 and *Y*3 positions (Phillips et al. 1987). The narrow peak at 59.90 ppm can be attributed to ^{IV}Al (Singh et al. 2005; Panikorovskii et al. 2017c). The relative integrated intensities of the peaks corresponding to ^{IV}Al and ^{VI}Al are 46% and 54%,

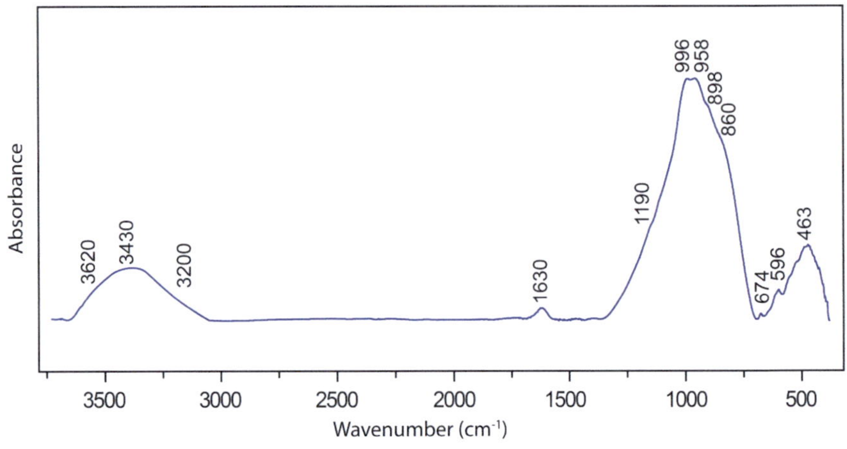

Fig. 3.77 Infrared spectrum of manaevite-(Ce)

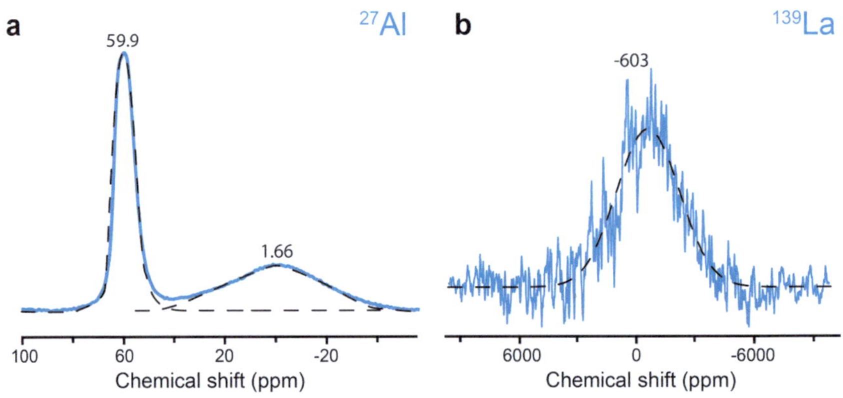

Fig. 3.78 Solid-state ^{27}Al (**a**) and ^{139}La (**b**) NMR spectra of manaevite-(Ce). The dashed lines show the approximating curves

respectively, which corresponds to ~2 Al *apfu* in tetrahedral coordination. A major part of ^{IV}Al may occur in metamict blocks.

The spectrum of ^{139}La (Fig. 3.78b) is represented by a broad peak with a chemical shift of –603 ppm, corresponding to La^{3+} occurring at the *X*3 site, according to the structural data.

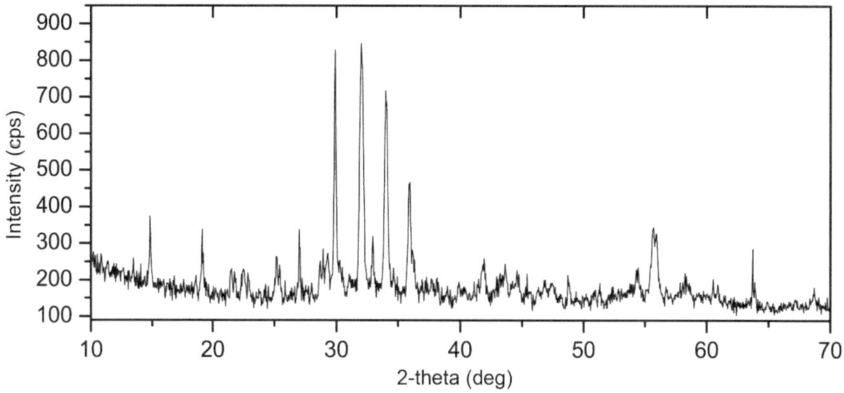

Fig. 3.79 Powder X-ray diffraction pattern of manaevite-(Ce)

3.9.7 Powder X-Ray Diffaction

The powder X-ray diffraction pattern of manaevite-(Ce) is shown in Fig. 3.79. The refined unit cell parameters ($a = 15.7957(5)$ Å and $c = 11.9180$ (5) Å) are less than those determined according to the single-crystal XRD data ($a = 15.925(1)$ and $c = 11.966$ (1) Å) due to the chemical zonality of the studied crystals. The increased values of the unit cell parameters of manaevite-(Ce) in comparison with the usual values for the VSMs ($a \sim 15.55$, $b \sim 11.80$ Å) are due to the occurrence of magnesium and titanium instead of aluminum in the octahedral *Y*2 and *Y*3 positions as well as a partial substitution of Ca with *REE* at the *X*3 and *X*4 sites. Data on the intensity, position of the lines and their *hkl* indexes are given in Table A.46.

3.9.8 Mössbauer Spectroscopy

The Mössbauer spectrum of manaevite-(Ce) (Fig. 3.80) can be described as a superposition of four symmetric doublets. The Mössbauer parameters of these doublets (isomer shift *IS*, quadrupole splitting *QS*, line width and areas under resonance doublets *S*) and their assignment based on crystal structure refinement are given in Table 3.17. In accordance with the data by Rusakov et al. (2006), Groat et al. (2013), all iron in manaevite-(Ce) occupies octahedral positions, *Y*2 and *Y*3. The $^{Y2}(Fe^{2+} + Fe^{3+})$:$^{Y3}(Fe^{2+} + Fe^{3+})$ ratio is 0.28:0.72. The absence of iron with five-fold coordination is in agreement with the single-crystal X-ray diffraction data.

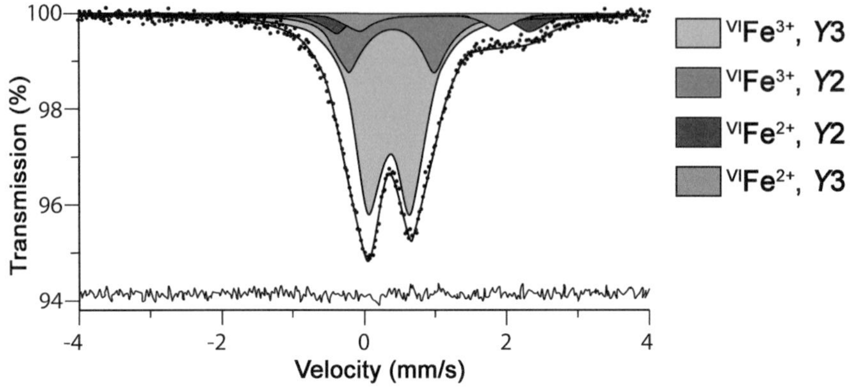

Fig. 3.80 Mössbauer spectrum of manaevite-(Ce)

Table 3.17 Hyperfine parameters of quadrupole doublets in the Mössbauer spectrum of manaevite-(Ce)

Doublet	Quadrupole splitting (mm/s)	Isomer shift (mm/s)	Line width (mm/s)	Area (%)	Assignment
1	0.366 ± 0.003	0.300 ± 0.003	0.487 ± 0.007	66.2 ± 1.0	$^{VI}Fe^{3+}$ at *Y*3
2	0.399 ± 0.006	0.613 ± 0.007		21.0 ± 0.9	$^{VI}Fe^{3+}$ at *Y*2
3	0.925 ± 0.026	0.984 ± 0.026		6.1 ± 0.5	$^{VI}Fe^{2+}$ at *Y*3
4	1.119 ± 0.005	1.380 ± 0.018		6.8 ± 0.6	$^{V}Fe^{2+}$ at *Y*2

3.9.9 Single-Crystal X-Ray Diffraction Analysis

The crystal structure of manaevite-(Ce) was refined in the *P*4/*nnc* space group with $R_1 = 0.035$ for 1527 independent reflections with the intensities $|F_o| \geq 4\sigma_F$ (Table A.47). Only one low-intense violating reflection for the *P*4/*nnc* space group was observed. Despite the increased unit cell parameters [$a = 15.9247(13)$, $b = 11.9661(10)$ Å], the structure of manaevite-(Ce) (Fig. 3.81) does not have significant distinctions from that of vesuvianite.

The crystal structure refinement with a fixed total occupancy of the *Z*-group cations lead to the R_1 value of 0.05 and unusually large displacement parameters for the *Z*1 and *Z*2 positions. The refined scattering factor for the *Z*1 position corresponds to 8.40 e^-, which, together with the significantly increased mean <*Z*1–O> bond length of 1.718 Å, indicates the presence of a significant fraction (about 40%) of vacant *Z*1 sites. According to chemical data, vacancies are coordinsted by the H_4O_4 clusters according to the $SiO_4^{4-} \leftrightarrow [H_4O_4]^{4-}$ substitution scheme (Galuskin 2005). As noted above, this type of substitution is common for low-temperature VSMs crystallized at 200–300 °C (Galuskin et al. 2003a, b). There is a lack of silicon also at the *Z*2 and *Z*3 sites whose scattering factors correspond to 10.50 and 12.04 e^-, respectively. The average <*Z*2–O> and <*Z*3–O> bond lengths are 1.691 and

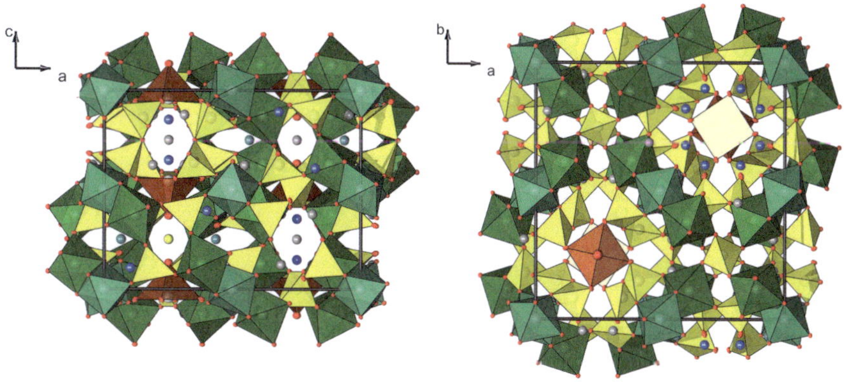

Fig. 3.81 The view of the crystal structure of manaevite-(Ce) along the *b* (left) and *c* (right) axes. The *Z* tetrahedra are yellow; the *Y*2 and *Y*3 octahedra are green, the five-coordinated *Y*1 polyhedron is orange. O atoms are highlighted in red, Ca atoms are gray (*X*1, *X*2), and the *REE*-dominant positions (*X*4 and *X*3) are shown with blue balls. Yellow and green balls indicate the low-populated positions of sulfur (*T*1) and aluminum (*T*2). The unit cell is outlined

1.644 Å, respectively, which is in a good agreement with the refined occupancies of these positions: $Si_{0.75}$ and $Si_{0.86}$. The hydrogarnet substitution, $SiO_4^{4-} \leftrightarrow [H_4O_4]^{4-}$ is impossible for nesosilicate sites. Low scattering of *Z*1 and *Z*2 sites may related to a partial metamictization or to the presence of B_2O_7 groups, to which the shoulder at 1190 cm^{-1} in the IR spectrum may belong.

The *X*1 и *X*2 positions with the coordination numbers 7 and 8 are occupied exclusively by Ca atoms, while the largest *X*3 polyhedron is *REE*-dominant, in accordance with (Fitzgerald et al. 1987; Hålenius et al. 2013). The refined occupancy of the *X*3 position is $Ce_{0.44}Ca_{0.36}\square_{0.20}$ (Fig. 3.82), in agreement with its scattering factor corresponding to 32.72 e^-, <*X*3–O> mean bond length of 2.584 Å, the total *REE* content in the mineral and the existence of an additional site which is close to *X*3 and is populated by H_2O on 20%. Rare-earth elements can also partially incorporate into the *X*4 position, the refined population of which is $Ca_{0.74}Ce_{0.26}$. The vacancy at the *X*3 position is associated with the incorporation of Th^{4+} and REE^{3+} cations into the manaevite-(Ce) structure in accordance with the substitution schemes $2Ca^{2+} \leftrightarrow Th^{4+} + \square$ and $3Ca^{2+} \leftrightarrow 2REE^{3+} + \square$.

In the crystal structure of manaevite-(Ce), Mg predominates at the *Y*1 and *Y*3 positions. The scattering factor of the tetragonal-pyramidal *Y*1 position corresponds to 6.0 e^-, i.e. half-population of this position by $Mg_{0.50}$ (Table A.48). The mean bond lengths in the *Y*2 and *Y*3 octahedra are significantly increased as compared with most other VSMs and are equal to 1.976 and 2.019 Å, respectively. For comparison, the lengths of analogous bonds in vesuvianite from the Vesuvius volcano (Panikorovskii et al. 2017b) lies within 1.895–1.896 Å for *Y*2–O and 1.961–1.966 Å for *Y*3–O (Table A.49). The refined scattering factors correspond to 14.66 and 15.78 e^- and the populations $[(Al,Mg)_{0.82}(Fe,Ti)_{0.18}]_{1.00}$ and $[(Mg,Al)_{0.73}(Fe,Ti)_{0.27}]_{1.00}$ for the *Y*2 and *Y*3 positions, respectively.

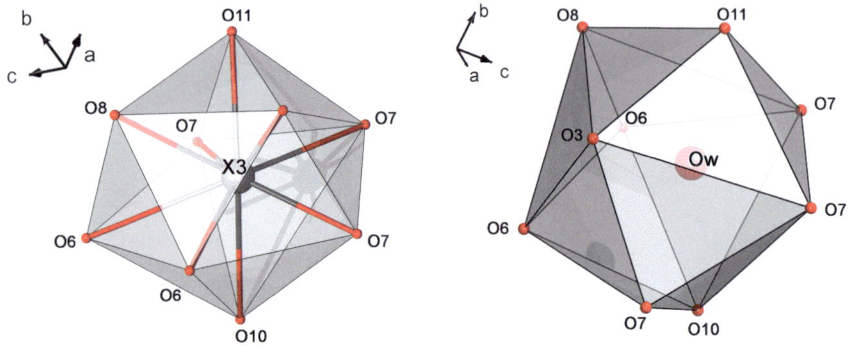

Fig. 3.82 Local environment of the $X3$ and Ow sites

The additional $T1$ and $T2$ positions are partially populated by Al and S atoms, respectively, and their refined occupancies are $Al_{0.08}$ and $S_{0.07}$. Incorporation of Al^{3+} into the $T1$ position is in agreement with the solid-state NMR data (Fig. 3.78a). According to the photoelectron spectroscopy data for the S2p level, sulfur in the vesuvianite structure occupies a position at the center of the $T2$ triangle.

The Ow site of molecular water was localized close to the $X3$ position (the $X3$–Ow distance is 0.35 Å) (Fig. 3.82), its refined population is $Ow_{0.20}$. Thus, the Ow position is populated when the $X3$ position is vacant. In fact, it is a split position whose total composition can be written (by combining $X3$ and Ow) as $[Ce_{0.44}Ca_{0.36}(H_2O)_{0.20}]_{1.00}$. The presence of molecular water in manaevite-(Ce) is confirmed by TG/DSC and CHN-analyzes and the band at 1630 cm^{-1} in the IR spectrum.

Taking into account the data of solid-state NMR and infrared spectroscopy, the crystal-chemical formula of manaevite-(Ce) can be written as follows:

$$^{X1}(Ca)_{\Sigma 2.00}\ ^{X2}(Ca)_{\Sigma 8.00}\ ^{X3}(Ce_{3.52}Ca_{2.88}\{H_2O\}_{1.60})_{\Sigma 8.00}$$
$$^{X4}(Ca_{0.74}Ce_{0.26})_{\Sigma 1.00}\ ^{Y1}Mg_{\Sigma 1.00}$$
$$^{Y2}[(Al,Mg)_{3.24}(Fe,Ti)_{0.76}]_{\Sigma 4.00}\ ^{Y3}[(Mg,Al)_{5.84}(Fe,Ti)_{2.16}]_{\Sigma 8.00}$$
$$^{Z1}[Si_{1.20}(H_4O_4)_{0.80}]_{\Sigma 2.00}$$
$$^{Z2}[Si_{6.00}(H_4O_4)_{2.00}]_{\Sigma 8.00}\ ^{Z3}(Si)_{\Sigma 8.00}\ ^{T1}(Al_{0.32}\square_{3.68})_{\Sigma 4.00}$$
$$^{T2}(S_{0.07}\square_{0.93})_{\Sigma 1.00}O_{69.00}(OH,O)_{\Sigma 9.60}.$$

3.10 Milanriederite

Incorporation of rare earth elements into the vesuvianite structure via the substitution $^{X3}Ca^{2+} + {}^{Y3}Al^{3+} \leftrightarrow {}^{X3}REE^{3+} + {}^{Y3}Mg^{2+}$ may lead to the predominance of Mg at the $Y3$ site, whereas $X3$ site still remains Ca-dominant (Chukanov et al. 2019).

Fig. 3.83 Crystal of milanriederite in a hausmannite-jacobsite ore, FOV width 4 mm (**a**) and separate crystal of milanriederite, FOV width 4 mm (**b**)

3.10.1 Occurrence and Physical Properties

Milanriederite occurs in a jacobsite-hausmannite ore of the Kombat deposit situated in the Grootfontein district, Otjozondjupa region, Northern Namibia. The mineral assemblage including milanriederite was formed by hydrothermal and metasomatic replacement and fracture-fill of dolostones of the Upper Proterozoic Hüttenberg Formation by polymineral aggregates containing Mn-rich spinels, calcite, dolomite, baryte, as well as various subordinate components including different lead oxychlorides (Innes and Chaplin 1986).

Milanriederite occurs as dark brownish red dipyramidal crystals up to 3 mm across embedded in granular baryte-calcite-jacobsite-hausmannite aggregates with minor glaucochroite and native copper (Fig. 3.83a, b). The major crystal form is {111} and the minor forms (observed on some crystals) are {001}, {110}, and {100}. Based on morphological features, one can conclude that milanriederite is the earliest mineral in this association, whereas native copper belongs to a late stage of mineral formation.

Milanriederite is brittle, has a Mohs' hardness of 6 and an uneven fracture. Neither cleavage nor parting is observed. The density measured by hydrostatic weighing is 3.53(2) g/cm^3. Density calculated using the empirical formula and unit-cell dimensions obtained from single-crystal X-ray diffraction data is 3.547 g/cm^3. Milanriederite is optically uniaxial (–), $\omega = 1.744(3)$, $\varepsilon = 1.737(3)$ (589 nm). The extinction is straight. Inspected with the polarizing microscope, the mineral is pleochroic from brownish-pink (*O*) to nearly colorless (*E*).

3.10.2 Chemical Composition

The chemical composition of milanriederite is given in Table 3.18. The empirical formula of milanriederite (based on 78 O atoms *pfu,* i.e. $Z = 2$) is $[Ca_{16.48}(La_{0.84}Y_{0.67}Ce_{0.16}Nd_{0.24}Pr_{0.06})Na_{0.47}]_{\Sigma 18.92}(Al_{5.92}Mg_{4.22}$

Table 3.18 Chemical composition of milanriederite after (Chukanov et al. 2019)

Component	Mean content	Ranges	*apfu*
Na_2O	0.47	0.42–0.57	0.47
MgO	5.49	5.07–6.72	4.22
CaO	29.86	28.34–31.08	16.48
Mn_2O_3	1.40	1.09–1.43	0.55
Al_2O_3	9.75	8.26–10.65	5.92
Fe_2O_3*	6.03	5.47–7.22	2.34
Y_2O_3	2.44	1.79–3.03	0.67
La_2O_3	4.43	3.43–5.30	0.84
Ce_2O_3	0.86	0.64–1.05	0.16
Pr_2O_3	0.32	0.27–0.42	0.06
Nd_2O_3	1.31	0.98–1.46	0.24
SiO_2	34.84	34.33–35.41	17.95
H_2O**	2.9(3)	0.42–0.57	9.96
Total	100.10		

* All Fe is trivalent according to Mössbauer data
** H_2O content measured by the Penfild method

$Fe^{3+}_{2.34}Mn^{3+}_{0.55})_{\Sigma 13.03}$ $Si_{17.95}O_{68.04}(OH)_{9.96}$. The simplified formula written in accordance with the structural data is (Ca, *REE*)$_{19}$$Fe^{3+}Al_4(Mg, Al, Fe^{3+})_8Si_{18}O_{68}(OH, O)_{10}$.

3.10.3 *Infrared Spectroscopy*

The IR spectrum of a milanriederite (Fig. 3.84) contains bands of O–H-stretching vibrations (in the range from 3000 to 3700 cm^{-1}), strong bands of Si–O-stretching vibrations of the groups SiO_4 and Si_2O_7 (in the range from 860 to 1080 cm^{-1}), as well as mixed lattice modes involving Si–O–Si and O–Si–O bending and *M*–O-stretching ($M = $ Al, Mg, Fe^{3+}) vibrations (below 660 cm^{-1}). The shoulder at 805 cm^{-1} is assigned to *M*···O–H bending vibrations. The absence of absorption bands in the range 1100–1500 cm^{-1} indicates the absence of boron-bearing groups.

The bands of O–H-stretching vibrations at 3670, 3629, 3567 and 3200 cm^{-1} correspond to the bands A, B, D and J, respectively, which are predominantly polarized along *c* and correspond to differently coordinated OH groups at the sites O11 (the bands A, B and D) and O10 (the band J) (Groat et al. 1995b). The band at 3320 cm^{-1} does not have analogues among O–H-stretching bands discussed by Groat et al. (1995) and is a specific feature of milanriederite.

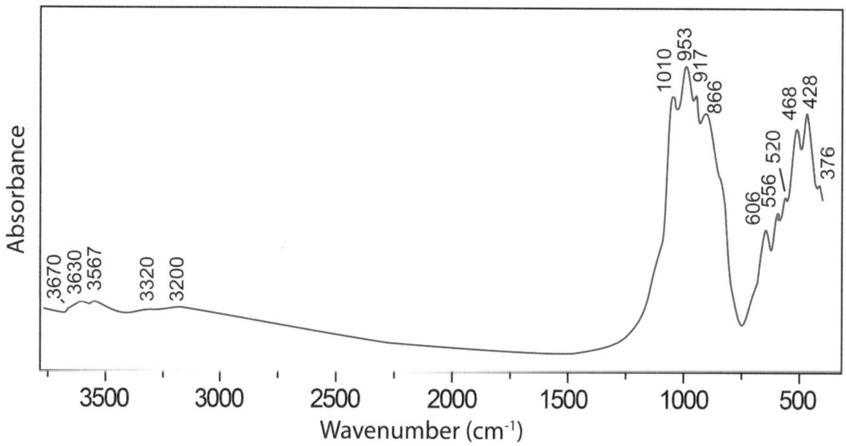

Fig. 3.84 Infrared spectrum of milanriederite

3.10.4 Mössbauer Spectroscopy

The milanriederite Mössbauer spectrum can be approximated by a symmetric doublet with *IS* = 0.363(3) mm/s, *QS* = 0.513(4) mm/s and *FWHM* = 0.461(8) mm/s (Fig. 3.85) which corresponds to trivalent iron. Unfortunately, a correct and unambiguous division of the Mössbauer spectrum into individual doublets is impossible because of their proximity.

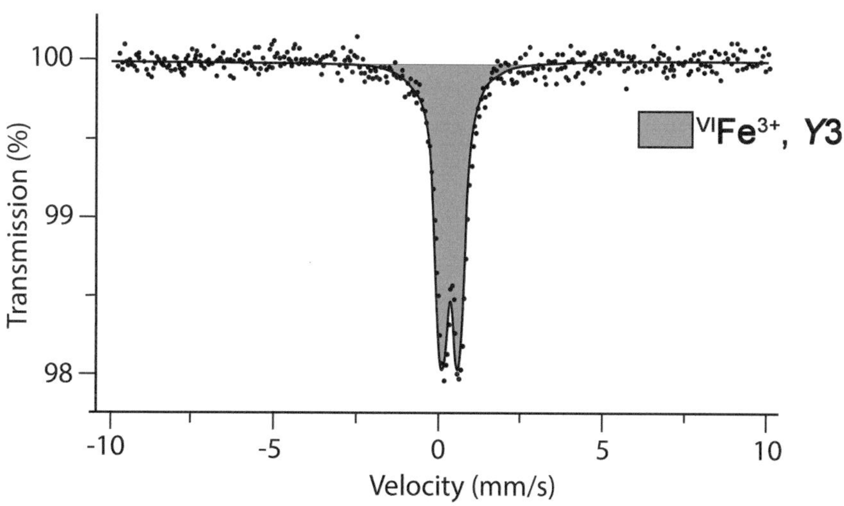

Fig. 3.85 Mössbauer spectrum of milanriederite

3.10.5 Powder X-Ray Diffraction

Data on intensities, line positions and their *hkl* indices for the powder X-ray diffraction pattern of milanriederite are given in Table A.50. The refined unit cell parameters of milanriederite are $a = 15.679(2)$, $c = 11.888(3)$ Å, $V = 2922(2)$ Å^3.

3.10.6 Single-Crystal X-Ray Diffraction Analysis

The crystal structure of milanriederite was solved based on single-crystal X-ray diffraction data for 1684 independent reflections with intensity $|F_o| \geq 4\sigma_F$ and refined to $R_1 = 0.027$ in the *P4/nnc* space group, because only 49 weak reflections violating absence conditions for this space group were observed (Table A.51). Final atom coordinates, displacement parameters and site occupancies are given in (Table A.52). Selected interatomic distances are reported in (Table A.53). According to the chemical data, milanriederite contains significant amounts of *REE* (both *Ln* and Y), and during the structure refinement their mean atomic number was accepted to be 51.

In milanriederite, tetrahedrally coordinated sites are occupied by Si atoms only, with mean <Si–O> distances in the range of 1.631–1.646 Å. The bond-valence sums (BVS) for the *Z*-sites of 3.88, 3.77 and 3.99 valence units indicate the absence of a hydrogarnet-type substitution. The *X*2 and *X*4 sites are occupied by Ca atoms only, while *X*1 (BVS 2.16) and *X*3 (BVS 2.20) sites contain admixtures of REE^{3+} which is typical for vesuvianite supergroup minerals (see Fitzgerald et al. 1987). The refined number of electrons (e_{ref}) at the *X*3 site is 26.82 e^-, and the mean <*X*3–O> distance is 2.593 Å, which is in accordance with the refined occupancy $Ca_{0.78}REE_{0.22}$. Rare earth elements and sodium also occur at the *X*1 site with the total refined occupancy $Ca_{0.75}Na_{0.15}REE_{0.10}$.

In the crystal structure of milanriederite, magnesium prevails over aluminum at the *Y*3 site, which is confirmed by the mean <*Y*3–O> distance of 2.015 Å. For comparison, corresponding mean distances for two samples of vesuvianite from its type locality, the Vesuvius volcano in Campania, Italy, are 1.966 and 1.961 Å (Panikorovskii et al. 2017b).

The total scattering factors of the *Y*2 (BVS 3.04) and *Y*3 sites are 13 and 14.94, which is in agreement with refined occupancies $[Al]_{1.00}$ for the *Y*2 site and $[Mg_{0.56}Al_{0.25}Fe^{3+}{}_{0.19}]_{1.00}$ for the *Y*3 site (Fig. 3.86). The refined populations of the *Y*2 and *Y*3 sites are also in agreement with the total chemical composition of milanriederite and BVS values for the mixed *Y*3 site (2.54 for Mg at *Y*3 and 2.27 for Al at *Y*3). The low BVS values of 1.79, 1.22 and 1.31 observed for O7, O10 and O11 sites, respectively, indicate their involvement in the hydrogen bonding system.

The refined crystal-chemical formula of milanriederite is:

$$\left[{}^{X2,4}(Ca_{8.80}Na_{0.20})_{\Sigma9.00}\,{}^{X3}(Ca_{6.24}REE_{1.76})_{\Sigma8.00}\,{}^{X1}(Ca_{1.50}Na_{0.30}REE_{0.20})_{\Sigma2.00}\right]^{Y1}$$
$$(Fe_{0.94}Mg_{0.06})_{\Sigma1.00}\,{}^{Y2}Al_{4.00}\,{}^{Y3}[Mg_{4.48}Al_{2.00}(Fe,Mn)_{1.52}]_{\Sigma8.00}[SiO_4]_{10.00}[Si_2O_7]_{4.00}(OH,O)_{10}.$$

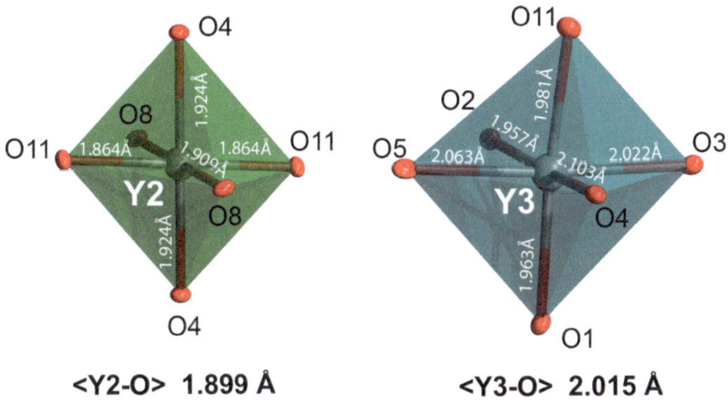

Fig. 3.86 Local coordination of the Mg-dominant *Y*3 site in the crystal structure of milanriederite

3.11 Ti-Rich Modraite

The Fe^{3+} cation is the most common component in the *Y*1 polyhedron. The presence of Fe^{2+} at the *Y*1 site in vesuvianite via Mössbauer spectroscopy was demonstrated by Manning and Tricker (1975). The first full description of Modraite (from Alchuri, Shigar Valley, Northern Areas, Pakistan), including complete chemical data, Mössbauer, MAS NMR and IR spectroscopy, powder and single-crystal XRD data, crystal structure refinement to $R_1 = 0.031$, optical data, was reported by Aksenov et al. (2016). This material (IMA 2018-164) was registered as 'ferrovesuvianite' with formula $Ca_{19}Fe^{2+}Al_4(Al, Ti, Fe)_8(SiO_4)_{10}(Si_2O_7)_4(OH, O)_{10}$, with Fe^{2+} at the *Y*1 site, however the approval of this mineral was suspended because of the absence of the VSMs nomenclature.

Despite this fact, a Fe^{2+}-dominant (at the *Y*1 site) analogue of vesuvianite was approved by the IMA CNMNC (the IMA proposal 2023-108a) with the name modraite and the idealized formula $Ca_{19}Fe^{2+}Al_4(Al_6Fe^{2+}{}_2)(\square_4)\square(SiO_4)_{10}(Si_2O_7)_4(OH)_{10}$ (Uher et al. 2024). However, the formula of modraite is to be modified to $Ca_{19}Fe^{2+}Al_4(Al_7R^{2+})(SiO_4)_{10}(Si_2O_7)_4[(OH)_9O]_{10}$ since subordinate divalent cations are not species-defining components in Y3 site (give only 25% of occupancy) and owing to short H···H distances which makes impossible the occurrence of 10 OH groups at the *W*1 and *W*2 sites. Material from Malé Karpaty is considered as a holotype of modraite and the Ti-rich variety from Shigar Valley as its co-type. The holotype material was collected a small unnamed valley in a forest of the MaléKarpaty (Little Carpathian Mountains), *ca.* 200 m to the southwest of Dolinkovský Hill, settlement of Harmónia, Modra, Pezinok District, Bratislava Region, Slovakia (48°2105000 N,17°1804300 E;) The complete data on the modraite holotype are still not published yet and in this chapter we report data on the Ti-rich variety of this mineral from Shigar Valley.

Fig. 3.87 Ti-rich modraite from Alchuri, Shigar Valley, Pakistan: parallel-prismatic agragates (**a**), druse (**b**), split crystal (**c**). 1—Ti-rich modraite, 2—calcite, 3—chloritite rock

3.11.1 Occurrence and Physical Properties

Ti-rich modraite originates from a coherent assemblage of the Alpine-cleft type hosted by chlorite grade metamorphosed metasediments outcropped near Alchuri, Shigar Valley. This mineral forms reddish-brown to almost black (with red reflections) blocky prismatic crystals (Fig. 3.87) in cavities. The main crystal forms are {100}, {001} and {110}; the subordinate form is {021}.

Associated minerals occurring in the host rock identified by means of electron microprobe analyses are quartz, clinochlore, albite, potassium feldspar, aegirine-augite, andradite, zoisite, calcite, titanite, fluorapatite, and zircon.

Modraite from Shigar Valley is optically uniaxial (–), with $\varepsilon = 1.740(4)$, $\omega = 1.749(2)$. Elongation is negative. Pleochroism is weak, from almost colourless on N_e to pale brown on N_o. Density measured by hydrostatic weighing is equal to 3.44(2) g cm^{-3}.

3.11.2 Chemical Composition

Modraite from Shigar Valley is anomaliously enriched in Ti and Na (with 5.85 wt% of TiO_2 and 0.91 wt% of Na_2O). The empirical formula based on five electron microprobe analyses and calculated on 50 cations is $(Ca_{17.73}Na_{1.38})_{\Sigma 19.11}$

$(Al_{7.82}Ti_{2.16}Fe^{3+}{}_{1.73}Fe^{2+}{}_{0.77}Mg_{0.61})_{\Sigma 13.09}Si_{17.79}O_{69.72}[(OH)_{7.21}F_{1.30}]_{\Sigma 8.51}$. The simplified formula written in accordance with the structural data is $(Ca, Na)_{19}Fe^{2+}Al_4(Al, Ti, Fe^{3+})_8Si_{18}O_{68}(OH,O)_{10}$. The chemical composition of modraite is given in Table 3.19.

3.11.3 Infrared Spectroscopy

The IR spectrum of Ti-rich modraite (Fig. 3.88) contains bands of O–H stretching vibrations (in the range from 3200 to 3600 cm^{-1}), strong bands of stretching vibrations of Si–O–Si in the groups Si_2O_7 (at 1021 cm^{-1}) and SiO_4 (at 906 cm^{-1}), as well as Si–O–Si and O–Si–O bending vibrations (below 630 cm^{-1}). The bands in the range 750–840 cm^{-1} correspond to bending vibrations of OH^- groups. The absence of absorption bands in the range 1100–1500 cm^{-1} indicates the absence of boron-bearing groups.

The bands of O–H-stretching vibrations at 3563, 3453 and 3220 cm^{-1} correspond to the D, F + G and J bands, respectively. These bands are predominantly polarized along *c* (Groat et al. 1995b). It is important to note that the presence of the F-band in the IR spectrum is a specific feature of Ti-rich vesuvianite samples (Groat et al. 1995b), and in the IR spectrum of Ti-rich vesuvianite from Alchuri this band has the highest intensity. Consequently, one can suppose that the band at 3453 cm^{-1} corresponds to OH groups coordinating Ti. The most probable configurations around

Table 3.19 Chemical composition of Ti-rich vesuvianite from Alchuri based on 5 electron microprobe analyses

Constituent	wt%	Range	*apfu*
Na_2O	0.91	0.81–1.00	0.88
CaO	33.66	33.42–33.92	18.11
MgO	0.84	0.76–0.95	0.63
FeO*	1.88	6.60–6.04**	0.79
Fe_2O_3*	4.67		1.77
Al_2O_3	13.50	13.19–13.75	7.99
TiO_2	5.85	5.67–5.98	2.21
SiO_2	36.20	34.76–35.42	17.62
F	0.84	0.74–0.95	1.33
–O=F	–0.35		
H_2O***	2.2		7.37
Total	100.20		

* Total iron content analysed as Fe_2O_3 is 6.76 wt% and is divided between FeO and Fe_2O_3 based on the Mössbauer data

** For total iron measured as Fe_2O_3

*** H_2O was determined by gas chromatography of the products of annealing at 1200 °C

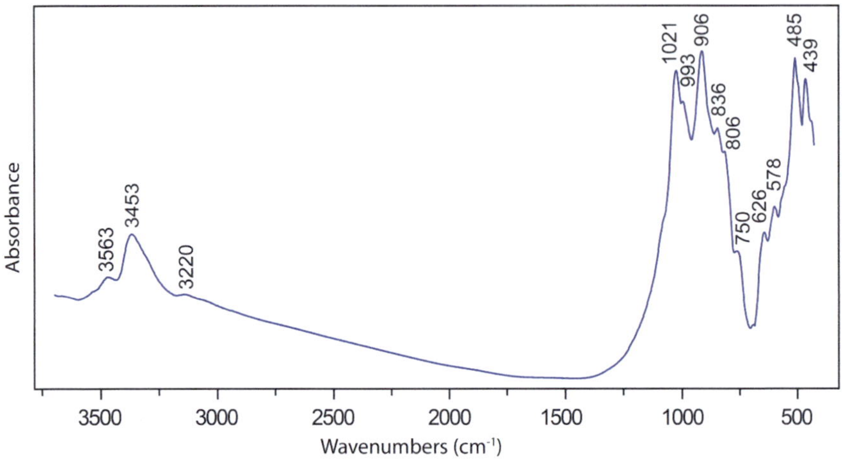

Fig. 3.88 Powder infrared absorption spectrum of modraite from Alchuri, Shigar Valley, Pakistan

the OH group are AlAlAl (the band at 3563 cm^{-1}), AlAlTi (3453 cm^{-1}) and $AlFe^{3+}Ti$ (3220 cm^{-1}).

3.11.4 Mössbauer Spectroscopy

The experimental spectrum of Ti-rich modraite can be described as a superposition of four symmetric doublets (Fig. 3.89). The Mössbauer parameters of these doublets (isomer shift *IS*, quadrupole splitting *QS*, line-width Γ and areas under resonance lines *S*) are listed in Table 3.20. It is important to note that the fitting of the experimental spectrum by three doublets is also possible (not shown here) but it leads to a higher value of goodness-of-fit parameter ($\chi^2 = 0.83$). In addition, three-doublet model contradicts structural data and rules out further consideration.

3.11.5 Solid-State Nuclear Magnetic Resonance

The solid-state ^{27}Al NMR spectrum of Ti-rich modraite (Fig. 3.90) contains one strong peak centered at –2.93 ppm that is in the range of chemical shifts for octahedrally coordinated aluminum (^{VI}Al) (Phillips et al. 1987; Olejniczak and Żabiński 1996). It is assigned to Al at the *Y*2 and *Y*3 positions. Broadening of this line is explained by the presence of different configurations around the *Y*2 and *Y*3 sites (AlAlAl, AlAlTi, $AlAlFe^{3+}$ and others: see the assignment of IR bands above). The absence of a peak near 39.4 ppm indicates the absence of Al at the *Y*1 site with the

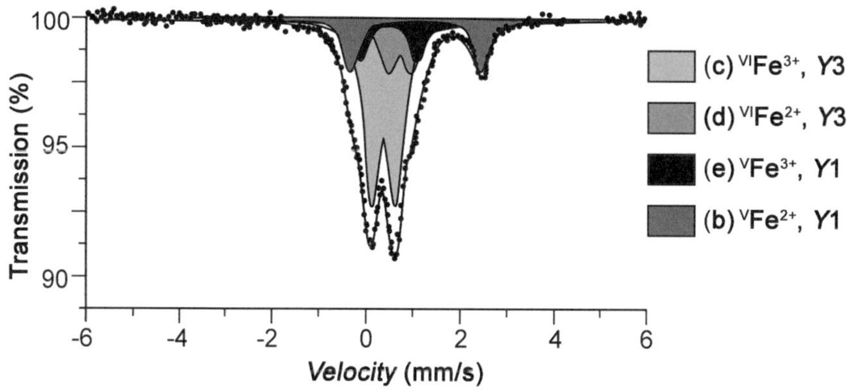

Fig. 3.89 Mössbaur spectra of Ti-rich modraite from Alchuri, Pakistan (N. V. Chukanov's sample)

Table 3.20 Hyperfine parameters of quadrupole doublets in the Mössbauer spectrum of Ti-rich modraite at 300 K

Quadrupole doublet*	Isomer shift, mm/s	Quadrupole splitting, mm/s	Line width, mm/s	Relative area, %	Assignment*
c	0.397 ± 0.002	0.523 ± 0.005	0.388 ± 0.006	53.7 ± 0.7	$^{VI}Fe^{3+}$ at *Y*3
e	0.518 ± 0.007	1.228 ± 0.014		13.3 ± 0.8	$^{V}Fe^{3+}$ at *Y*1b
d	0.754 ± 0.009	0.506 ± 0.014		15.7 ± 0.6	$^{VI}Fe^{2+}$ at *Y*3
b	1.096 ± 0.004	2.796 ± 0.008		17.2 ± 0.3	$^{V}Fe^{2+}$ at *Y*1a

* The symbols and the assignment of the quadrupole doublets are given in accordance with the data from (Rusakov et al. 2006) which are based on the analysis of Mössbauer spectra of 17 vesuvianite samples from different localities. Taking into account Fe–O distances and site multiplicities, quadrupole doublet *D* could not be assigned to Fe at distorted *Y*1 sites

five-fold coordination. The broad symmetric peaks observed at ±130 ppm belong to spinning sidebands.

3.11.6 Powder X-Ray Diffraction

Data on intensities, line positions and their *hkl* indices in the powder X-ray diffraction pattern of Ti-rich modraite are given in (Table A.54). The tetragonal unit-cell parameters refined from the powder data are: $a = 15.5655(6)$, $c = 11.8330(8)$ Å, $V = 2867.0(2)$ Å^3.

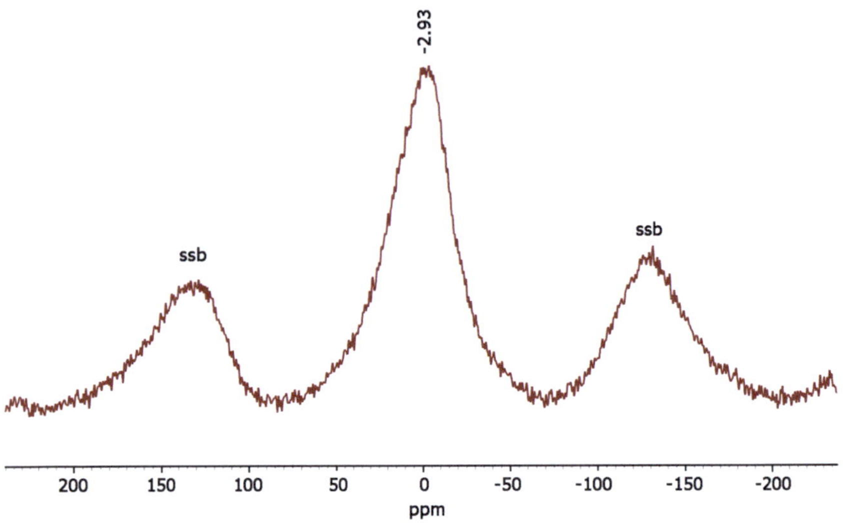

Fig. 3.90 The ^{27}Al MAS NMR spectrum of Ti-rich modraite

3.11.7 Single-Crystal X-Ray Diffraction Analysis

The crystal structure of Ti-rich modraite was refined at *P4/nnc* space group with $R_1 = 3.12$, $wR_2 = 5.65$, GooF = 1.02 for 11247 $I > 2\sigma(I)$ (Table A.55). The highest peak and the deepest hole in the final residual electron density map are 1.04 and −0.68 e Å^{-3}, respectively.

In the crystal structure of the studied sample, Na selectively occupies the *X*1 and *X*4 positions (Panikorovskii et al. 2016b). Based on the values of the refined site-occupancy factor and bond distances, the occupancy $Ca_{0.55}Na_{0.45}$ corresponding to 0.90 Na *apfu* is attributed to the *X*1 site with the mean *X*1–O distance of 2.442 Å. The *X*3 site is usually considered as eightfold coordinated, however in Ti-rich modraite *X*3-coordination including a long *X*3–O distance is ninefold, with <*X*3–O> = 2.551 Å (Table A.560. The refined occupancy of the *X*4 site is $Ca_{0.50}$. The *Z* positions are occupied by Si atoms.

The *Y*2- and *Y*3-centered octahedra are combined via common edges in linear trimers. Like almost all other investigated samples of vesuvianite-group minerals, the smallest $Y2O_6$ octahedron of Ti-rich modraite (with the mean <*Y*2–O> distance of 1.896 Å) is occupied predominantly by Al, possibly with a minor amount of Mg. The *Y*3 site is also predominantly occupied by Al. In accordance with the Mössbauer spectrum and chemical composition, this site contains Fe^{2+}, Fe^{3+} and Ti (Fig. 3.91). The key site defining the studied sample as modraite is *Y*1 with predominant occupancy by Fe^{2+}. During refinement, elongation of thermal ellipsoid of the *Y*1 site along *c* axis was observed. In the final structure model, the *Y*1 site is split into the *Y*1a and *Y*1b subsites (Fig. 3.91) with the *Y*1a–*Y*1b distance of 0.381 Å. The *Y*1a polyhedron is occupied by $Fe^{2+}{}_{0.15}Mg_{0.11}$ while the smaller *Y*1b

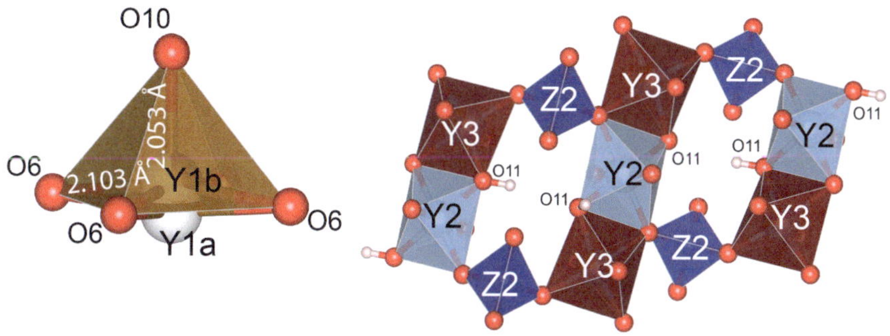

Fig. 3.91 Splitting of the *Y*1 site and the arrangement of the Ti-bearing *Y*3 octahedra in the crystal structure of Ti-rich modraite

polyhedron has the occupancy $Fe^{3+}{}_{0.17}Fe^{2+}{}_{0.07}$. The distribution of ferric and ferrous iron between the *Y*1-sites is confirmed by Mössbauer spectroscopy.

In accordance with the structural and Mössbauer spectroscopy data, a major part of Fe^{3+} and all Ti occur in the *Y*3 position. This fact may be related to the affinity of this position to amphoteric cations with high Lewis acid strengths. Indeed, the Lewis acid strengths is equal to 0.528 for Fe^{3+}, 0.583 for Al^{3+} and 0.676 for Ti^{4+} where for all bivalent cations this parameter is below 0.4 (Gagné and Hawthorne 2017).

Based on the refined site-scattering factors, the crystal-chemical formula of Ti-rich modraite can be written as follows ($Z = 2$):

$$^{X1}(Ca_{1.10}Na_{0.90})\ {}^{X2}Ca_{8.00}\ {}^{X3}Ca_{8.00}\ {}^{X4}Ca_{1.00}\ {}^{Y1}[Fe^{3+}{}_{0.34}Fe^{2+}{}_{0.30}Mg_{0.22}Fe^{2+}_{0.14}]_{1.00}$$
$$^{Y2}(Al_{3.58}Mg_{0.42})_{4.00}\ {}^{Y3}(Al_{4.04}Ti_{2.20}Fe^{3+}{}_{1.36}Fe^{2+}_{0.40})_{8.00}\ {}^{Z1}(Si)_{2.00}$$
$$^{Z2}(Si)_{8.00}\ {}^{Z3}(Si)_{8.00}(O)_{68.00}\ {}^{W}[(OH)_{5.86}O_{2.8}F_{1.34}]_{10.00}.$$

Chapter 4
New Data on the Crystal Chemistry of VSMs and Their Application in Geology

4.1 Garnet-Vesuvianite Relationships

In terms of the modular approach (Ferraris et al. 2014), the basis of the crystal structures of VSMs can be described as a combination of relatively rigid modules (rods) of the grossular structure $[Ca_{10}(Al_6Mg_2)_8(SiO_4)_{10}]^{2+}$ and $[Ca_9Fe^{3+}(Si_2O_7)_4(OH)O]^{6-}$ modules containing Si_2O_7 diorthogroups (below, "vesuvianite" modules), which are connected with each other via common vertices of the *Y*2 $[Al(OH)_2]_4{}^{4+}$ octahedra (Fig. 4.1). The ordering of cations in channel positions of "vesuvianite" modules in low-symmetric VSMs varieties leads to changes in volumes of corresponding polyhedra and is compensated by cooperative crystal-chemical adaptation including changes in configurations of polyhedra belonging to the garnet module.

Minerals of the garnet supergroup are characterized by wide variations in chemical composition. Their general formula is $^{[12]}X_3{}^{[6]}Y_2(^{[4]}Z\text{Ø}_4)_3$ where X = Mg, Ca, Mn^{2+}, Fe^{2+}, Y; Y = Mg, Al, V^{3+}, Cr^{3+}, Mn^{3+}, Fe^{3+}, Sc^{3+}, Si, Ti, Zr, Sn, Sb^{5+}, Te^{6+}, U^{6+}; Z = Li, Zn, Al, Si, V^{5+}, As^{5+}, □ (□ is vacancy), Ø = O, OH, F; the numbers in square brackets indicate the coordination number.

Various mechanisms of the adaptation in the VSMs crystal structures caused by changes in the chemical composition is the most important factor determining their structural stability. An important consequence of this compensation mechanism is the presence of a large number of complex substitution schemes, which are discussed in detail in this chapter. The change in the chemical composition and, as a result, the in the VSMs structures is reflected in a regular change in their physical properties (IR and NMR spectra, TG/DSC, optical characteristics etc.). The information on the features of crystal chemistry of VSMs presented in this chapter makes it possible to relate the nature of the structural diversity of these minerals with the factor of structural stability.

T. L. Panikorovskii and N. V. Chukanov, *Vesuvianite-Supergroup Minerals*, Springer Mineralogy, https://doi.org/10.1007/978-3-032-06970-2_4

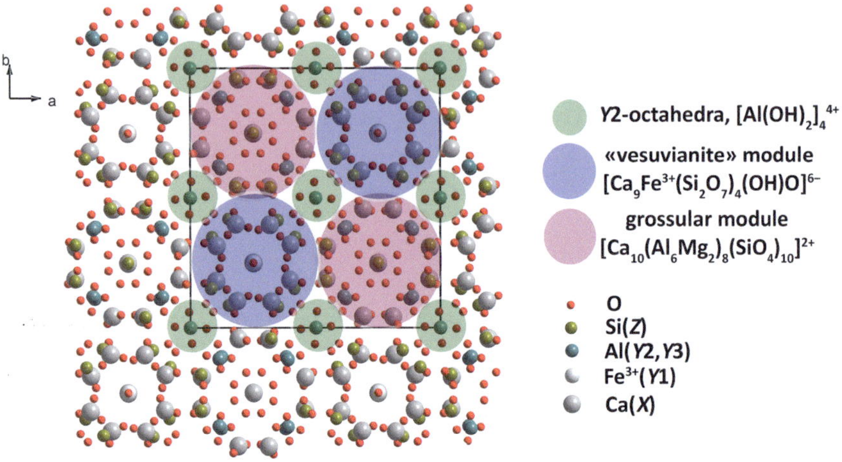

Fig. 4.1 The projection of the magnesiovesuvianite crystal structure along the *c* axis

4.1.1 Garnet-Vesuvianite Oriented Intergrowths

Grossular and vesuvianite intergrowths are common in nature, as in samples from Sierra de Cruces, Sierra Mojada Municipality, Coahuila, Mexico. This deposit is better known for its raspberry-colored grossular, but only yellowish-green grossular intergrows with vesuvianite (Geiger et al. 1999). Vesuvianite and grossular intergrowths are also common in contact-metasomatic rocks on the bank of the Vilyui River in Yakutia (Fig. 4.2a). Visual examination of the samples revealed that some of the grossular crystals are regularly arranged (Fig. 4.2b). Along with our experimental studies, we modeled the crystal arrangement for an ideal case in which all intergrowths are regularly oriented relative to the garnet block (Fig. 4.2c).

Oriented (uniaxial) intergrowths of grossular and wiluite were studied by the X-ray diffraction analysis. The intergrowths array of data was processed in CrysAlisPro (Agilent Technologies 2012) as a multicomponent crystal. The reciprocal orientation of the unit cell is shown in Fig. 4.3a. The presence of unidimensional grossular modules in the structure of the VSM was the determining factor in the appearance of such uniaxial intergrowths (Panikorovskii et al. 2014). Attention to the orientation of identical fragments of the structure in the VSMs is shown in a number of works (Rucklidge et al. 1975; Allen and Burnham 1992; Kurazhkovskaya et al. 2005). However, in the last two papers, an additional orthosilicate group was included incorrectly in the repeating fragment of the structure. The correct arrangement of the same fragment of the structure of grossular and wiluite is given in Fig. 4.3b.

It is suggested that regular intergrowths are formed by the joint growth of crystals, whereas irregular intergrowths were formed when earlier garnet was trapped during the rapid growth of wiluite crystals to form poikilitic crystals (Panikorovskii 2025b).

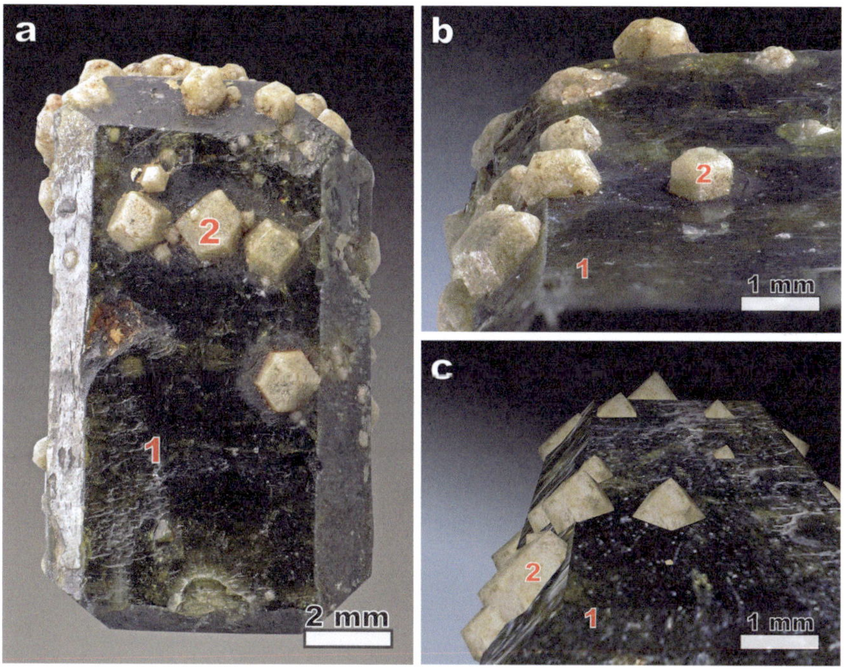

Fig. 4.2 Oriented intergrowths of wiluite and grossular from the Vilyuy River basin, Russia: images of crystals (**a**, **b**) and a theoretical model with a regular arrangement of all grossular crystals (**c**). 1—wiluite, 2—grossular

4.1.2 Hydrogarnet-Type Substitution in the Vesuvianite Structure

The hydrogarnet-type substitution, $(SiO_4)^{4-} \leftrightarrow (H_4O_4)^{4-}$, plays a significant role in the VSMs crystal chemistry. However, in the latest crystal chemistry review on the VSMs (Groat et al. 1992a) this matter is not discussed despite the mentioned work presents chemical data of vesuvianite from Sevard Peninsula (Alaska, USA) with a significant Si deficiency and an excess of H_2O. We have studied the mechanism of incorporation of the $[H_4O_4]^{4-}$ clusters into the VSMs structures using XRD, TGA/DSC, solid-state NMR, IR spectroscopy and chemical analysis. The obtained data are in agreement with the results of the studies by E. V. Galuskin and co-authors (Galuskin et al. 2003b; Galuskin 2005). This type of VSMs is usually called "hydrovesuvianite".

The samples of Si-deficient B-bearing vesuvianite were studied in detail. They form fractured white or yellowish crusts on wiluite (Fig. 3.10b) and crystals on pseudomorphs of $[H_4O_4]^{4-}$-bearing grossular ("hibschite") after a wadalite-like mineral (Fig. 1.7b). The theoretical limit of the replacement $(SiO_4)^{4-} \leftrightarrow (H_4O_4)^{4-}$

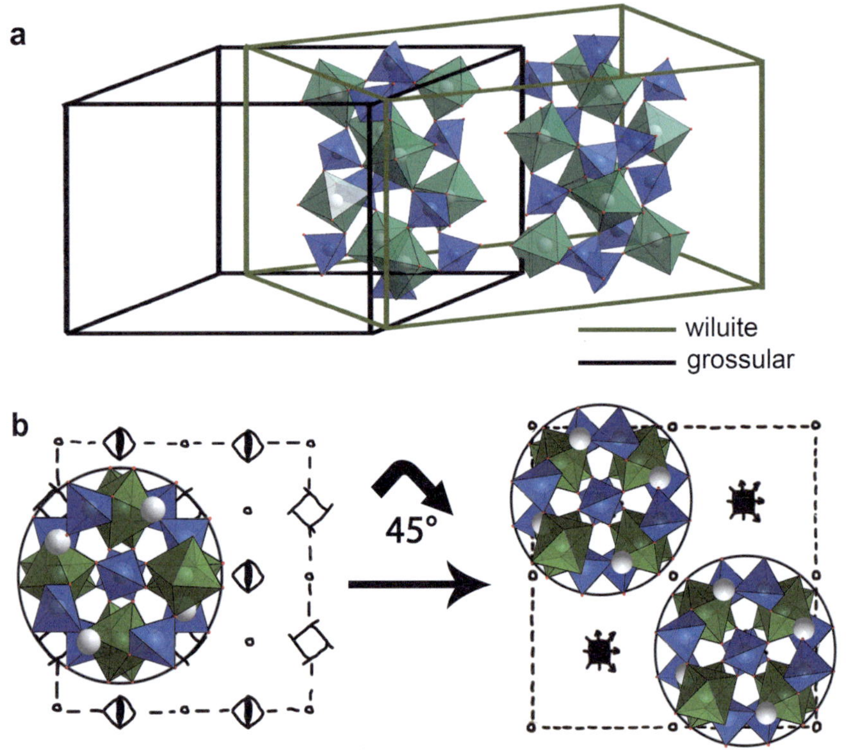

Fig. 4.3 Mutual arrangement of the unit cells and blocks of the structure of grossular with the layout of the garnet block (**a**) and the diagram showing (001) projections of the garnet block in the structures of grossular and VSM (the latter is rotated by 45° and shifted along the *z* axis by 1/8 with respect to the garnet block of grossular) (**b**). The *Y*3 octahedra are green, the *Z*1 and *Z*2 tetrahedra are blue and the *X* sites are gray

in the structure of vesuvianite was determined as the complete replacement of the orthosilicate groups.

The average lengths of the Z–O bonds in the VSMs with 100% populated *Z*1 and *Z*2 (orthosilicate) sites are in the ranges of 1.63–1.64 and 1.64–1.65 Å, respectively (Groat et al. 1992a; Ohkawa 1994; Armbruster and Gnos 2000b, c; Galuskin et al. 2003b; Bellatreccia et al. 2005b). For Si-deficient vesuvianite (Galuskin et al. 2007a), the <*Z*1–O> and <*Z*2–O> mean bond lengths are equal to 1.687 (with the occupancy factor of 0.75) and 1.660 Å (with the occupancy factor of 0.84), respectively.

Among the eight structurally characterized Si-deficient VSMs, the maximum silicon deficiency (up to 40%) was observed for the *Z*1 and *Z*2 positions (Fig. 4.4), and the mean <*Z*1,2–O> bond lengths for the sorosilicate positions reached 1.710 Å. At the same time, the refined population of the positions of nesosilicate groups (Z3) was close to 100%, and the maximum values of the mean <Z3–O> bond length did not exceed 1.640 Å. The calculation of the local balance of valences for the sample

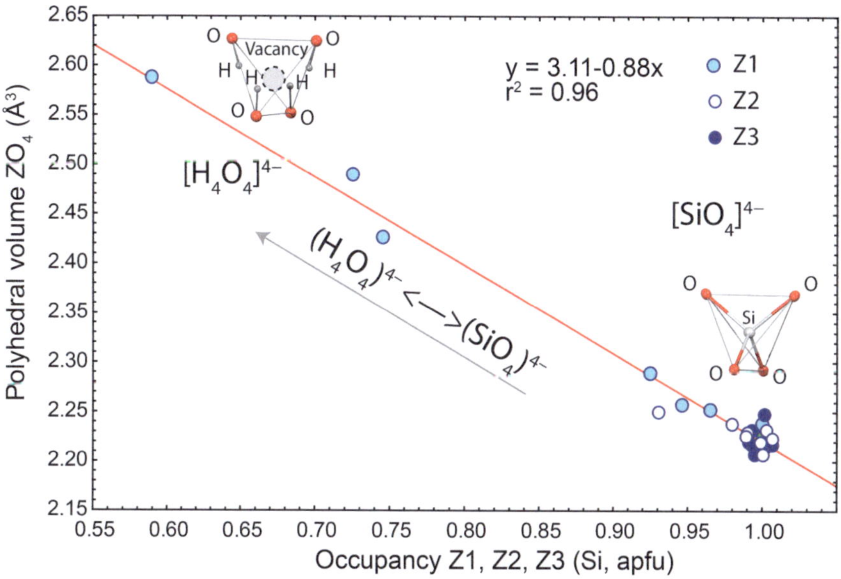

Fig. 4.4 A plot of volume (Å^3) sorosilicate *Z*1 and nesosilicate *Z*2 tetrahedra in VSMs versus the degree of their occupancy

with 27% (for *Z*1) and 10% (for *Z*2) vacancies are 3.28 and 3.69 *v.u.*, respectively (Panikorovskii et al. 2016a).

According to the thermogravimetric studies, "hydrovesuvianite" from the Wiluy River basin demonstrates two-stage weight loss of 4.30 wt% during the calcination sample to 1200 °C (Fig. 4.5a). The first stage of weight loss (2.08 wt%) corresponds to the disappearing of the $(H_4O_4)^{4-}$ clusters, and the second weight loss stage, accompanied by an endothermic peak with a minimum at 1014 °C (2.22 wt%), corresponds to the loss of OH groups at the O10 and O11 sites accompanied by subsequent melting of the mineral. The total H_2O content in this sample corresponds to 13.81 OH^-, which exceeds the maximum content of OH groups at the O10 and O11 sites (9.00 OH^- groups in total). In the ^1H MAS NMR spectrum of "hydrovesuvianite", compared with the analogous spectrum of magnesiovesuvianite, an additional peak is observed (see Sect. 3.1.6), which corresponds to the appearance of an additional hydrogen position (Fig. 4.5).

A significant number of vacancies at the *Z*1 and *Z*2 sites accompanied by the increase of corresponding ZO_4 tetrahedral volume does not have a significant influence on the unit cell parameters because the expansion of the *Z*1 tetrahedron is compensated by a compression of the *X*1 polyhedron (the Ca-bearing dodecahedron) along the *c* axis (Fig. 4.6). The height of the *Z*1 tetrahedron along the *c* axis in fluorine-rich vesuvianite is equal to 2.02 Å (Galuskin et al. 2003a) and the corresponding height of the *X*1 dodecahedron is 3.87 Å. In the structure of "hydrovesuvianite" from the Vilyuy River basin, the average height of the *Z*1 tetrahedron is 2.089 Å, and

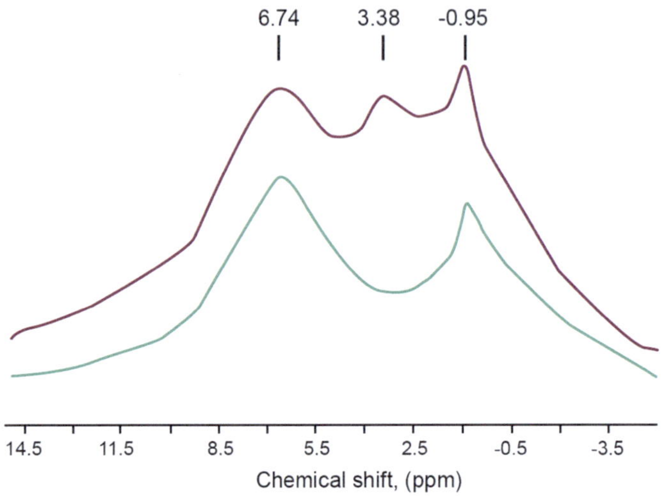

Fig. 4.5 The ^{1}H NMR spectrum of "hydrovesuvianite" (red curve) and magnesiovesuvianite with 9.11 OH groups per formula unit (green curve)

corresponding height of the *X*1 dodecahedron is 3.812 Å. Thus, the size of the repeat unit of the cationic chains of polyhedra ...*Z*1–*X*1–*Z*1–*X*1–*Z*1... remains almost unchanged on a partial substitution in accordance with the hydrogarnet scheme.

Thus, the theoretical degree of the substitution $(SiO_4)^{4-} \leftrightarrow (H_4O_4)^{4-}$ in VSMs is limited by the number of orthosilicate groups, since the incorporation of the $[H_4O_4]$ complexes into the *Z*3 sorosilicate sites would be accompanied by the formation of vacancies in the *X*4 channel positions (Panikorovskii et al. 2016a). On the other hand, in the Si–O–Si fragment containing two *Z*3 sites coordinated by a common O

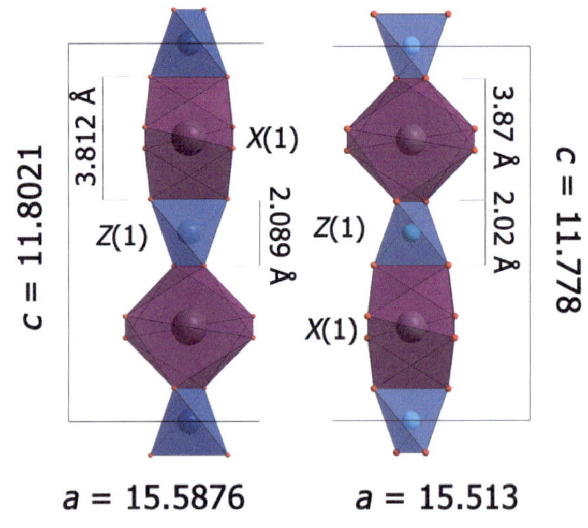

Fig. 4.6 Geometric characteristics of the *Z*1–*X*1–*Z*1–*X*1–*Z*1 rods in 'hydrovesuvianite' (left) and in fluorine-rich vesuvianite (right), after (Galuskin et al. 2003a)

atom, the Si–O bonds are covalent, and protonation of this oxygen at a vacancy in the position of one Si atom would lead to the cleavage of the Si–O bond, which is energetically unfavourable. In addition, the transformation of the Si_2O_7 group into $SiO_3(OH)$ with an acidic OH group is not observed in IR spectra.

In other words, the substitution of the hydrogarnet type in vesuvianite is limited only to the garnet block and occurs only in orthosilicate positions.

4.1.3 Relationships Between Chemical Compositions of Wiluite and Ti-Rich Grossular

Vesuvianite-supergroup minerals are widespread in nature (Groat et al. 1992a; Gnos and Armbruster 2006). The stability range of garnet is wider compared to the VSMs (Galuskina et al. 1998b; Grew et al. 2013). The composition of andradite-grossular garnets is close to the composition of VSMs, so these minerals are often associated with each other (Dey et al. 2019; Butek et al. 2021). As a result of changes in crystallization conditions (temperature and chemical composition of the environment), the equilibrium shifts and either garnet, or vesuvianite, or simultaneously both minerals are formed (Galuskin 2005). The chemical composition of vesuvianite and associated garnet can be a sensitive tool for reconstructing the conditions of mineral formation.

In diagnostics by electron microprobe analysis, especially with poor polishing preparations and using EDS-mode analyses, VSMs are very often confused with garnets. Our goal is to determine the differences and common features of the chemical composition of garnet and vesuvianite and to determine reliable criteria that will allow distinguishing minerals of these two supergroups, as well as the stability factors of the structural types of garnet and vesuvianite depending on the chemical composition. Let us consider several examples: zonal grossular, inclusions of earlier zonal grossular in wiluite and the case of the formation of regular intergrowths of vesuvianite and garnet described in Sect. 4.1.1.

Grossular from various localities often exhibits distinct chemical zoning (Antao 2021). Zoned grossular crystals associated with wiluite were found in akhtarandite rocks from the Vilyui River basin (Fig. 4.7a). The cores of these grossular crystals (Fig. 4.7b) are enriched in Zr, Cr, Ti, Fe, and Mg, are opaque, and dark brown. The intermediate zone is also enriched in Ti, Fe, and Mg, but is transparent and light brown, and the transparent margins are close in composition to ideal grossular with minor impurities of Fe and Mg (Panikorovskii 2025b).

Such zoned crystals of grossular, the chemical composition profile of which is shown in Fig. 4.8, are captured during the growth of later wilute crystals. It should be noted that this grossular has a relatively high MgO content (up to 1 wt%). The Ti content in some zones correlates with the Fe content, in some others with Mg or simultaneously with Mg and Fe. Most likely, Ti is included in the grossular structure according to the schemes $2Al^{3+} \leftrightarrow Ti^{4+} + Mg^{2+}$ and $2Al^{3+} \leftrightarrow Ti^{4+} + Fe^{2+}$.

Fig. 4.7 Zoned grossular crystals on akhtarandite: photograph of the sample (**a**) and BSE image of zoned grossular crystals (**b**). 1—achtarandite, 2—grossular

During the formation of oriented intergrowths of wiluite and grossular (Fig. 4.9), wiluite formed first. Then (after the formation of Ti-enriched zones in wiluite: analysis 2 in Table 4.1) and with a further increase in the Ti content, a transition to the garnet structural type occurs and Ti-enriched grossular is formed (analysis 3 in Table 4.1). Thereafter, alternating of zones enriched in Fe and Ti (analyses 4 and 5 in Table 4.1) and zones close in composition to pure grossular are formed.

It should be noted that there are inclusions of prismatic pseudomorphs of montmorillonite (most likely, after gehlenite) in wiluite and grossular, which may indicate a smooth change in the chemical composition of the system during the transition from wiluite to grossular. This may also indicate the preservation of the temperature of the system, with a change in the chemical composition. Thus, there is the change in the chemical composition of the system (a decrease in the Mg content and an increase in the Ti content) that leads to a change in the structural type from vesuvianite to garnet.

The general regularities in the compositions of garnet and wiluite from the Vilyui River basin, Russia are as follows. The Mg content in wiluite is always much higher (3.5–7.0 wt% of MgO), while the maximum MgO content in grossular rarely exceeds 1.0 wt%. The maximum Fe and Ti contents in grossular are higher than in VSMs. Despite the closeness of the structures, the main factor is the absence of a significant isomorphism between pyralspite and ugrandite garnet series, therefore, with significant Mg contents, the VSM structure is more stable than garnet one whereas with small ones (<1 wt% of MgO), the garnet structure is more stable.

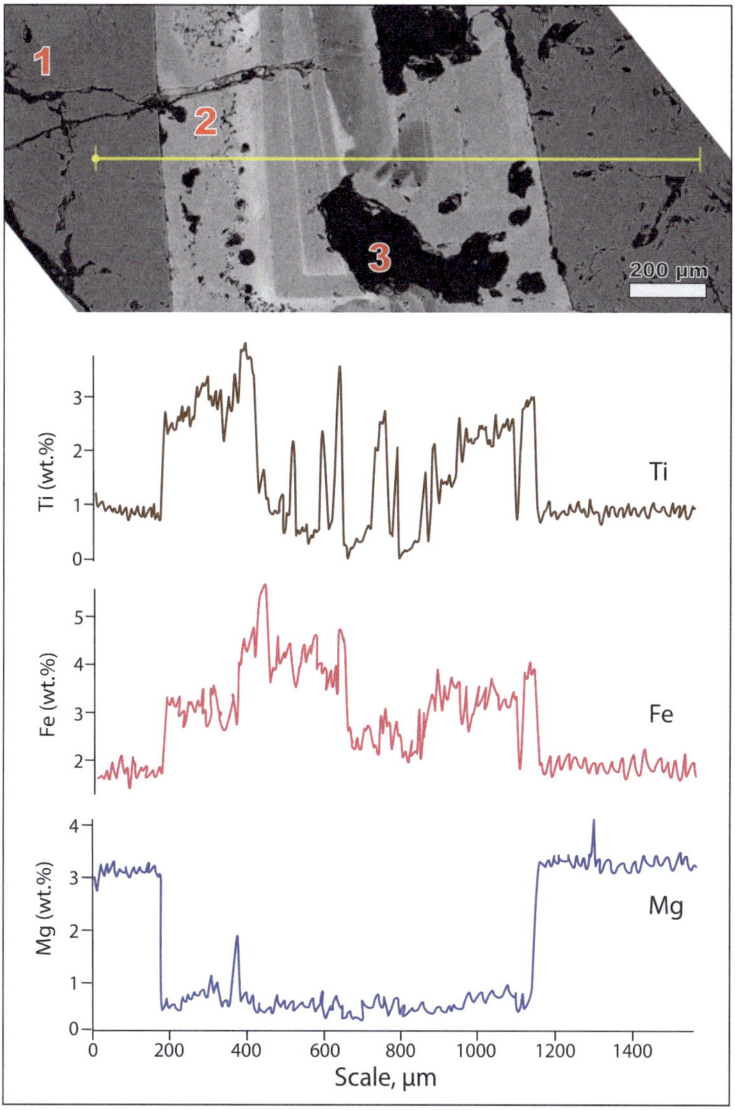

Fig. 4.8 Profile of Ti, Fe and Mg contents in a zonal inclusion of grossular in wiluite from the Vilyuy River basin. 1—wiluite, 2—grossular, 3—montmorillonite

4.1.4 Compositional Relationships Between Cr-Bearing Garnet and Vesuvianite

Emerald-green vesuvianite (Fig. 1.27) containing significant Cr^{3+} amounts occurs in rodingite deposits in association with chromite ores (Fitzgerald et al. 1986a;

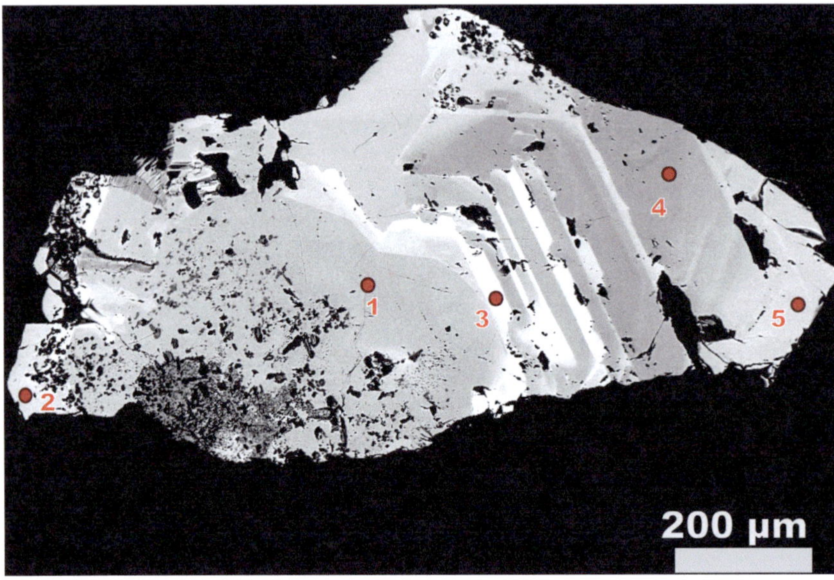

Fig. 4.9 Oriented intergrowths of wiluite (1, 2) and zoned Ti-bearing grossular (3–5) in a sample from the Vilyui River basin, Russia. Numbers correspond to analyses in Table 4.1

Platonov et al. 1993b; Hålenius 2000; Rogkala et al. 2022b). We previously investigated Cr-bearing low-symmetry vesuvianite from Monetnaya Dacha, Middle Urals (Krivovichev et al. 2013), but its association did not contain garnet group minerals. However, we encountered altered samples of chromitite ores containing both garnet group minerals and vesuvianite (Fig. 4.10) in the Bazhenovskoe deposit, Middle Urals.

Many chromitite bodies found in serpentinites of the Bazhenovskoe deposit contain chlorite-vesuvianite-garnet cement, i.e. minerals of typical rodingite paragenesis are developed in them. It is likely that chromitites served as a geochemical barrier on which excess Ca and Al were precipitated during serpentinization of peridotites (Erokhin 2017b). The mineral associations of chromite ore xenoliths in the rodingite matrix are usually represented by a wide variety of minerals, including uvarovite, vesuvianite, diopside, and clinochlore (Antonov 2003b). Massive chromite ores contain admixtures of Al_2O_3 (1–2 wt%) and MgO (1–2 wt%). The most typical mineral in such rocks is Cr-bearing clinochlore of pink or reddish color (Fig. 4.10). Clinochlore common in composition was found in rims during the formation of pseudomorphs of montmorillonite after an unknown mineral (presumably, after monticellite or melilite).

Veins filled with Cr-rich grossular-uvarovite (Fig. 4.11a) can be found directly in the chromite ore in paragenesis with Cr-bearing chlorite (with ~1.0 wt% of Cr_2O_3), containing inclusions of uvarovite (Fig. 4.10). Vesuvianite is found either as individual crystals growing on clinochlore in rock cavities (Fig. 4.10) or as inclusions in

Table 4.1 Chemical composition of oriented intergrowths of wiluite (1, 2) and grossular (3–5) in samples from the Vilyuy River basin. Analysis points are marked in Fig. 4.9

Sample	1	2	3	4	5
MgO	4.56	4.60	0.99	0.74	0.72
Al_2O_3	12.31	10.44	10.96	19.09	15.97
SiO_2	36.52	35.00	35.70	39.78	40.17
CaO	36.02	34.69	35.73	36.07	36.72
TiO_2	1.50	4.20	7.28	0.25	0.37
MnO	0.13	0.12	–	0.08	0.07
FeO	1.86	2.74	6.37	2.65	4.09
B_2O_3	3.67	3.67	–	–	–
H_2O	2.67	1.74	–	–	–
Total	99.22	97.20	97.03	98.66	98.11
apfu					
Mg^{2+}	3.35	3.51	0.08	0.08	0.12
Al^{3+}	7.14	6.29	1.72	1.46	1.04
Si^{4+}	17.98	17.89	3.05	3.11	2.88
Ca^{2+}	19.00	19.00	2.96	3.05	3.09
Ti^{4+}	0.56	1.62	0.01	0.02	0.44
Mn^{2+}	0.05	0.05	–	–	–
Fe^{2+}	0.77	1.17	0.17	0.27	0.43
B^{3+}	3.12	3.24	–	–	–
H^{+}	8.72	5.91	–	–	–

Note All iron is considered as Fe^{2+}. The H_2O content was calculated from the charge balance requirement

clinochlore (Fig. 4.11b). Vesuvianite crystals are zonal (Fig. 4.11c), which is due to variable contents of Cr^{3+}, Fe^{3+} и Al^{3+}: analyses 1–3 in Table 4.2.

In the garnets, the Cr^{3+} content is nearly constant, but the amounts of Fe^{3+} and Al^{3+} are variable. Depending on the Fe^{3+} and Al^{3+} contents, garnet group minerals, can fall into the composition field of grossular or uvarovite analyses 4–6 in Table 4.2. It should be noted that the garnet group minerals in our samples are practically free of Mg, the maximum MgO content is 0.80 wt%, whereas in the VSMs the amount of magnesium is higher and is almost constant (~3.0 wt% of MgO or 2 apfu).

VSMs and garnets in rodingitized chromitites of the Bazhenovskoe deposit generally show similar zoning in terms of Al^{3+}–Fe^{3+}–Cr^{3+}, but the isomorphic capacity of garnet is much wider for Fe^{3+} and Cr^{3+} than for vesuvianite, while vesuvianite is characterized by a constant admixture of MgO of about 3.0 wt%. It is likely that vesuvianite and garnet were formed according to the following scheme (Panikorovskii and Shilovskikh 2025):

Fig. 4.10 Cr-bearing vesuvianite and uvarovite from the Bazhenovskoe deposit, Middle Urals (sample of A. A. Antonov). 1—Cr-bearing vesuvianite, 2—Cr-bearing clinochlore, 3—uvarovite, 4—clinochlore, 5—montmorillonite, 6—chromite

$$6\mathrm{Chr} + 17\mathrm{Di} + 7\mathrm{An} + 17\mathrm{Brc} + 8\mathrm{H_2O} + 8\mathrm{H_2O} \rightarrow 5\mathrm{Clc} + \mathrm{Ves} \\ + \mathrm{Uv} + 2\mathrm{Wo} + 10\mathrm{En} + 2\mathrm{Mgc} + \mathrm{O_2}\uparrow$$

$$6\mathrm{FeCr_2O_4} + 17\mathrm{CaMgSi_2O_6} + 7\mathrm{CaAl_2Si_2O_8} \\ + 17\mathrm{Mg(OH)_2} + 8\mathrm{H_2O} \rightarrow 5\left(\mathrm{Mg_4Fe^{2+}}\right)\mathrm{Cr[Si_3Al]O_{10}(OH)_8} \\ + \mathrm{Ca_{19}Fe^{2+}Mg_2(Al_9Cr)Si_{18}O_{68}(OH)_{10}} + \mathrm{Ca_3Cr_2Si_3O_{12}} \\ + 2\mathrm{CaSiO_3} + 10\mathrm{MgSiO_3} + 2\mathrm{MgCr_2O_3} + \mathrm{O_2}\uparrow.$$

In the process of rodingitization of chromite ores, the composition of chromite plays an important role. If there is practically no Mg impurity in it, as in the Saranovskoye deposit, then uvarovite is formed. If there is Mg in chromite, then mainly vesuvianite is formed, as in the Bazhenovskoye deposit (Antonov 2003b). In rodingites, the formation of garnet, clinochlore and vesuvianite occurred in the same temperature range of 350–470 °C; the shift in the equilibrium between Ves/(Grt + Chl) occurs depending on the proportion of carbon dioxide in the fluid, X_{CO2}. Vesuvianite crystallizes at low values of this parameter, while the Grt + Chl pair is formed

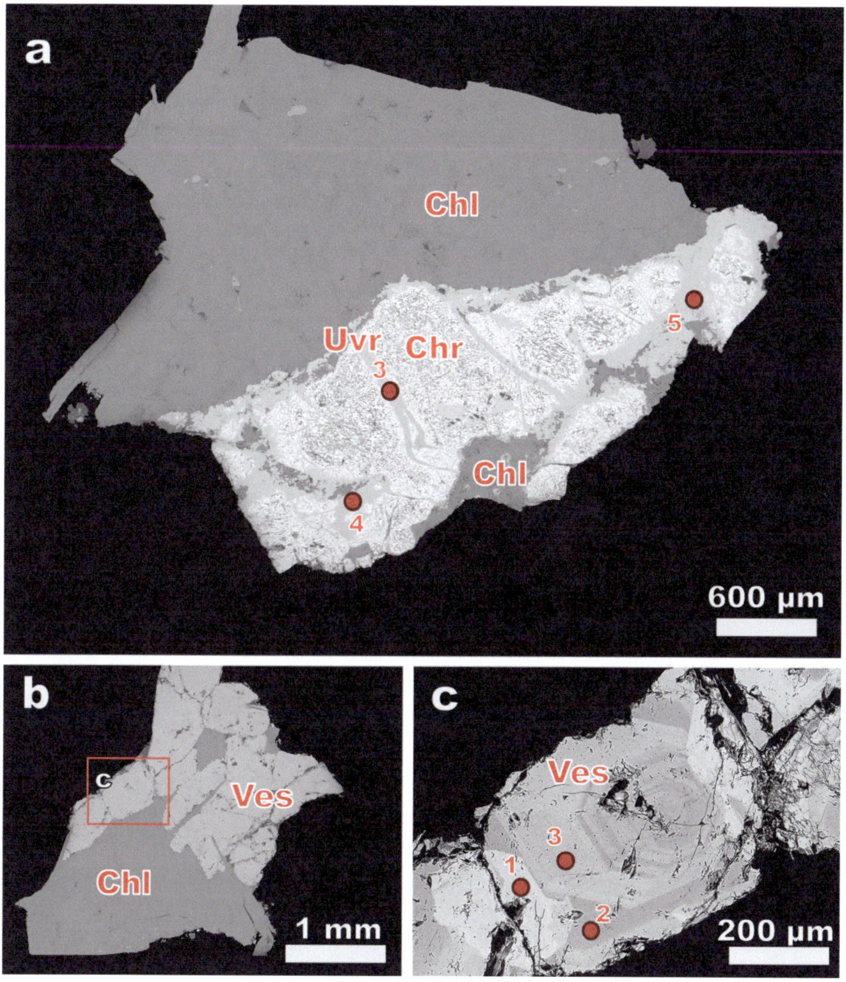

Fig. 4.11 BSE images of chromite-chlorite-vesuvianite rock from rodingites of the Bazhenovskoe deposit, Urals. The spots 1–6 correspond to the analyses of VSMs and garnets given in Table 4.2

at its higher values (Murzin et al. 2020). From our point of view, the key factor in the formation of the VSMs is the presence of Mg in the system, and in the Ves/(Grt + Chl) equilibrium, the main role is played by the proportion of carbon dioxide in the fluid.

Table 4.2 Chemical composition of Cr-containing vesuvianite (1–3), grossular (3–5) and grossular-uvarovite (4–6) in samples from the Bazhenovskoe deposit

Sample →	1	2	3	4	5	6
Content, wt%						
SiO_2	36.55	36.99	37.03	36.29	36.90	36.61
Al_2O_3	14.8	17.99	16.07	6.99	9.63	9.65
Cr_2O_3	1.49	0.47	1.38	11.36	11.98	11.85
Fe_2O_3	4.9	1.63	3.03	10.14	5.63	6.31
MgO	2.95	2.96	3.01	0.66	0.83	0.34
CaO	36.1	36.75	36.44	34.45	34.90	34.99
H_2O	2.75	2.90	2.76			
Total	99.54	99.69	99.72	99.89	99.87	99.75
Formula coefficients, apfu						
Si^{4+}	17.95	17.85	18.02	2.95	2.95	2.94
Al^{3+}	8.57	10.23	9.22	0.67	0.91	0.91
Cr^{3+}	0.58	0.18	0.53	0.73	0.76	0.75
Fe^{3+}	1.81	0.59	1.11	0.62	0.34	0.38
Mg^{2+}	2.16	2.13	2.18	0.08	0.10	0.04
Ca^{2+}	19.00	19.00	19.00	3.00	2.99	3.01
H^+	8.99	9.34	8.98			

4.1.5 Indicatory Significance of the Association of Henritermierite with Mg-Rich VSMs in Mn Ores. An Example of Mn and Mg Separation

Garnet with the composition $Ca_3Mn^{3+}{}_2(SiO_4)_3$ is unknown, despite the widespread occurrence of its constituent components. Most likely, this fact results from crystal chemical factors and is associated with relatively large ionic radii of Ca и Mn^{3+}. However, the "hydrogarnet" henritermierite, $Ca_3Mn^{3+}{}_2(SiO_4)_2[\square(OH)_4]$, is known. This mineral is very rare and was found only on the territory of the African continent, in the metasomatic deposits of manganese in the Kalahari Manganese Field, South Africa (Wessels Mine, Hotazel and N'Chwaning I, II Mines: (Kuleshov 2012)) and Zagora Province, Morocco (Tachgagalt Mine: Gaudefroy et al. 1969). All finds of henritermierite belong to mineral associations including Mn^{3+} oxides and Mg-rich minerals of the vesuvianite group. Unlike other representatives of the garnet supergroup, which, with the exception of holtstamite, $Ca_3Al_2(SiO_4)_2[\square(OH)_4]$, have cubic symmetry, henritermierite is tetragonal (Aubry et al. 1969) and nikmelnikovite is trigonal (Krivovichev et al. 2021).

A new find of henritermierite in the Combat deposit, Namibia was described by Chukanov et al. (2023a). The mineral was discovered as a component of manganese ore of metasomatic origin from, where it is associated with

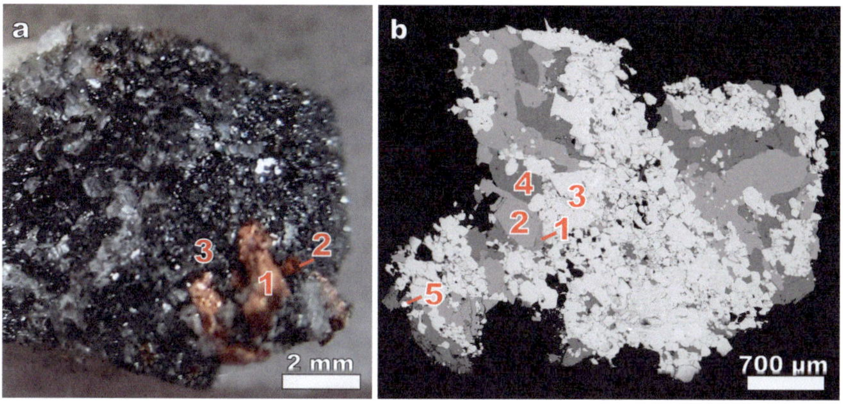

Fig. 4.12 Photo (**a**) and BSE-image of Mg-rich vesuvianite (1), henritermierite (2), hausmannite (3), calcite (4), and liebigite (5) in a specimen from the Combat deposit. BSE image of a polished section (**b**)

hausmannite, calcite, native copper, Mg-rich vesuvianite and late liebigite (Fig. 4.12). The compositions (Table 4.3) of the associated mineral of the vesuvianite-milanriederite solid-solution series correspond to the crystal chemical formulae $(Ca_{17.61}Mn_{0.43}La_{0.38}Ce_{0.19}Pr_{0.19}Nd_{0.10}Y_{0.10})_{\Sigma 19.00}$ $^{Y1}Fe^{3+}$ $^{Y2}Al_4$ $^{Y3}(Al_{4.19}Mg_{3.43}Fe^{3+}_{0.24}Mn^{3+}_{0.14})_{\Sigma 8.00}$ $Si_{18.00}[O_{68.53}(OH)_{0.47}]_{\Sigma 69.00}(OH)_9$ (Sample 1) and $(Ca_{18.36}Sr_{0.28}Y_{0.19}Sm_{0.04}Gd_{0.04}Nd_{0.03}Eu_{0.02}Dy_{0.02}Pr_{0.01}Yb_{0.01})_{\Sigma 19.00}$ $^{Y1}(Fe^{3+}_{0.59}Cu^{2+}_{0.41})_{1.00}$ $^{Y2}Al_4$ $^{Y3}(Mg_{2.90}Al_{2.71}Mn^{3+}_{1.41}Fe^{3+}_{0.59}Sc^{3+}_{0.18}Ti_{0.11}Cr_{0.10})_{\Sigma 8.00}$ $^{T}Al_{1.00}$ $Si_{18.00}$ $O_{69.00}$ $[(OH)_{6.84}O_{2.16}]_{9.00}$ (Sample 2). For comparison: in milanriederite, the $Y3$ site is Mg-dominant and the content of $Y3$ cations is $[Mg_{4.48}Al_{2.00}(Fe,Mn)_{1.52}]_{\Sigma 8.00}$ (Chukanov et al. 2019). The presence of copper in Sample 2 is noteworthy (by analogy with cyprine from the Kalahari manganese deposits, we assume that the ion Cu^{2+} occurs at the $Y1$ site).

The results obtained allow us to draw a number of conclusions. The presence of the rare garnet henritermierite and its association with high-magnesium minerals of the vesuvianite group is a characteristic feature of specific manganese ores of metasomatic genesis, formed at relatively low oxygen fugacity and containing manganese in relatively low-valent states (Mn^{3+}, Mn^{2+}). Despite a number of analogies between the crystalline structures of garnet and vesuvianite, in ores of this type, the Mn^{3+} and Al^{3+} cations exhibit a pronounced affinity for henritermierite and minerals of the vesuvianite group, respectively.

It should be noted that the incorporation of a significant amount of Mg into the structure of garnets of the ugrandite series was observed for the first time. We assume that this is a feature of very low-temperature and low-symmetry varieties of garnet-group minerals, which are relatively rare in nature. Therefore, in routine EDS analysis, we recommend distinguishing VSMs from garnets by the Mg content.

Table 4.3 Chemical composition of henritermierite

	Kombat mine (our data)	Wessels mine (our data)	Tachgagalt mine (Gaudefroy et al. 1969)
Content (wt%)			
CaO	31.45 (30.51–31.88)	36.07	35.45
MgO	5.32 (5.00–5.68)	0.13	–
Mn_2O_3	28.41 (27.22–29.73)	31.00	24.90
Fe_2O_3	0.20 (0–0.56)	0.90	0.95
Al_2O_3	0.14 (0–0.30)	1.19	5.95
SiO_2	30.74 (30.27–31.21)	25.76	24.65
TiO_2	0.02 (0–0.08)	–	–
H_2O	No data	No data	7.93
F	1.01 (0.23–2.34)	–	–
$-O{=}F_2$	0.43 (0.10–0.99)	–	–
Total	96.86	95.05	99.83
Formula coefficients calculated on 5 (X + Y) cations			
Ca	2.54 (2.50–2.56)	3.00	2.94
Mg	0.61 (0.57–0.64)	0.02	–
Mn^{3+}	1.81 (1.75–1.92)	1.83	1.47
Fe^{3+}	0.01 (0.00–0.03)	0.05	0.06
Al	0.01 (0.00–0.03)	0.11	0.54
Si	2.32 (2.30–2.35)	2.00	1.91
F	0.25 (0.06 – 0.55)	–	–

4.2 Substitutions and Mechanisms of Cooperative Crystal-Chemical Adaptation in the Vesuvianite Structure

Recently, we have introduced a next step of development a novel method for reconstructing the petrogenesis of magmatic and postmagmatic systems based on the typomorphic properties, known as the topo-crystal-chemical approach. This method involves analyzing the spatial distribution of crystallographic and crystal-chemical characteristics of rock-forming and major ore minerals. This approach has been successfully implemented in studying pyrochlore, forsterite, and sulfides in the Kovdor massif, as well as minerals of the eudialyte group in the Lovozero massif (Mikhailova et al. 2018; Ivanyuk et al. 2018a, b; Panikorovskii et al. 2021). VSMs from a single deposit can differ in chemical composition, the degree of cation ordering in the channels, the ordering of 2–3 valence cations in octahedra, and whether or not they form merohedral twins. All these factors play a significant role in determining the PT conditions for VSM crystallization, as well as the characteristics of the chemical composition of the crystallization environment.

4.2.1 Substitution Schemes

The structural and crystal-chemical diversity of VSMs crystallized under various conditions in various geological settings (skarns, achtaragdite rocks, rodingites, altered alkaline rocks, calcareous schists, volcanogenic-sedimentary rocks of the trap formation) results in a significant number of different homovalent and heterovalent substitutions schemes, given in Table 4.4. The inclusion of cations with different ionic radii into the VSMs structures causes various crystal-chemical mechanisms of structure adaptation associated with changes in the volumes of coordination polyhedra.

Table 4.4 Structural substitution schemes in VSM

Substitution scheme	References
$^{T1}B^{3+} + {}^{Y}Mg^{2+} + 2{}^{O10,11}O^{2-} \leftrightarrow {}^{T1}\square + {}^{Y}Al^{3+} + 2{}^{O10,11}(OH)^{-}$	Groat et al. (1992a)
$^{T2}B^{3+} + 2{}^{O10,11}O^{2-} \leftrightarrow {}^{T2}\square + {}^{O10,11}(OH)^{-}$	Groat et al. (1992a)
$^{Z1,2}(SiO_4)^{4-} \leftrightarrow {}^{Z1,2}(H_4O_4)^{4-}$	Galuskin et al. (2003b)
$^{Y2}Mg^{2+} + {}^{Y3}Ti^{4+} \leftrightarrow {}^{Y2}Al^{3+} + {}^{Y3}Al^{3+}$	Aksenov et al. (2016)
$^{X1,4}Ca^{2+} + {}^{Y2,3}Al^{3+} \leftrightarrow {}^{X1,4}Na^{+} + {}^{Y2,3}Ti^{4+}$	Panikorovskii et al. (2016b)
$^{X3}Ca^{2+} + {}^{Y2,3}Al^{3+} \leftrightarrow {}^{X3}REE^{3+} + {}^{Y2,3}Mg^{2+}$	Fitzgerald et al. (1987)
$^{O10,11}OH^{-} \leftrightarrow {}^{O10,11}F^{-}$	Groat et al. (1992b)
$^{Y1}Mn^{3+} \leftrightarrow {}^{Y1}Fe^{3+}$	Armbruster et al. (2002)
$^{O10}OH^{-} \leftrightarrow {}^{O10}Cl^{-}$	Galuskin et al. (2005)
$^{Y1}Cu^{2+} + {}^{Y2,3}Mn^{3+} \leftrightarrow {}^{Y1}Fe^{3+} + {}^{Y2,3}Mg^{2+}$	Panikorovskii et al. (2017e)
$^{Y3}Mg^{2+} + {}^{Y2}Al^{3+} \leftrightarrow {}^{Y3}Fe^{3+} + {}^{Y2}Mn^{2+}$	Panikorovskii et al. (2017e)
$^{Y3}Al^{3+} \leftrightarrow {}^{Y3}Mn^{3+}$	Panikorovskii et al. (2017e)
$^{Y3}Cr^{3+} \leftrightarrow {}^{Y3}Fe^{3+}$	Hålenius (2000)
$^{X3}Bi^{3+} + {}^{Y2,3}Mg^{2+} \leftrightarrow {}^{X3}Ca^{2+} + {}^{Y2,3}Al^{3+}$	Hålenius et al. (2013)
$^{X3}Pb^{2+} \leftrightarrow {}^{X3}Ca^{2+}$	Hålenius et al. (2013)
$^{Y1}Al^{3+} \leftrightarrow {}^{Y1}Fe^{3+}$	Fitzgerald et al. (1986a)
$^{Y1}Mn^{2+} \leftrightarrow {}^{Y1}Fe^{2+}$	Groat and Evans (2012)
$^{X3}Fe^{2+} \leftrightarrow {}^{X3}Ca^{2+}$	Groat et al. (2013)
$^{Z2}As^{5+} + {}^{Y3}Mg^{2+} \leftrightarrow {}^{Z2}Si^{4+} + {}^{Y3}Al^{3+}$	Groat et al. (2013)
$^{Y1}Mg^{2+} + {}^{Y3}Al^{3+} \leftrightarrow {}^{Y1}Fe^{3+} + {}^{Y3}Mg^{2+}$	Panikorovskii et al. (2017d)
$^{T1}Al^{3+} + {}^{Y1-3}Mg^{2+} + {}^{H1}2\square \leftrightarrow 2H^{+} + {}^{Y1-3}Fe^{3+} + {}^{T1}\square$	Groat et al. (1994b)
$^{X}U,Th^{4+} + {}^{Y}Fe^{2+} \leftrightarrow {}^{X}Ca^{2+} + {}^{Y}Ti^{4+}$	Kononova (1960), Himmelberg and Miller (1980)
$^{Y}Ti^{4+} + {}^{O10,11}O^{2-} \leftrightarrow {}^{Y}Fe^{3+} + {}^{O10,11}OH^{-}$	Kurazhkovskaya et al. (2005)
$^{Z1}Si^{4+} + {}^{X3}Ca^{2+} \leftrightarrow {}^{Z1}Al^{3+} + {}^{X3}REE^{3+}$	This work

4.2.2 Cation Diversity in the 5-Coordinated Pyramid

In the VSMs, the most diverse cation occupancy is observed in the 5-coordinated tetragonal pyramid *Y*1 (Ohkawa et al. 1992). This is probably due to the fact that the coordination number 5 is transitional between tetrahedral and octahedral coordination and is therefore relatively rare. A similar cation diversity is observed for the 5-coordinated pyramid in litidionite group minerals (Chukanov et al. 2015; Balassone et al. 2022b, 2024).

It is assumed that different temperatures of VSMs formation (from 250 to 800 °C) lead to different ratios of structural rods with different symmetry and, as a result, to different general symmetry of these minerals ($P4nc \rightarrow P4/n \rightarrow P4/nnc$). This most strongly affects the sizes of the *Y*1A,B polyhedra (Fig. 4.13) and is the main cause of the diversity of cations at this site. The *Y*1 site can host different cations, including, in order of increasing ionic radius, (Fig. 4.14): Al^{3+}, Fe^{3+}, Mn^{3+}, Cu^{2+}, Mg^{2+}, Fe^{2+}, Mn^{2+} (Fitzgerald et al. 1986b; Phillips et al. 1987; Ohkawa et al. 1992; Armbruster et al. 2002; Panikorovskii et al. 2017a, d, e). Depending on the population of the *Y*1 site, the polyhedral volume variations can exceed 30%.

Eight mineral species of the eleven vesuvianite-supergroup minerals approved by the IMA CNMNC are characterized by different predominant cations in the *Y*1A,B positions. In these minerals, these positions are occupied by Fe^{3+} (vesuvianite, milanriederite), Fe^{2+} (fluorvesuvianite, hongheite and modraite), Mg^{2+} (magnesiovesuvianite, wiluite, and manaevite-(Ce)), Mn^{3+} (manganivesuvianite), Al^{3+} (alumovesuvianite) and Cu^{2+} (cyprine). This confirms the role of the *Y*1A,B positions as the key sites determining the diversity of VSMs. However, some other sites also play a species-defining role in the VSMs.

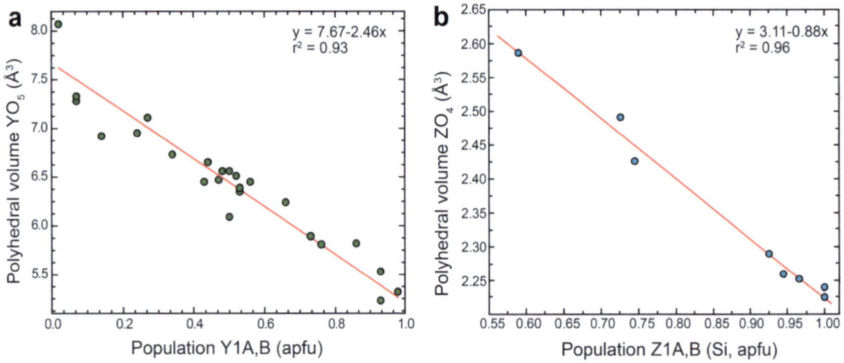

Fig. 4.13 The dependencies of the volumes of the five-vertex pyramid and $Z1O_4$ tetrahedra in the VSMs structures on the degree of their filling according to the works (Krivovichev et al. 2013; Panikorovskii et al. 2016c, 2017a, d, e) and X-ray data for vesuvianite samples from the Akhmatovskaya pit (sample 28975 FM), wiluite (E. V. Galuskin's sample), Kovdor *REE*-bearing vesuvianite (M. M. Moiseev's sample), Kovdor Fe-rich vesuvianite (V. N. Yakovenchuk's sample), rock-forming vesuvianite from Pitkäranta (sample 189/12409 SM) and from Shishimskaya pit (sample 180/12292 SM)

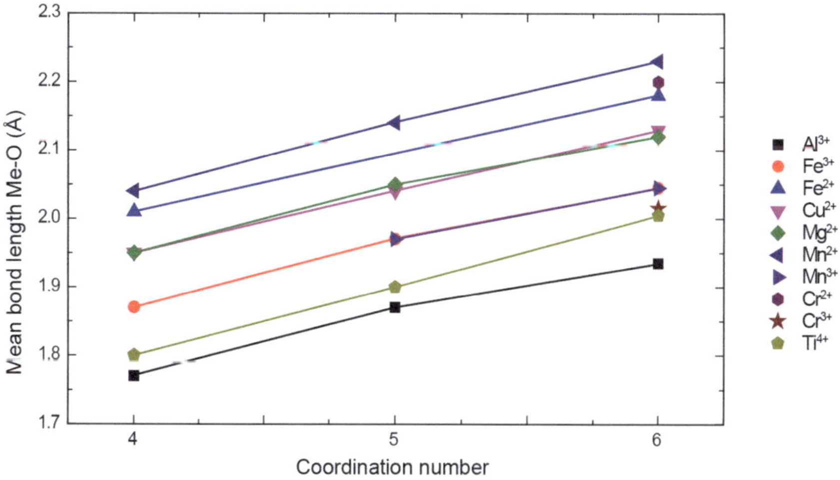

Fig. 4.14 The relationship between the average Me–O bond length and the coordination number of *Y*-cations (Shannon 1976; Hawthorne and Gagné 2024)

4.2.3 *Cooperative Adaptation of the* X4A,B *and* Y3A,B *Polyhedra*

In the VSMs structures, symmetry changes according to the *P4*/*nnc* → *P*4/*n* or *P*4/*nnc* → *P*4*nc* schemes are due to the ordering of the cations in the tetragonal antiprism *X*4A,B and the tetragonal pyramid *Y*1A,B positions (Fig. 1.41). These positions are located in the channels of the structure in the sequence *Y*1–*X*4–*X*4–*Y*1. In the *P4*/*nnc* structure, the occupancy of each of these positions is 50% because of a short distance between neighboring *X*4 and *Y*1 sites (about 1.3 Å). In the ordered VSMs (*P4*/*n* or *P*4*nc*), the population of the *Y*1A and *X*4B positions is equal, as is the population of the *X*4A and *Y*1B positions. The total occupancy of the pairs of positions *Y*1A + *Y*1B and *X*4A + *X*4B is 100%. Depending on the degree of occupancy of ordered rod sites, their polyhedral volume can vary significantly.

Increasing the size of the *X*4A square antiprism with the refined occupancy $Ca_{0.66}$ and volume $V = 24.43$ Å^3 in the structure of cyprine (as compared to the *X*4A polyhedron in vesuvianite) leads to an increase in stresses in the polyhedral layer (Fig. 4.15a) and causes compression of the *Y*3B octahedron with the refined occupancy $Al_{0.75}Mn_{0.25}$ and $V = 9.73$ Å^3. The relative compression of the *X*4B polyhedron (Fig. 4.15b) with a refined occupancy of $Ca_{0.34}$ and $V = 24.05$ Å is compensated by an increase of the size of the *Y*3A octahedron with the refined occupancy of $Al_{0.52}Mn_{0.48}$ and $V = 10.11$ Å^3 due to the substitution $^{Y3A,B}Al^{3+} \leftrightarrow {}^{Y3A,B}Mn^{3+}$.

A similar redistribution of cations was observed in the *P*4*nc* vesuvianite modification. In the crystal structure of vesuvianite from Jordanow (Poland), the refined occupancy of the channel positions *X*4A and *X*4B are $Ca_{0.87}$ and $Ca_{0.13}$, and the polyhedral volumes are 24.91 and 23.86 Å^3, respectively. The population and volumes of

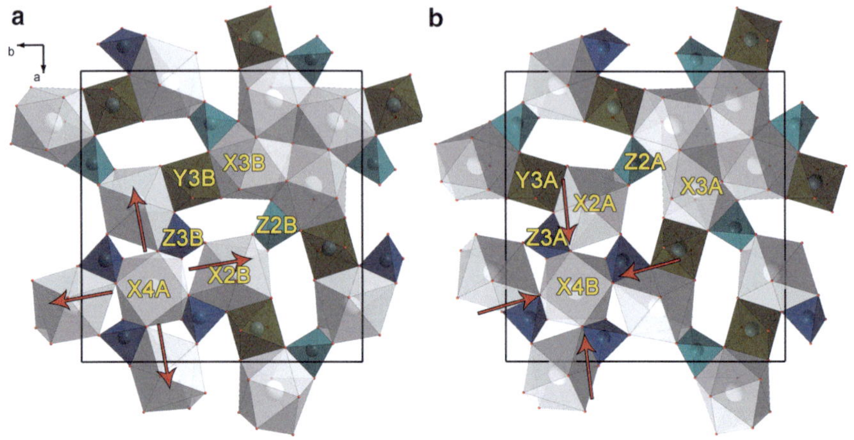

Fig. 4.15 Polyhedral sheet in the crystal structure of the zone-2 cyprine grain at different levels z along the c axis: **a** $z = 0.85$; **b** $z = 0.65$. Red arrows show the directions of tension arising as a result of different occupations of the Ca4A,B positions

Table 4.5 Polyhedral volumes of the X4A,B and Y3A,B sites (Å^3) in VSMs structures

Sample	Locality	X4B	X4A	Y3A	Y3B
198/14298 SM	Akhmatovskaya pit, South Urals	23.57	24.86	9.65	10.44
28975 FM	Akhmatovskaya pit, South Urals	25.05	28.61	9.56	10.64
13988 FM	Ala di Stura, Lanzo Valleys	23.59	24.88	9.74	10.38
17342 FM	Kluchevskiy mine	24.49	23.69	10.13	9.86
A. A. Antonov's sample	Bazhenovskoe deposit	24.89	23.78	10.32	9.64
70026 FM	Asbestos mine	24.45	22.94	10.12	9.38

the octahedral Y3A and Y3B positions are $Al_{0.94}Fe_{0.06}$ and $Al_{0.97}Fe_{0.03}$ and 10.40 and 9.71 Å^3, respectively. The average bond lengths <Y3A–O> and <Y3B–O> are significantly different and are equal to 1.990 and 1.947 Å, respectively. Thus, an increase in the joint volume of the pair of polyhedra X4A + Y3A results in a decrease of the joint volume of the X4B + Y3B pair (Table 4.5).

4.2.4 Mechanism of the Formation of Merohedral Twins

Statistical analysis for 150 structure refinements of VSMs showed that the mean value of the R-factor of the structure refinement for samples with the $P4/n$ symmetry ($R_{1\mathrm{mean}} = 0.053$) significantly exceeds that for samples with other space groups ($R_{1\mathrm{mean}} = 0.025$ for $P4nc$ and $R_{1\mathrm{mean}} = 0.026$ for $P4/nnc$). A significant part of the studied VSMs with the $P4/n$ symmetry (Fig. 4.16) was presented by samples twinned on the two-fold axis along the [110] direction (Fig. 4.17). After the application of corresponding twinning matrix (010 100 00-1), the mean R-factor decreased to $R_{1\text{mean}} = 0.030$. It is remarkable that all the merohedral twins of the VSMs have a high degree of cation ordering.

As a quantitatively measure characterizing the degree of order in the channels of the structure, the ordering coefficient, $K_O = |\delta(X4A) - \delta(X4B)| \cdot 100\%$, can be used (here $\delta(X4A)$ and $\delta(X4B)$ are the occupancy values of the corresponding positions). The graph of the distribution of the values of the order factor for 80 samples of the VSM samples is shown at Fig. 4.16. The K_O values > 68% correspond to low-symmetry VSMs modifications (space group *P*4/*n* or *P*4*nc*), the first of which is characterized by the presence of merohedral twins.

With a gradual increase of the cation ordering in the channels of the structure, two mechanisms of reducing the overall symmetry of the VSMs are realized: *P*4/*nnc* → *P*4/*n* → *P*4*nc* and *P*4/*nnc* → *P*4/*n* → *P*4/*n* (merohedral twin). An increase

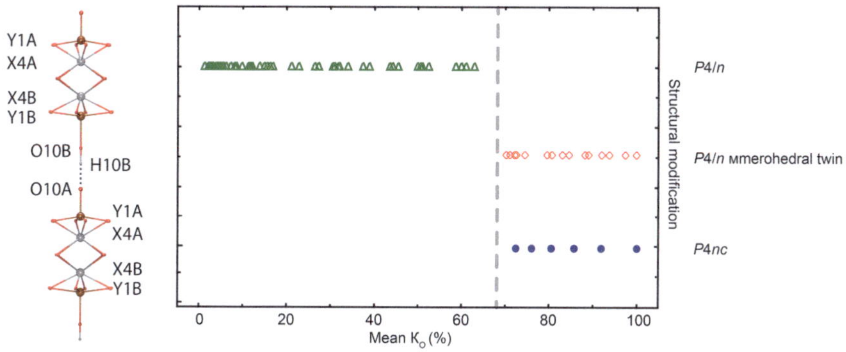

Fig. 4.16 Projection of channel *Y*1–*X*4–*X*4–*Y*1 postions in the VSM structure along the *c* axis (left) and VSMs symmetry versus ordering coefficient for cations in the structural channels (right)

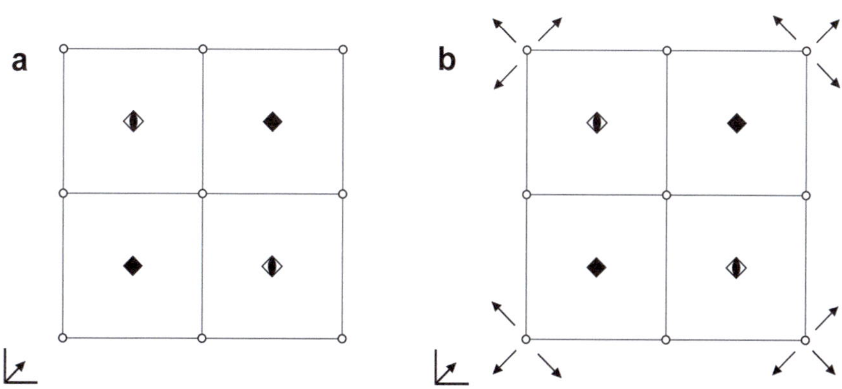

Fig. 4.17 Symmetry elements of the space group *P*4/*n* projected along the *c* axis (**a**) with additional merohedral twinning axes (**b**)

in the degree of ordering is accompanied by an increase in the number of violating reflections for the space group *P4/nnc* (Fig. 1.45a).

The obtained distribution of the K_O values, as well as the connection of this coefficient with the formation of various VSMs modifications is in agreement with the data of geothermometry (Gnos and Armbruster 2006) and the model of structural evolution of the vesuvianite from the Kovdor massif (Panikorovskii et al. 2017f). According to these models, the *P4/nnc* modification is formed during normal kinetics of crystal growth in the temperature range of 400–800 °C, the *P4/n* modification crystallizes in the range of 300–400 °C, whereas for the formation of the *P4nc* modification and *P4/n* merohedral twins temperatures below 300 °C are required (Fig. 1.45b). Consequently, these data are of practical importance for the development of the VSMs geothermometer.

4.3 New Crystal Chemistry Data: Implications for the Nomenclature of the Vesuvianite-Supergroup Minerals

By now, many questions have accumulated regarding the writing of formulas and names of VSMs, as well as the rules for identifying mineral species within this supergroup. The data on the crystal chemistry of VSMs accumulated over the past 30 years make it possible to understand regularities of the entry of cations into various positions of the structure, which makes it possible to propose a universal method for calculating formulas of these minerals. Based on this study, several important conclusions were made about the relationships between the unit cell parameters of VSMs and the chemical composition and the lack of a relationship between the unit cell parameters and the symmetry of VSMs.

4.3.1 Main Trends in Unit Cell Parameters. Division of the Supergroup into Groups

As shown in Sect. 1.4.4 using data by Gnos and Armbruster (2006), there are several trends related to unit cell parameters in the VSMs, which are clearly visible in the UCP diagram (Fig. 1.51). We constructed a similar diagram, but with a much larger number of samples, expanding the intervals for the parameter a from 15.47–15.74 Å to 15.47–15.94 Å, while for the parameter *c* the interval remained the same (11.67–11.97 Å). The abundances of the main components of the chemical composition were plotted on this diagram as isolines. Thus, it was possible to establish a relationship between the chemical composition, structural isomorphism schemes and physical properties and to propose a division of the vesuvianite supergroup into groups.

In the UCP diagram, two trends can be distinguished in general. Trend 1 (Fig. 4.18) shows a decrease in the *c* parameter with a simultaneous increase in the *a* parameter. This trend was first noted by Groat (1988). Trend 1 is clearly associated with variations in the Mg^{2+} content (Fig. 4.18a), as well as B^{3+} content (Fig. 4.18b). This trend corresponds to the isomorphism schemes $^{T1}B^{3+} + {}^{Y}Mg^{2+} + 2^{O11}O^{2-} \leftrightarrow {}^{T1}\square + {}^{Y}Al^{3+} + 2^{O11}(OH)^{-}$ (Fig. 4.19a, b) and $^{T2}B^{3+} + 2^{O10}O^{2-} \leftrightarrow {}^{T2}\square + {}^{O10,11}(OH)^{-}$ (Fig. 4.19c, d). The incorporation of boron into the *T*2 site with triangular coordination causes the substitution of the O10–H1···O10 group of atoms by the BO_3 triangle with the ^{T2}B–O10 and ^{T2}B–O12 bonds, which causes a significant reduction in the *c* parameter from 11.80–11.90 Å in B-free VSMs to 11.66–11.75 Å in wiluite. At the same time, the incorporation of boron into the tetrahedral *T*1 site leads to an increase in the *a* parameter from 15.55 to 15.75 Å. Such a significant change in the unit cell parameters leads to a change in the optical sign from "−" in B-deficient VSMs to "+" in wiluite (Fig. 1.33). The change in optical sign occurs at B^{3+} contents of about 2.5 *apfu*, which corresponds to the vesuvianite-wiluite boundary. Therefore, we consider it appropriate to distinguish a wiluite group in which the change in the UCP parameters (Trend 1) is associated with the occupation of additional positions *T*1 and *T*2.

Trend 2 in the UCP diagram (Figs. 4.18a and 4.20a, b) is associated with a simultaneous increase in both the *a* and *c* parameters. In the work by Gnos and Armbruster (2006), this regularity is interpreted as a manifestation of two separate trends, one of which is related to the hydrogarnet-type isomorphism, $^{Z1,2}(SiO_4)^{4-} \leftrightarrow {}^{Z1,2}(H_4O_4)^{4-}$, and the second one to the incorporation of manganese according to the schemes $^{Y3}Al^{3+} \leftrightarrow {}^{Y3}Mn^{3+}$ and $^{Y1}Fe^{2+} \leftrightarrow {}^{Y1}Mn^{2+}$. We have found that by expanding the number of studied samples, these two trends can be combined into Trend 2, which in general can be reduced to substitutions in *X*, *Y* and *Z* positions of vesuvianite with cations with a larger radius, which leads to a uniform increase in the volume of coordination polyhedra and the total volume of the unit cell.

In most VSMs, the *X* sites are occupied by Ca with insignificant admixtures of other components. However, structural changes associated with the incorporation of cations with different charges and radii into the *X* group of sites are of particular interest because these cations are important geochemical markers.

Among studied VSMs, manaevite-(Ce) from the Kovdor massif is characterized by the maximum content of rare-earth elements (>4 *apfu*), minimum Al content (<4 *apfu*) and a significant Mg content (>4 *apfu*). Another *REE*-rich VSM, milanriederite, containing about 2 *REE apfu*, is also Mg-rich (with 4.24 Mg *apfu*). In VSMs, rare-earth elements occur at the *X*3 and *X*4 sites. These and numerous other examples show that the increase of the *REE* content in the VSMs is accompanied by an increase amount of magnesium in the *Y*2 and *Y*3 octahedral positions. These regularities correspond to the following heterovalent substitution scheme: $^{Y1,2}Al^{3+} + {}^{X3,4}Ca^{2+} \leftrightarrow {}^{Y2,3}Mg^{2+} + {}^{X3,4}REE^{3+}$.

The unit cell parameters of the VSMs linearly increase with the increase of the *REE* content because in this case smaller cations (Ca^{2+} and Al^{3+}) are substituted by larger ones (REE^{3+} and Mg^{2+}, respectively). As a result, the lanthanide-enriched

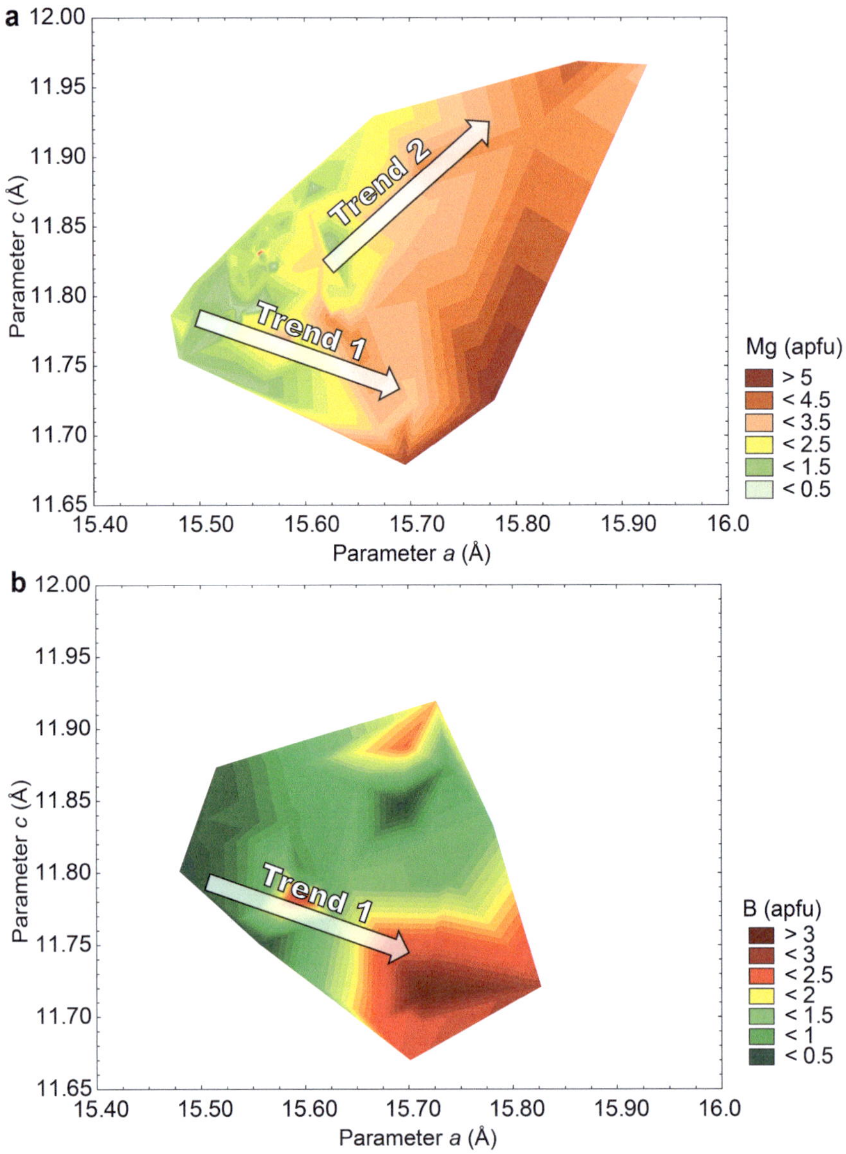

Fig. 4.18 Diagrams of the dependences of the VSMs unit cell parameters on the content of magnesium (**a**), and boron (**b**)

vesuvianite samples (in particular, those from the Kovdor massif) show significant deviations from the mean values of the *a* and *c* parameters of the VSMs (Fig. 4.20a).

According to Groat et al. (1992a), Al does not replace Si in VSMs. However, refinement of the crystal structure of VSMs samples with different *REE* contents from

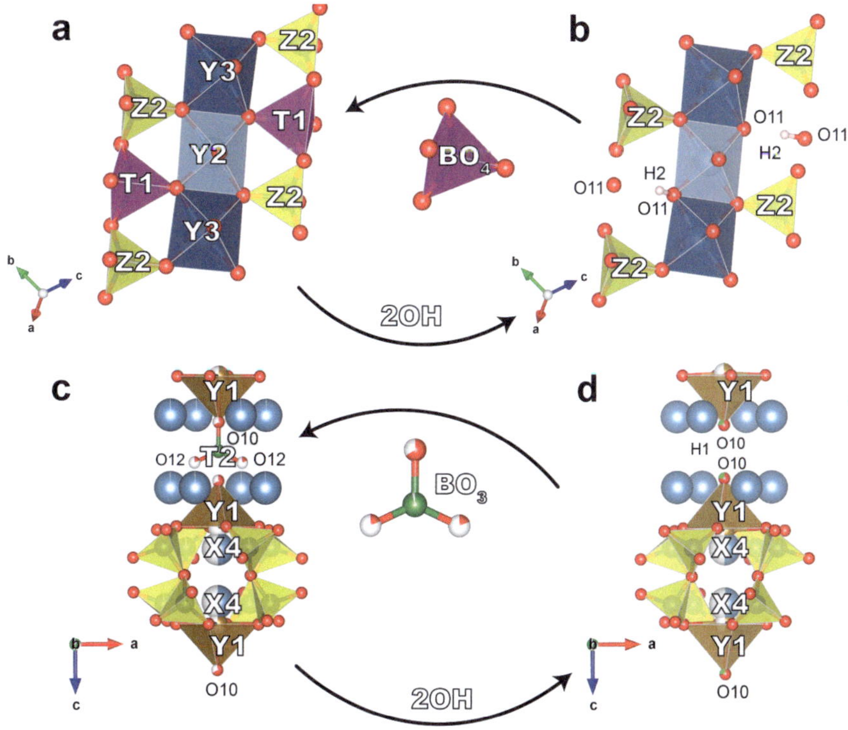

Fig. 4.19 Mechanism of boron incorporation into the crystal structure of vesuvianite, corresponding to Trend 1. The location of BO_4 in the *T*1 position in wiluite (**a**); the local environment of the *T*1 position in a B-free VSM (**b**), the location of boron in the channels of the VSM structure in the *T*2 position (**c**), the location of hydroxyl groups in the vesuvianite structure in the case of a vacancy in the *T*2 position (**d**)

the Kovdor massif showed an increase of the polyhedral volume of the sorosilicate groups from 2.21 to 2.45 Å^3, caused by an increase in the proportion of $(H_4O_4)^{4-}$ or Al^{3+} in the *Z*1 and *Z*2 positions.

As noted above, the presence of aluminum with tetrahedral coordination in the VSMs structures was confirmed by the presence of the corresponding peak at 60 ppm in the ^{27}Al NMR spectrum (Fig. 3.78a). In other words, the replacement of Ca by *REE* can be compensated by the introduction of Al in the tetrahedron of the position according to the scheme $^{Z1,2}Si^{4+} + {}^{X3,4}Ca^{2+} \leftrightarrow {}^{Z1,2}Al^{3+} + {}^{X3,4}REE^{3+}$ accompanied by the substitution of the SiO_4^{4-} groups by the $(H_4O_4)^{4-}$ clusters, which also leads to an increase in the volume of the ZO_4 tetrahedra (Panikorovskii et al. 2016a).

The increase of *REE* content (and related increase of Mg content) in the VSMs composition results in an increase of the <*Y*2,3–O> mean bond lengths and volumes of the *Y*2- and *Y*3-octahedra (Fig. 4.21).

Summarizing the above, it should be noted that Trend 2 generally relates to the formation of the potential manaevite group, as well as "hydrovesuvianite". The

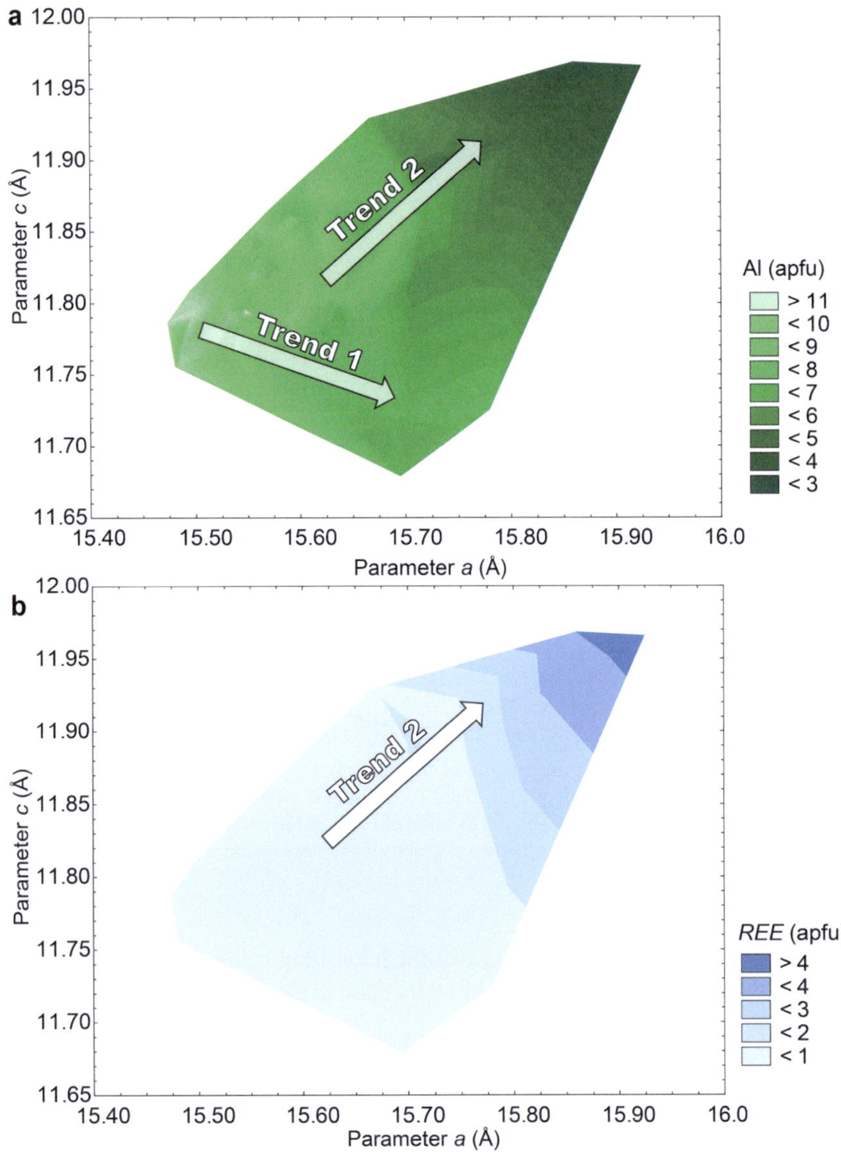

Fig. 4.20 Diagrams of the dependences of the VSMs unit cell parameters on the content of aluminum (**a**) and rare earth elements (**b**)

symbate increase in the *a* and *c* parameters is easily recorded by PXRD data, so X-ray diffraction can be recommended as a method for express assignment of VSMs (together with EDS and WDS data) to a particular group in the vesuvianite super-group. The simultaneous increase in UCP parameters leads to the preservation of the

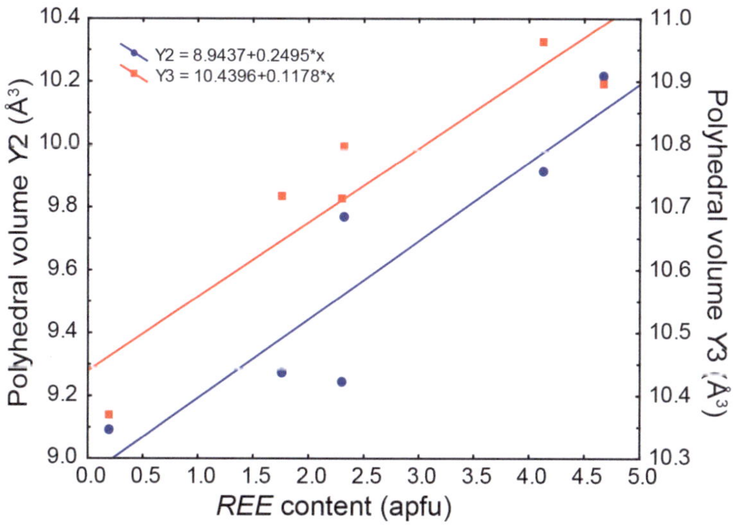

Fig. 4.21 Volumes of the $Y2O_6$ and $Y3O_6$ octahedra *versus REE* content in VSMs

optical sign "–", which can also be distinguished as a diagnostic feature. The UCP parameters for the studied samples and the area of potential vesuvianite groups are shown in Fig. 4.22.

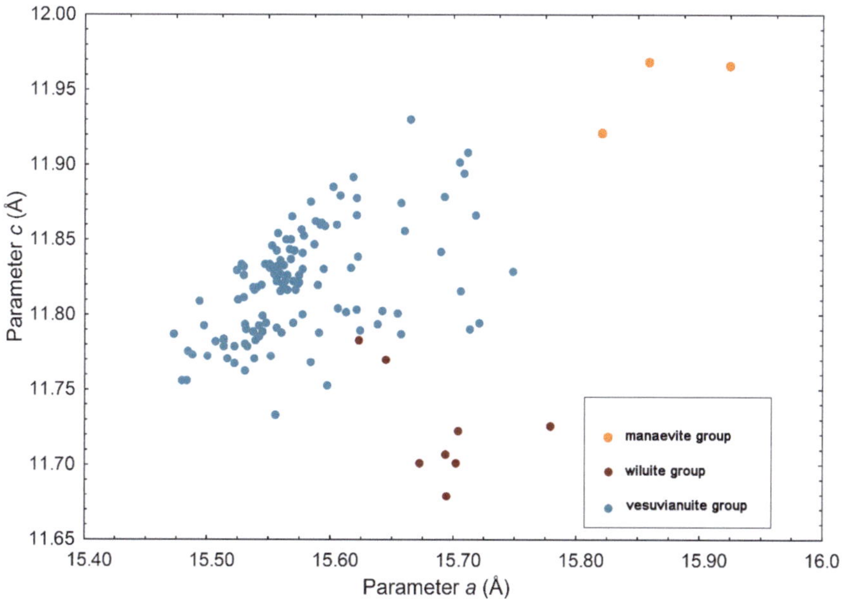

Fig. 4.22 UCP parameters of different VSMs and their separation between the groups

4.3.2 Substitutions at the Y3 Octahedral Site

Most of the samples studied in this work belong to vesuvianite itself, whose ideal formula is $Ca_{19}Fe^{3+}Al_4(Al_6Mg_2)(SiO_4)_{10}(Si_2O_7)_4[(OH)_9O]$. One of the main nomenclatural problems related to this formula is the presence of two mixed-occupied positions, although only one mixed-occupied position is allowed for end-member formulas of minerals (Hawthorne 2002). Which are these positions? In vesuvianite, the O10 position is half occupied by OH^- and half occupied by O^{2-} (Panikorovskii et al. 2017b). Simultaneous occupancy of this site by O^{2-} is possible, as e.g. in wiluite, but the occurrence of two $^{O10}OH^-$ groups per unit cell is impossible due to geometric factors (this is discussed in more detail in Sect. 4.3.3). The second mixed-occupied position is the *Y*3 site, which usually contains $Al^{3+}{}_6Me^{2+}{}_2$ *apfu*. The divalent *Y*3-cations are usually Mg^{2+}, Fe^{2+} and/or Mn^{2+}. Statistics confirm our observation: most experimental points lie near the theoretical occupancy of this position by 25% with Me^{2+} and 75% with Me^{3+} (Fig. 4.23).

In order to avoid difficulties in derivation the end-member formulas of VSMs, for example, when incorporation of minor amounts of Ti^{4+} takes place, we propose to use the following end-member formula for vesuvianite: $Ca_{19}Fe^{3+}Al_4(Al_6Me^{2+}{}_2)(SiO_4)_{10}(Si_2O_7)_4[OH_9O]$. This will avoid an unnecessary increase in the number of mineral species in the vesuvianite group by approximately 2 times and not violate the basic 50% rule (Nickel and Grice 1998), according to which a new mineral species can be defined if at least one of the crystallographic positions is filled by more than 50% in a different way.

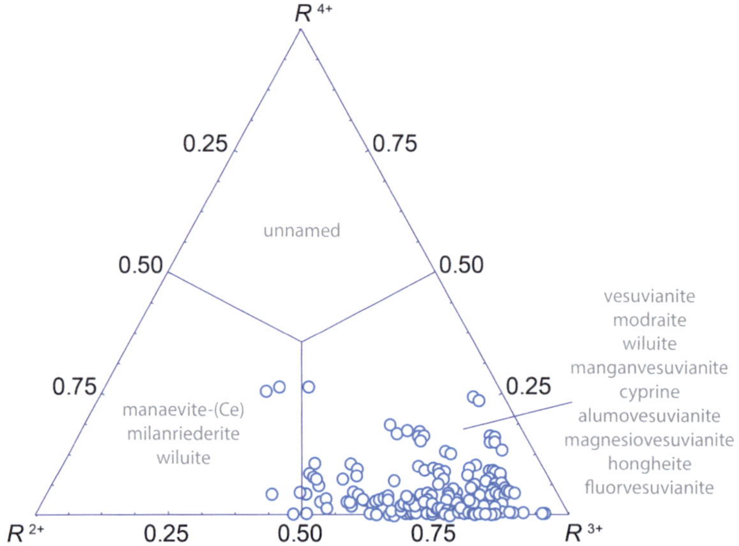

Fig. 4.23 Triangular diagram showing population of the *Y*3 site by cations with different oxidation states

In addition, not only divalent but also significant amounts of tetravalent cations, in particular Ti^{4+}, can be incorporated into the *Y*3 position. The highest TiO_2 contents reported by (Groat et al. 1995c) for vesuvianite are 2.7–2.8 wt% which corresponds to 1 Ti *apfu*. In the IR spectra of all samples with Ti contents more than 0.35 apfu, studied in the cited paper. However, according to our data, Ti-rich VSMs are characterized by the Ti content up to 2.20 Ti *apfu* (in a sample from Alchuri, Shigar Valley, Northern Areas, Pakistan: Chukanov et al. 2018a). In milanriederite, incorporation of a significant amount of Mg in the *Y*3 octahedron is compensated by the incorporation of *REE* in the *X* sites. However, as noted above, the main heterovalent substitutions in the VSMs involving only Y-cations occur in accordance with the general scheme $2M^{3+} \leftrightarrow Ti^{4+} + M^{2+}$where $M^{3+} = Al^{3+}$, Fe^{3+} or Mn^{3+} and $M^{2+} = Mg^{2+}$, Fe^{2+} or Mn^{2+}. Substitutions of this kind involving Fe^{2+} are the most common case.

Thus, the component that predominates among the non-trivalent cations at the *Y*3 site should be considered as a charge-balancing component in the end-member formula. It does not matter how many of these major charge-balancing cations there are.

4.3.3 Why the VSM Formula Cannot Have 10 OH Groups

The hydrogen bonding network in VGMs channels involves the O10 (2 apfu) sites. These sites can be occupied by either oxygen atoms from hydroxyl (OH) groups or fluoride (F^-) anions. Typically, the hydrogen site H2 (1 *apfu*), participates in the hydrogen bonding O10–H10···O10, with bond distances ranging from 2.80 to 3.05 Å (Groat et al. 1996b; Lager et al. 1999b; Galuskin et al. 2007a). Due to a short hydrogen–hydrogen contact with an interatomic distance of approximately 0.70–1.0 Å, the H10 site in the *P*4/*nnc* structure is half-populated (Fig. 4.24a), allowing only one of the adjacent H10A or H10B sites to be occupied in the low-symmetry *P*4/*n* and *P*4*nc* modifications (Pavese et al. 1998b; Panikorovskii et al. 2016b).

Cyprine, $Ca_{19}Cu^{2+}Al_4(Al_7R^{2+})(SiO_4)_{10}(Si_2O_7)_4O(OH)_9$, was initially approved with the formula $Ca_{19}Cu^{2+}(Al,Mg)_{12}(SiO_4)_{10}(Si_2O_7)_4(OH)_{10}$ (Panikorovskii et al. 2017e), which differs from the former formula in the OH content. The formula of cyprine was modified since the complete occupancy of H sites in the channels (the H10A, H10B sites) is unlikely due to the short H···H distances of ~1 Å (Fig. 4.24a). In cyprine, as in aluminovesuviane, the space group *P*4/*n* is realized. In this case, the OH group is associated with the vacant *Y*1B site, and O^{2-} occurs at the fully occupied *Y*1A site (Fig. 4.24b).

Thus, if all Si positions are fully occupied and there is no hydrogarnet-type isomorphism, the coexistence of 10 OH groups in vesuvianite is impossible since there is no sufficient free space in the channels for two fully occupied H10 positions. At the same time, there are no steric restrictions for the introduction of two fluoride anions or two oxygen atoms into the O10 position.

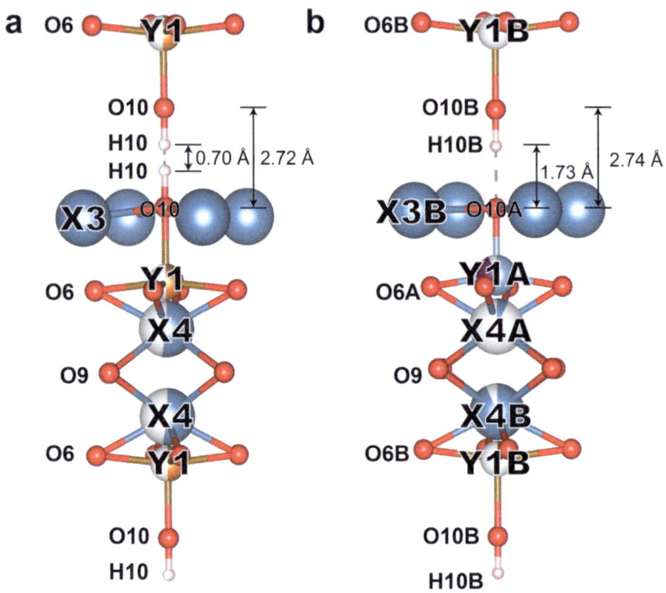

Fig. 4.24 Arrangement of hydrogen bonding in *P*4/nnc vesuvianite based on data (Lager et al. 1999b) (**a**) and alumovesuvianite channels (**b**)

4.3.4 Na Incorporation into the Vesuvianite Structure. Potential Na-Dominant VSM

Sodium is a ***minor*** component in VSMs. Significant amounts of Na, together with Ti and *REE*, were noted in VSMs from altered alkaline, basic or ultrabasic rocks. In particular, Na-bearing VSMs occur in jadeite aggregates from Myanmar (Nyunt et al. 2009) and alpine type alkaline veins from the Alchuri valley, Pakistan (Aksenov et al. 2016).

The works by Chumakov et al. (1948), Batieva and Bel'kov (1984), Batieva et al. (1987) mentioned Na-containing vesuvianite from the contact zone of the gneiss-amphibolite rocks of the Keivskaya series, alkaline granites of the West Keivy mounts and nepheline syenites of the Sakharjok massif. The crystal chemistry of these samples has been studied by us in detail. The refinement of the crystal structures of three samples of Na-containing vesuvianite (with 0.5, 0.6 and 1.3 Na *apfu*, Table 1.11, analysis 2) showed that sodium selectively occupies positions with the coordination number of 8 (Fig. 4.25). In samples from the West Keivy massif, the *X*2 (coordination number 7) and *X*3 (coordination number 9) positions are completely occupied by Ca atoms. The calculated population of the *X*1 position is $Ca_{0.96}Na_{0.04}$ for the sample 979 BM and $Ca_{0.93}Na_{0.07}$ for the sample 50578 FM. The *X*4 position located in the structural channels has a significant amount of sodium: $Ca_{0.54}Na_{0.46}$ (for the sample 979 BM) and $Ca_{0.68}Na_{0.32}$ (for the sample 50578 FM); the corresponding scattering factors correspond to 15.4 e^{-} and 16.8 e^{-}. In the sample from

Alchuri (Pakistan), all Na is localized at the $X1$ site whose population is $Ca_{0.55}Na_{0.45}$. The refined scattering factor for this position corresponds to 16 e^-. $X1$ site has 4 common edges with $Y3$ octahedra (Fig. 4.25c), it seems that the isomorphism $Ca^{2+} + Al^{3+} \leftrightarrow Na^+ + Ti^{4+}$ is realized also at the local level.

The data of solid-state ^{23}Na NMR (3.15a) confirm the results of the single-crystal X-ray structure analysis. The observed peak at –15.8 ppm corresponds to Na with the coordination number of 8–9 (Xue and Stebbins 1993; Fechtelkord et al. 2001).

The lack of positive charge caused by the replacement of Ca by Na in the $X1$ and $X4$ positions is compensated by Ti incorporation in the Al-dominant octahedra (mainly, $Y3$). According to the available statistics, all samples with more than 0.1 Na *apfu* contain a significant amount of Ti, i.e. sodium is incorporated in the VSMs structures in accordance with the scheme $Ca^{2+} + Al^{3+} \leftrightarrow Na^+ + Ti^{4+}$ (Fig. 4.26, trend #1). In turn, the incorporation of Ti into the VSMs structures may be realized via an alternative non-sodium scheme: $Ti^{4+} + Mg^{2+} \leftrightarrow 2Al^{3+}$ (Fig. 4.26, trend #2).

Due to a high content of Na and F in alkaline fluids, a part of calcium in the vesuvianite structure is replaced by sodium according to the scheme $Ca^{2+} + Al^{3+} \leftrightarrow Na^+ + Ti^{4+}$, and a part of the OH groups is replaced by fluorine. Sodium is localized in the intrachannel $X1$ and $X4$ positions of the mineral structure (Fig. 4.25b), mainly concentrating at $X4$. This is also indicated by the increased <$X4$–O> average bond length, equal to 2.470 Å, which is due to the excess of the ionic radius of Na^+ over the Ca^{2+} radius (for comparison, the average <$X4$–O> bond length in the Cu-containing Norwegian vesuvianite, in which the $X4$ position is fully populated with calcium, is 2.456–2.458 Å (Panikorovskii et al. 2016b). The interesting feature of the studied samples is the dominance of Fe^{2+} over Fe^{3+} in the $Y1$ position, unlike most VSMs samples from other localities, studied by us.

The process of vesuvianite formation in such rocks can be realized via the following reaction:

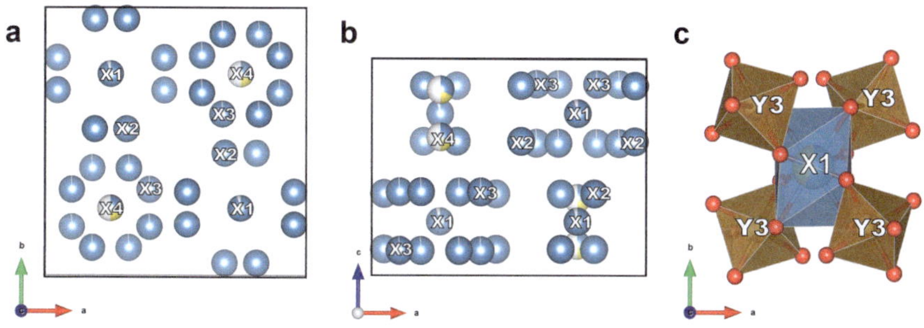

Fig. 4.25 Crystal structure of Na-bearing vesuvianite: projection of X positions on the (001) plane (**a**), projection of X positions on the (010) axis (**b**) Local coordination of Ti and Na at edge-connected $Y3$ and $X1$ polyhedra

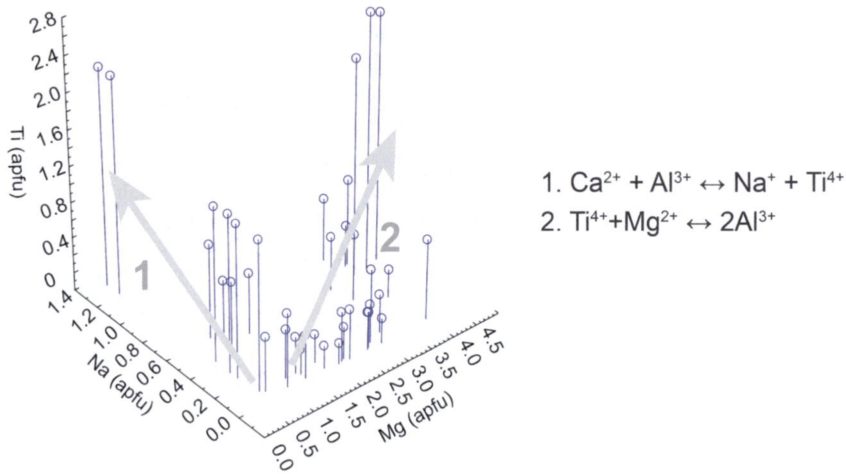

Fig. 4.26 Na, Mg and Ti contents in VSMs

$$\underset{\text{anorthite}}{16CaAl_2Si_2O_8} + \underset{\text{diopside}}{3CaMgSi_2O_6} + 5H_2O + 10Na_2O \rightarrow \underset{\text{vesuvianite}}{Ca_{19}Mg_3Al_{10}Si_{18}O_{68}(OH)_{10}}$$
$$+ \underset{\text{“nepheline”}}{20NaAlSiO_4} + Al_2O_3.$$

The data obtained indicate a high probability of the discovery of “natrovesuvianite and “natrofluorvesuvianite”, the formation of which is quite possible under the conditions of fenitization of high calcium aluminous rocks. This situation is typical for both exocontacts of the West Keivy alkaline granitic massif and fenitized basalt xenoliths in nepheline syenites and foidolites of the Khibiny and Lovozero massifs.

4.3.5 Incorporation of Sulfur into the Vesuvianite Structure

Depending on the local coordination, the binding energy of tetra- and hexavalent sulfur can vary in the range of 160–171 eV (Lindberg et al. 1970; Kelemen et al. 1990). Therefore, powders of hannebachite $(CaSO_3)_2 \cdot H_2O$ and ettringite $Ca_6Al_2(SO_4)_3(OH)_{12} \cdot 26H_2O$ were used as standards of S^{4+} and S^{6+}. Based on the published data, a broad intense unseparated 2p (1/2 and 3/2) sulfur peak at 168.7 eV in the spectrum of lemon-yellow vesuvianite from Monzoni, Italy (Fig. 1.31) corresponds to sulfate sulfur whereas the additional signal in the range of 164–166 eV is due to sulfide sulfur (Sapozhnikov et al. 2023) (Fig. 4.27).

The nature of sulfide sulfur in yellow vesuvianite was studied by a complex of spectroscopic methods (Shendrik et al. 2025).

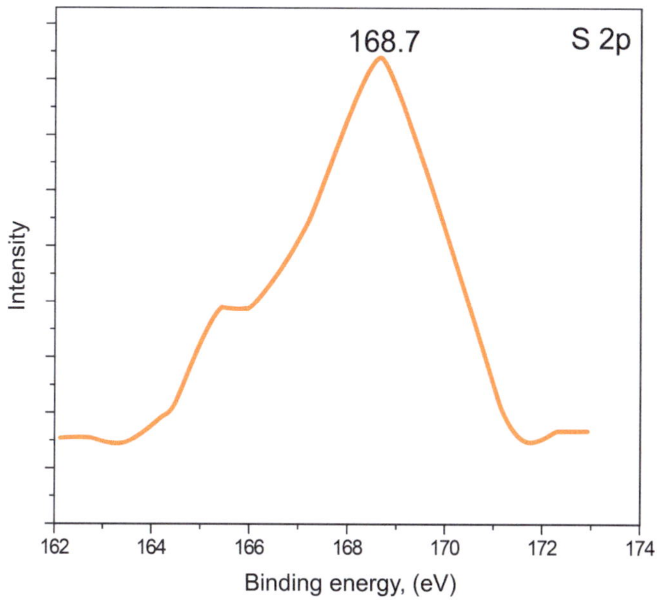

Fig. 4.27 Photoelectron spectrum of vesuvianite specimen from Monzoni, Italy (sample 56-3057 SM)

The luminescence spectrum of yellow S-containing vesuvianite from Monzoni, Italy, excited at 450 nm, shows a broad luminescence peak at 550 nm with a weak vibrational structure with a period about 560 cm^{-1} (Fig. 4.28). The electronic absorption spectrum (optical spectrum in the visible, near IR and UV regions) of this sample (Fig. 4.28) shows a shoulder at about 450 nm. This vibrational structure is similar to that often observed in the luminescence spectra of S-containing microporous aluminosilicates and beryllosilicates and is caused by the presence of the $S_2^{\cdot-}$ radical anions in them (Steudel and Chivers 2019; Chukanov et al. 2020, 2021, 2022a, b, c).

Unlike vesuvianite, in minerals of the sodalite and cancrinite groups the luminescence band is shifted to the red region of the spectrum. This is probably due to the fact that the $S_2^{\cdot-}$ radical anion in vesuvianite with a relatively dense structure is in a stronger crystalline field compared to these anion radicals in microporous minerals.

A complex ESR spectrum is observed for a sample of yellow S-bearing vesuvianite from the Talnakh deposit (Taimyr) (Fig. 4.29). The strongest 6-component signal with a *g*-factor of 2.002 is associated with Mn^{2+} ions in the high-spin state. Weaker EPR signals with $g_1 = 2.38$, $g_2 = 2.23$ and $g_3 = 1.82$ are also observed, similar to the previously observed $S_2^{\cdot-}$ ESR spectra in alkali halide salt crystals (Vannotti and Morton 1967).

Raman spectra of S-bearing vesuvianite show a strong luminescence caused by the presence of $S_2^{\cdot-}$ and Fe^{3+} (Fig. 4.30). The bands at 549 and 595 cm^{-1} (Fig. 4.31) correspond to stretching and combination (stretching + libration) modes of $S_2^{\cdot-}$, respectively. The band at 974 cm^{-1} is probably due to SiO_4^{2-} symmetric stretching

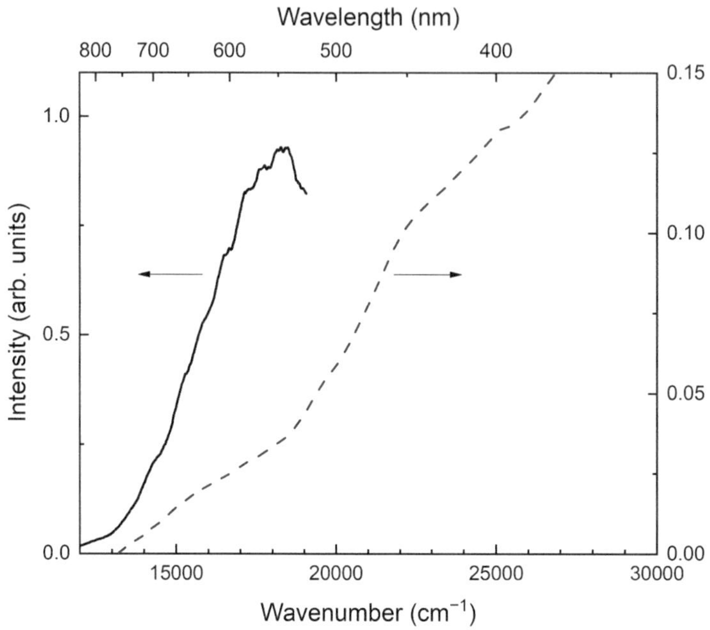

Fig. 4.28 Luminescence (solid line) and absorption (dashed curve) spectra of yellow vesuvianite from Monzoni

Fig. 4.29 ESR spectrum of yellow S-bearing vesuvianite from the Talnakh

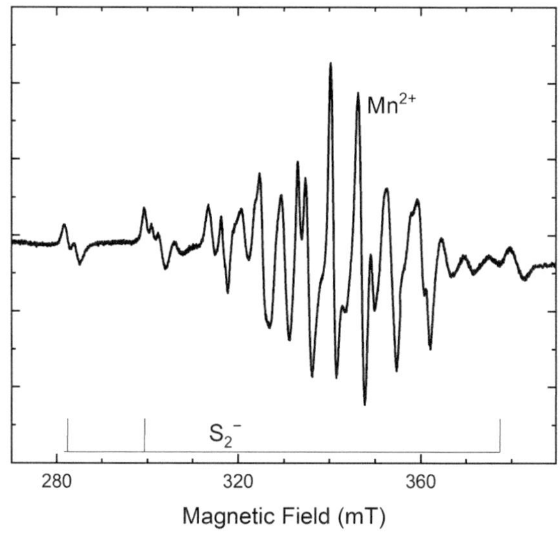

vibrations [$A_1(\nu_1)$ mode] or asymmetric stretching vibrations of Si–O bonds. Other Raman bands in the range of 350–1050 cm^{-1} correspond to vibrations of the SiO_4^{4-}

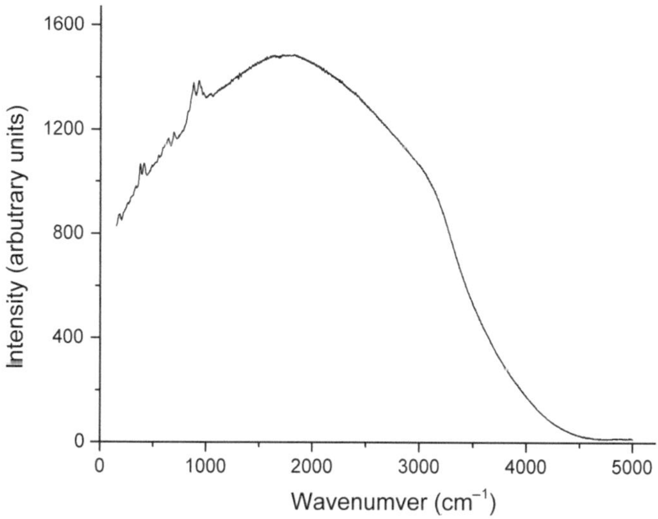

Fig. 4.30 Luminescence in the Raman spectrum of yellow S-bearing vesuvianite from Monzoni

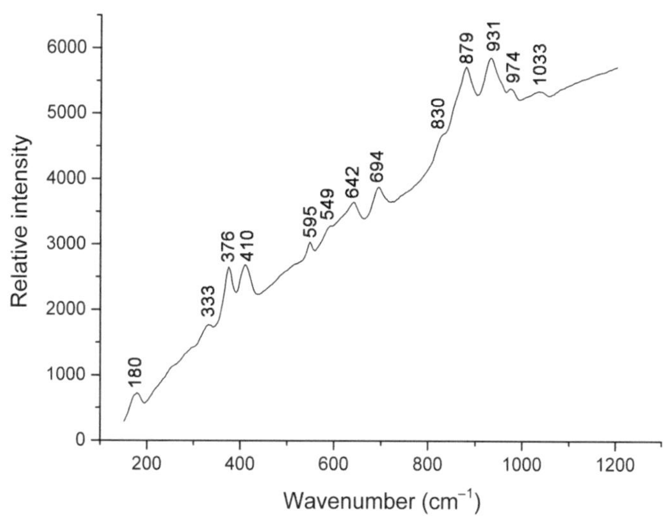

Fig. 4.31 Raman spectrum of yellow S-bearing vesuvianite from Monzoni in the low-frequency region

and $Si_2O_7^{6-}$ groups, partly combined with M···O–H bending vibrations. Weak bands of lattice modes are observed below 350 cm^{-1}.

The most plausible structural model corresponds to the S^{2-} radical anion, which is incorporated according to the scheme $^{T2}S^{2-} \leftrightarrow {}^{10}O^{2-}$ (Fig. 4.32a). It is quite likely that approximately 80–90% of sulfur is included in the structure of vesuvianite according

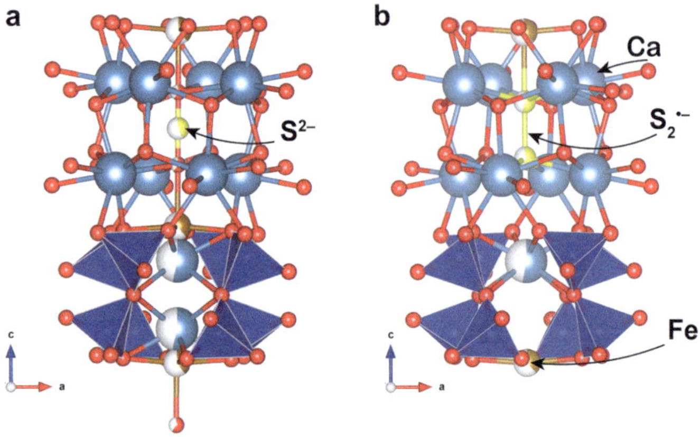

Fig. 4.32 Mechanism of incorporation of S^{2-} into the vesuvianite structure (**a**) and a fragment of the vesuvianite structure containing the $S_2^{\cdot-}$ radical anion showing local coordination of $S_2^{\cdot-}$ (**b**)

to this scheme. At the same time, the S^{2-} anion is undetectable by means of ESR and vibrational spectroscopy. Its presence is indirectly indicated by the vacancy in the O10 position and yellow color of S-bearing vesuvianite. Since the S–O10 distance is equal to 1.28 Å, two anions cannot be occupied simultaneously. The coordination of S in the *T*2 position excludes triangular and tetrahedral coordination, since there are no peaks of electron density that would relate to additional O atoms. Since the sulfur content correlates with the intensity of the yellow color of vesuvianite, and all spectroscopic methods indicate the presence of the $S_2^{\cdot-}$ radical anion, we assume that the remaining 10–20% of sulfide sulfur occur in the channels of the structure as the $S_2^{\cdot-}$ radical anion. In this case, the O10 position is vacant, and $S_2^{\cdot-}$ is included in the coordination sphere of Fe^{3+} (Fig. 4.32b).

The incorporation of sulfur into the vesuvianite structure occurs in the *T*2 position. The main substitution mechanism is ${}^{T2}S^{2-} + {}^{O10}\square \leftrightarrow {}^{10}O^{2-} + {}^{T2}\square$. A potential new mineral formed as a result of the introduction of sulfur would have the formula $Ca_{19}Fe^{2+}Al_4(Al_6Mg_2)(SiO_4)_{10}(Si_2O_7)_4S^{2-}(OH)_8$.

4.3.6 Recommendation on Formula Calculation

We recommend a standardized procedure for allocating cations in all VSMs based on complete chemical data normalized to 19 *X* cations. Other normalization methods has certain disadvantages. For instance, normalizing to 50 cations is ineffective when there is a significant presence of boron in the VSM (i.e., >1 wt% B_2O_3). Additionally, normalization based on 18 Si atoms can be ambiguous due to potential silicon deficiency (see Sect. 4.1.2). Renormalization using 13 *Y* cations is not advisable because high-temperature samples often exhibit an excess of *Y*-elements (or, more accurately, up to 0.5 atoms per formula unit (apfu) of 0.5 apfu Al + Fe^{3+} at the *T* sites) (Groat et al. 1994a, b).

While normalization to 19 X is not ideal, it represents the best compromise available. In rare cases, such as manaevite-(Ce) (Moiseev et al. 2020), cation vacancies at the X sites are substituted by H_2O molecules. Normalization to 18 Si at the Z sites is not preferred due to the potential for hydrogarnet-like substitutions at *Z*1 and *Z*2. Furthermore, normalization to 13 cations at the *Y* sites is problematic if there is an excess of Al at the *T*1 site. Normalizing based on anion charges is also inadequate due to variable OH/F substitutions at certain O sites, particularly those influenced by B incorporation at *T*2. The incorporation of B and S leads to the emergence of an additional oxygen atom (O12) at the *W*3 site.

In most cases, researchers have access only to EDS or WDS chemical data, and typically VSMs do not contain significant amounts of B^{3+} and REE^{3+}. The total expected cationic charge is +147, with the sum of ($Y1 + Y2 + Y3$) typically yielding a total charge of +37 in 95% of all published data. In 88% of cases, $T1,2 = 0$ and Ca = 19, making normalization to either 18 Si or 50 cations appropriate. For these instances, we propose a calculation scheme based on 50 cations (excluding H).

4.4 Geological Applications

The VSMs are excellent indicators of mineral formation conditions. Once formed, VSMs retain their symmetry until complete destruction and can be used as a key tool for reconstructing mineral formation conditions (Justiniano et al. 2025a).

Regarding the *P*–*T* conditions under which vesuvianite-group minerals form, these conditions can vary significantly. In rodingites, which belong to the prehnite-pumpellyite–greenschist facies, crystallization occurs at temperatures ranging from 200 to 400 °C and pressures between 150 and 350 MPa (Peters 1961, 1965; Stalder and Aufdenblatten 1995b; Gnos and Armbruster 2006). In contrast, vesuvianite in metacarbonate and calc-silicate rocks, associated with the amphibolite–granulite facies, typically forms at higher temperatures of 650 to 800 °C and pressures of 450 to 750 MPa (Trommsdorff 1968b; Gnos and Armbruster 2006). Vesuvianite related to skarns, found in contact aureoles, generally forms at elevated temperatures (>600 °C) and lower pressures (<100 to 300 MPa) (Gnos and Armbruster 2006). Additionally, vesuvianite can be discovered in high-pressure environments, reaching up to 2 GPa, such as in serpentinized peridotites from the Raspas Complex (Halama et al. 2013).

In addition to reconstructing PT formation conditions, the role of VSMs for dating contact metamorphic rocks is increasing (Romer 1992b; Stifeeva et al. 2024b; Han et al. 2025b). In this section, we discuss, which of the VSMs samples can be used for this purpose and which cannot. In addition, we propose using a method for estimating the formation temperature of vesuvianite based on the number of reflections that violate the extinction rules for the *P*4/*nnc* space group (Panikorovskii et al. 2024b). Information on the relationship between the IR spectra of vesuvianite and the formation temperature may also be very useful for geologists.

4.4.1 Vesuvianite-Based Diffraction Geothermometer

During diffraction studies, we noted that along with the increase in the ordering coefficient K_O, there was also an increase in the number of reflections violating the extinction rules for the space group *P*4/*nnc*. Usually, the most ordered VSMs with the space groups *P*4/*nnc* and *P*4*nc* have the lowest crystallization temperatures. We decided to test this assumption for a series of samples from the Kovdor massif in accordance with the previously determined dependence of the crystallization temperature of vesuvianite on the number of violating reflections (Panikorovskii et al. 2024b).

The diversity of VSMs in skarnoids of the Kovdor massif is described in detail in Sect. 1.1.5. The evolution of VSMs here is very clearly traced from early apourtite pargasite-vesuvianite rocks containing apple-green vesuvianite (Fig. 4.33a) to calcite-glagolevite-vesuvianite rocks (Moiseev et al. 2020, Fig. 4.33b), in which the *REE* content increases with time, up to the formation of manaevite-(Ce).

We selected six samples of VSMs for which the crystallization order could be clearly traced. All samples were examined using X-ray diffraction analysis. Information on the space group and the number of violating reflections for the space group *P*4/*nnc* is given in Table 4.6. Sample 1 was found in skarnoids, samples 2–6 formed later and their formation is associated with the intrusion of late calcite-carbonatite dikes. The crystallization temperature of the VSMs is determined in accordance with the graph of the dependence of the number of violating reflections on the crystallization temperature shown in Fig. 1.48a. The formation temperature of manaevite-(Ce), estimated as 280 °C, is discussed in detail by Moiseev et al. (2020). The resulting symmetry for manaevite-(Ce) is *P*4/*nnc* despite the low-temperature genesis is related to the equal number of the *P*4/*n* and *P*4*nc* domains in its structure (Table 1.5). Similar cases of resulting symmetry for low-temperature samples are also discussed in the works by Allen and Burnham (1992), Galuskin et al. (2003a), Britvin et al. (2003).

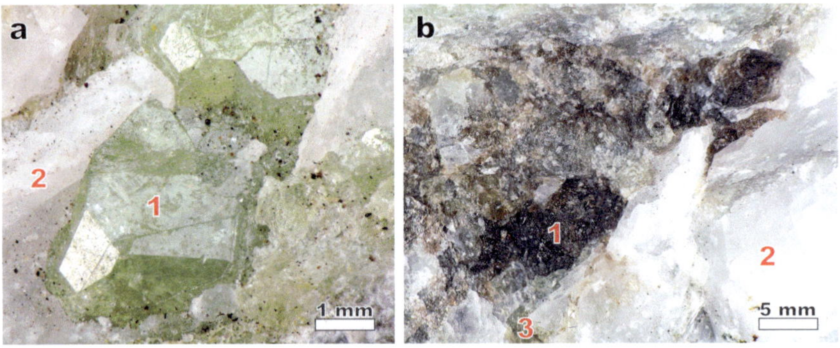

Fig. 4.33 Images of two main types of VSMs from the Kovdor massif: apple-green vesuvianite common in skarnoids (**a**) and brown REE-rich vesuvianite genetically related to the intrusion of calcite-carbonatite dikes (**b**). 1—vesuvianite, 2—calcite, 3—glagolevite

Table 4.6 Number of violating reflections for the *P4/nnc* space group, unit-cell parameters and *REE* content in VSMs from the Kovdor massif

Sample No	Type of VSM	Number of violating reflections	Space group	*a* (Å)	*c* (Å)	*REE (apfu)*
1	Early apple-green	274	*P4nc*	15.6051	11.8595	0.20
2	Core of brown *REE*-bearing vesuvianite from the vesuvianite-glagolevite rock	304	*P4/n*	15.7051	11.9015	1.76
3	Rim of brown *REE*-bearing vesuvianite from the vesuvianite-glagolevite rock	194	*P4/n*	15.7092	11.8940	2.30
4	Core of brown *REE*-bearing vesuvianite in calcite matrix	110	*P4/n*	15.8213	11.9209	2.32
5	Orange manaevite-(Ce) crusts	19	*P4/n*	15.8591	11.9686	4.13
6	Latest orange manaevite-(Ce) crusts	1	*P4/nnc*	15.9247	11.9661	4.68

The crystal structure refinement of VSMs from Kovdor showed a decreasing degree of ordering in the cationic rods *Y*1–*X*4–*X*4–*Y*1 with sample numbers from 1 to 6, according to their crystallization sequence, which indicates an increase in symmetry with the time. Simultaneously with the symmetrization (due to an increase in temperature) the admixture of rare-earth elements, accompanied by a partial substitution of Al with Mg, causes a linear increase in the unit cell parameters *a* and *c* of the minerals (Fig. 4.34).

The formation temperature of earlier green vesuvianite is estimated as 370 °C, which is in good agreement with both the general symmetry of the samples and the formation temperature of skarnoids according to the magnetite-ilmenite geothermometer (Fig. 4.35a). The symmetry increase (Fig. 4.35b) of the Kovdor samples and increase of the *REE* content in their composition with time of their formation is probably due to the alteration of skarnoids by late carbonatite melts, which caused both the rare-earth specialization of the VSMs and their relatively high-temperature crystallization (300–470 °C). The lowering in temperature during the formation of late low-temperature manaevite-(Ce) to 280 °C is associated with the end of the influence of late carbonatite melts and their cooling.

The increase in symmetry of the Kovdor VSMs and the increase in the content of *REE* in their composition with the time of their formation are probably related to the alteration of scarnoids under the influence of late carbonatite melts, which caused

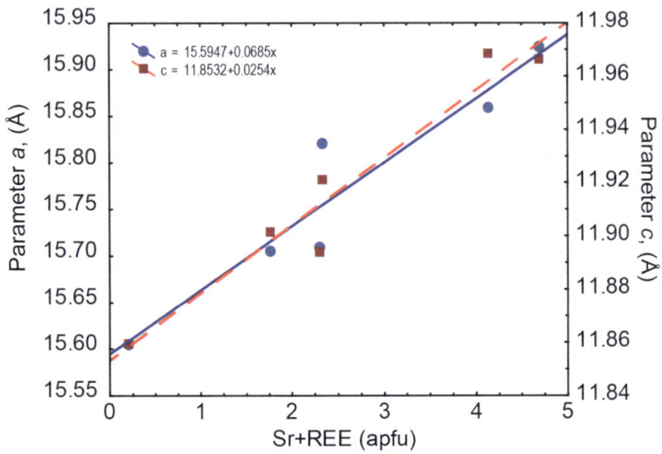

Fig. 4.34 Unit cell parameters *versus REE* content in VSMs from Kovdor

both the rare earth specialization of the VSMs and their relatively high-temperature crystallization (at 300–480 °C), with a consequent drop in temperature up to 280 °C during the formation of late low-temperature manaevite-(Ce), in contrast to the lower temperature of 370 °C for the earlier green vesuvianite.

Thus, the diffraction method for the estimation of the formation temperature of VSMs can be used for a more detailed study of geological processes. Despite the fact that the average temperature of formation of calcite-carbonatite dykes in the Kovdor massif, according to the magnetite-ilmenite thermometer, is 360 °C, in different evolution periods the temperature of rock formation varied widely, up to 480 °C.

4.4.2 *Geological Dating*

Contact-metasomatic rocks are widely distributed across all continents and hold significant industrial importance. They are associated with some of the largest deposits of Fe, W, Pb, Zn, and other metals. However, "direct" dating of skarns presents a complex challenge due to the scarcity of a reliable primary geochronometer minerals. Geochronological studies have predominantly focused on dating the ore minerals found within skarns, including molybdenite, titanite, wolframite, pyrrhotite, retrograde minerals (such as phlogopite and muscovite), rare accessory minerals like allanite and vesuvianite, as well as hydrothermal minerals such as zircon and baddeleyite. It is important to note that sulfide mineralization, in contrast to magnetite, is often classified as a superimposed type, indicating that the processes of mineralization are temporally distinct from those involved in skarn formation. Nonetheless, alongside pyroxene, B-bearing vesuvianite serves as the primary rock-forming mineral in ore skarns (Stifeeva et al. 2024b).

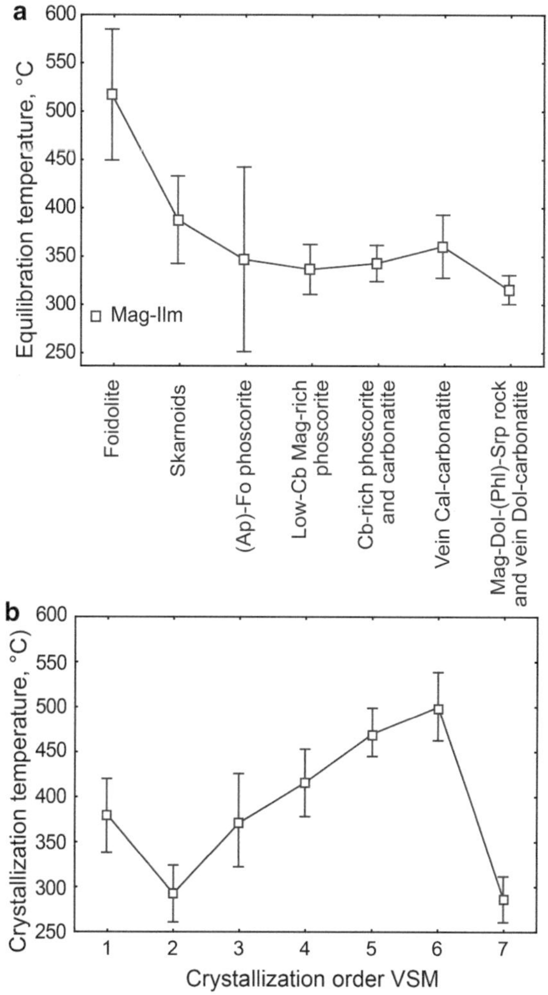

Fig. 4.35 The formation temperature of various rocks of the Kovdor massif according to the magnetite-ilmenite geothermometer (Ivanyuk et al. 2017) and the formation temperature of VSMs according to the dependence from the work by Panikorovskii et al. (2024b)

The use of VSMs as U–Pb geochronometers has become widespread in recent years (Zhang et al. 2022b; Han et al. 2025b). It is obvious that the presence of excessive amounts of common lead significantly complicates the use of VSMs for U–Pb geochronological studies, and existing correction methods cannot always be applied to the results obtained using local geochronological methods. It is also important to note that one of the key conditions ensuring the reliability and correctness of geochronological data obtained using local U–Pb research methods is the use of a certified single-matrix standard. At present, there is no generally applicable and well-studied vesuvianite standard for such studies.

We have carried out 30 U–Pb dating experiments using VSMs from different deposits and it turned out that low-temperature VSMs, for example, manaevite-(Ce) (~280 °C) and rodingite samples from the Ural mines (300–400 °C), are characterized by a high proportion of common lead. Suitable parameters for dating (the highest values of the proportion of inherited lead) were noted for VSMs with a relatively high content of B_2O_3 (> 1 wt%) and wiluite (with B_2O_3 > 2.4 wt%). Datings were obtained for skarns of Templtone (Canada), Yulia (Khakasia, Russia), Pitkäranta (Karelia, Russia), as well as Vilyui River basin and Titovskoe deposit (both in Yakutia, Russia).

Being a relatively early mineral, wiluite exhibits a high degree of stability during superimposed processes. It is typical of rocks in distal exoskarn zones, and provides direct estimates of the age of skarn formation, similar to how zirconium minerals provide information on the age of magmatic processes. Based on the results of detailed crystallographic and geochronological studies, it can be concluded that, due to its crystal chemical properties, wiluite has the greatest potential as a U–Pb geochronometer among the minerals of the vesuvianite group.

4.4.3 Reconstruction of Mineral Formation Conditions

Three types of vesuvianite (Fig. 4.36), differing in color, habit, or mineral assemblage, have been identified in the Loma Marcelo skarn, a group of polymetamorphosed carbonate xenoliths hosted by the Ediacaran calc-alkaline granites of the Sauce Chico Complex (Ventania System, Argentina). The three types correspond to high-F VSMs and show slight compositional differences, probably related to local variations in the protolith composition. The crystal structures of two samples (LM-9 and LM-10), with a predominance of Fe^{2+} at the *Y*1 site, were refined in the *P*4/*nnc* space group. Based on the number of reflections violating the extinction rules for the *P*4/*nnc* space group, the presence of domains with the *P*4/*n* symmetry, and the absence of hydrogarnet-type substitutions, crystallization temperatures (450 °C and 550 °C, respectively) were estimated for these samples. Although the third sample (LM-13) could not be structurally characterized, its Raman spectrum indicates low formation temperature (300–400 °C).

The three vesuvianite samples studied apparently reflect a retrograde sequence related to Early Cambrian contact metamorphism, according to the following sequence: LM-10 (type b, *ca.* 550 °C) → LM-9 (type a, *ca.* 450 °C) → LM-13 (type c, 300–400 °C). During the subsequent long-term Permian tectonometamorphic event, secondary fluid inclusions were entrapped, which contributed to the formation of stable isotopes. For two-phase fluid inclusions (Fig. 4.37) hosted in vesuvianite crystals, entrapment temperatures are between 270 and 339 °C at entrapment pressures of ca 250 MPa. Determination of stable H and O isotopes in vesuvianite concentrates shows that interaction with hydrous metamorphic fluids contributed to the change of $^{2}H/^{1}H$ ratios in hydroxyl groups at W-sites, while $^{18}O/^{16}O$ ratios in silicate groups did not change significantly (Justiniano et al. 2025a).

According to our results, symmetry does not affect the unit cell parameters of VSMs, as it was previously thought (Fig. 4.38). The Loma Marcelo skarn samples exhibit *a*/*c* ratios similar to those of *P*4*nc*, *P*4/*n* Mn-bearing and *P*4/*nnc* F-bearing

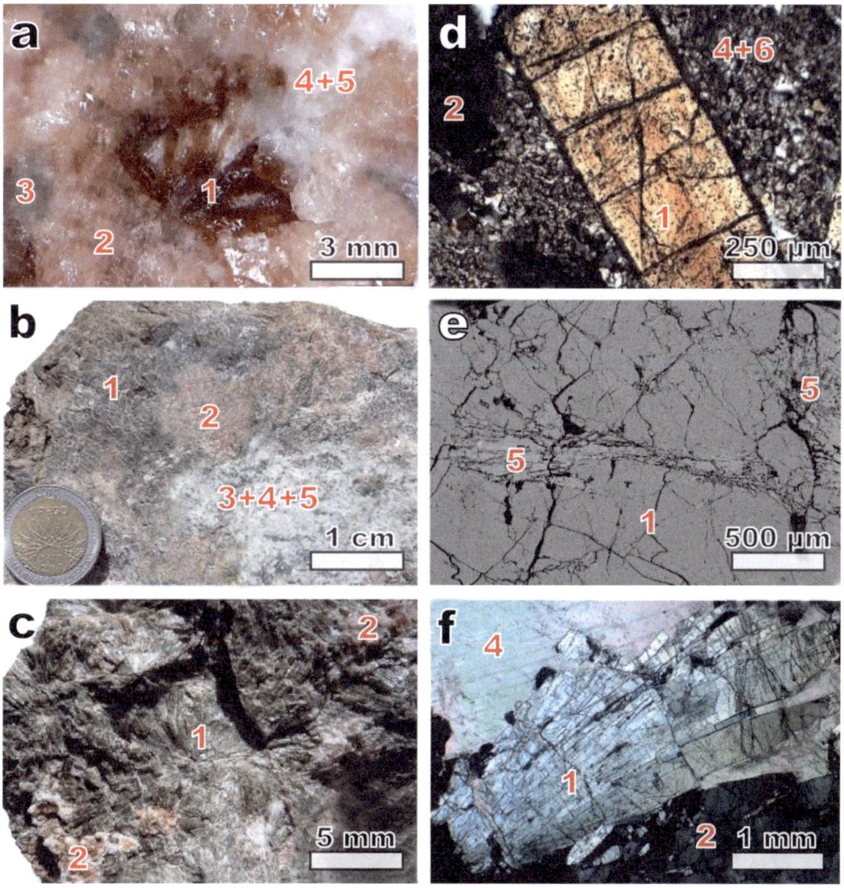

Fig. 4.36 Vesuvianite types from the Loma Marcelo skarn. **a** Reddish brown fan spherulite of LM-9 vesuvianite (hand specimen). **b** Brownish green irregular mass of LM-10 vesuvianite (hand specimen). The yellow dashed line shows the boundary between vesuvianite and a previous paragenesis composed of garnet (Grt), clinopyroxene (Cpx), wollastonite (Wo), and calcite (Cal). Coin diameter = 2.3 cm. **c** Green spherical spherulites of LM-13 vesuvianite (hand specimen). **d** Zoned and fractured prismatic crystal of type a vesuvianite (analyzed polarized light). **e** Wollastonite (Wo) relict hosted in a fractured crystal of type b vesuvianite (back-scattered electron image). **f** Fan spherulite of type c vesuvianite (analyzed polarized light). 1—vesuvianite, 2—grossular-andradite, 3—clinopyroxene, 4—calcite, 5—wollastonite, 6—quartz

vesuvianites, indicating the absence of symmetry influence. Finally, it should be noted that the Loma Marcelo skarn vesuvianite is highly stable under different conditions, with the structure fixed during its formation and unchanged thereafter, as evidenced by fluid inclusions and stable isotopes from a later long-term regional metamorphic event that did not affect its structure. This suggests that vesuvianite group minerals may serve as reliable X-ray diffraction-based thermometers and do not undergo structural changes after formation.

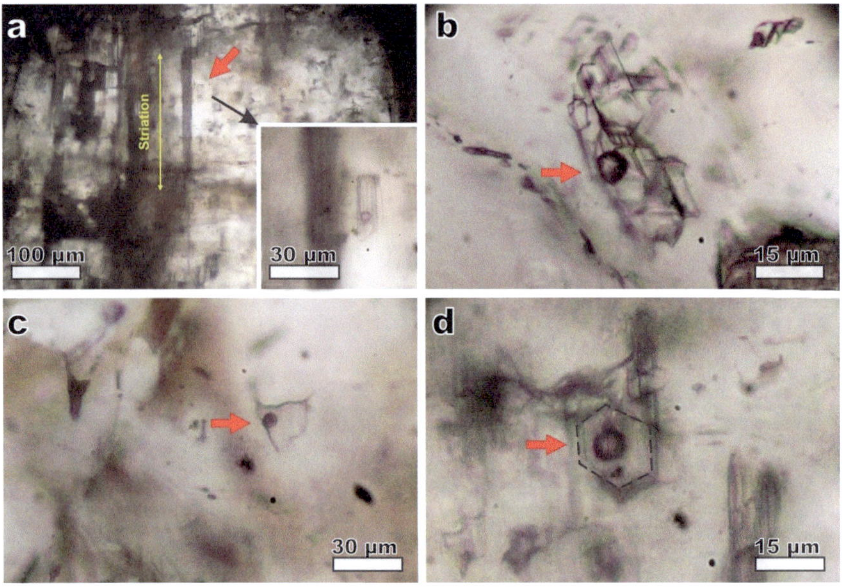

Fig. 4.37 Fluid inclusions in vesuvianite crystal from the Loma Marcelo skarn (type c, LM-13). **a** Tabular fluid inclusion parallel to the crystal striation. **b** Tabular fluid inclusion. **c** Irregular fluid inclusion. **d** Regular fluid inclusion (negative crystal)

4.4.4 Rapid Estimation of the Formation Temperature from the IR Spectrum

Traditionally, VSMs are divided into low-temperature, with crystallization temperatures of 250–450 °C and space groups *P*4/*n* and *P*4*nc*, and high-temperature, with crystallization temperatures of 450–800 °C and space group *P*4/*nnc* (Allen and Burnham 1992). We propose a new methodology that allows VSMs to be classified as high-temperature or low-temperature depending on their IR spectra.

Previous studies of the difference between IR spectra of low-symmetry (*P*4/*n* и *P*4*nc*) and high-symmetry (*P*4/*nnc*) VSMs in the region of stretching and bending vibrations of $(SiO_4)^{4-}$ and $(Si_2O_7)^{6-}$ groups, as well as in the stretching region of O–H bonds (Kurazhkovskaya et al. 2005; Borovikova and Kurazhkovskaya 2006) were partially confirmed by our subsequent studies (Chukanov et al. 2018a). The presence of a band about 3670 cm^{-1} in the IR spectra of low-symmetry VSMs noted in Borovikova (2005) is associated with an increase in the number of independent positions in them.

According to the Libowitzky (1999) correlations, the bands in the region of 3590–3670 cm^{-1} can be attributed to the weakest hydrogen bonds. The bands A–C observed in this range are polarized, with a small angle with respect to the *c* axis and probably belong to OH groups in the O11 position (Fig. 4.39). As a rule, in the IR spectra of high-symmetry VSMs, these bands have relatively low intensities. The band at

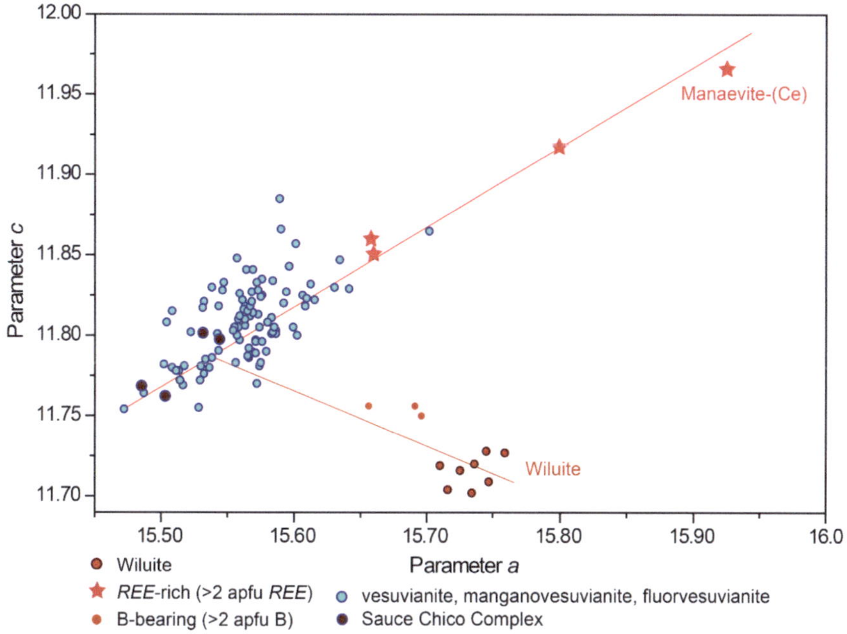

Fig. 4.38 VSMs unit cell parameters discriminated by the rock type and space group: *a*-axis length vs. *c*-axis length. The samples studied in this work are shown as yellow circles with their respective numbers. The unit cell parameters are compiled from Armbruster and Gnos (2000a, 2000b), Balassone et al. (2011), Britvin et al. (2003), Chukanov et al. (2019), Coda et al. (1970), Fitzgerald et al. (1986a, 1986b, 1987), Galuskin et al. (2003, 2007a), Giuseppetti and Mazzi (1983), Groat et al. (1992, 1994a, 1994b, 1996), Hålenius et al. (2013), Kurazhkovskaya et al. (2005), Lager et al. (1999), Ohkawa (1994), Ohkawa et al. (1992), Pavese et al. (1998), Rucklidge et al. (1975), and Warren and Modell (1931)

3628–3632 cm^{-1} (B band) is characteristic of low-symmetry VSMs spectra. The increase of intensities of the A–C bands in the spectra of low-symmetry samples can be explained by the enhancement of polarization of the O11–H1…O7 hydrogen bond compared to the high-symmetry VSMs (Table 4.7).

According to neutron diffraction data (Lager et al. 1999a), the ∠(DHA) angle for the O11–H1···O7 hydrogen bond in high-symmetry VSMs varieties is within 105–109°, i.e. less than the minimum angle value 120° for the formation of a weak hydrogen bond (Desiraju and Steiner 1999). In low-symmetric VSMs, the O11–H1···O7 bond strength increases with the ∠(O11H1O7) angle, reaching 148° in the *P*4*nc* vesuvianite structure from the Karmankul'skii cordon in South Urals (Panikorovskii et al. 2016c) and 142° in alumovesuvianite (Panikorovskii et al. 2017a).

Thus, in the IR spectra of low-temperature VSMs (250–450 °C) there are quite intense bands in the range of 3628–3690 cm^{-1}, whereas in the spectra of high-temperature VSMs (450–800 °C) these bands are not observed.

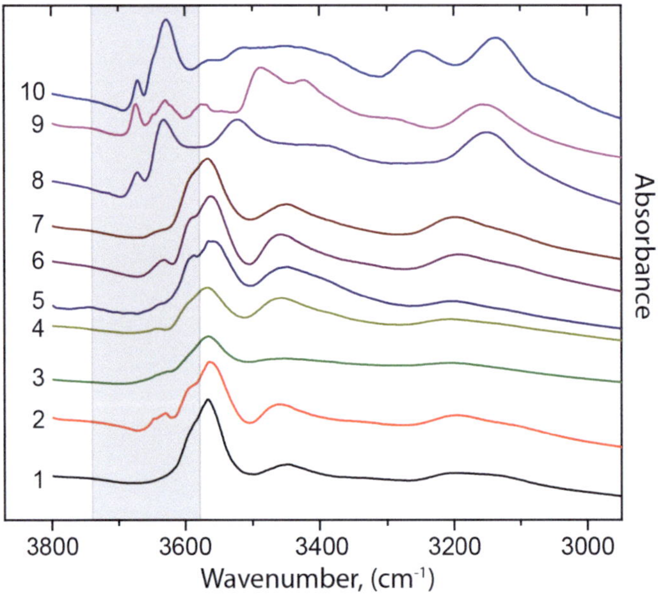

Fig. 4.39 IR spectra of VSMs samples with different symmetry in the region of O–H stretching vibrations: the curves 1–7 correspond to VSMs with the *P*4/*nnc* symmetry and the curves 8–10 to VSMs with the space group *P*4/*n*. The area of A–C bands is highlighted in gray. The symmetry of the samples was determined from the single-crystal XRD data

Table 4.7 Hydrogen bond parameters for *P*4/*nnc* vesuvianite according to Lager et al. (1999b)

D···A (Å)	∠(DHA) (°)	Bond
2.784(4)	109(3)	O11–H11…O7
2.720(5)	180	O10–H10…O10
3.038(4)	169(4)	O11–H11…O11

4.5 Conclusions

The role of VSMs as indicators of the crystallization temperature of metasomatic rocks has been confirmed. Indeed, once crystallized, vesuvianite retains information about the age of the rocks, the crystallization temperature, while inclusions carry information about pressure and the isotopic composition of fluids. We have demonstrated that wiluite or B-bearing vesuvianite is most suitable for U–Pb dating among VSMs. Based on the number of reflections violating extinction rules for the space group *P*4/*nnc*, a method for diffraction geothermometry has been developed and tested. An express method for determining the approximate crystallization temperature of metasomatic rocks has been developed based on IR spectroscopy data. Among the main trends, the changes of the VSMs unit cell parameterd have been identified, together with the proposed mechanism formula calculation, can be used in the new nomenclature of this supergroup.

Detailed crystal chemical studies of about 170 VSMs samples made it possible to determine the mechanisms of incorporation cations in structural positions and propose a sequence of cations in order of preferability of their occurrence at various sites. Sodium cations selectively occupy the $X1$ and $X2$, positions, whereas trivalent cations, REE^{3+}, Bi^{3+} and Sn^{3+} incorporate into the $X3$ position. Rare-earth elements can also partially incorporate into the $X4$ position. A *REE* analogue of vesuvianite has been synthesized, which confirms the patterns of *REE* entry into the $X3$ and $X4$ positions. The remaining X positions are populated by Ca atoms. The octahedral $Y2$ position is populated in the sequence $Al^{3+} \rightarrow Mn^{2+} \rightarrow Mg^{2+}$. The octahedral $Y3$ position is mainly filled by trivalent cations in the sequence $Al^{3+} \rightarrow Fe^{3+} \rightarrow Mn^{3+} \rightarrow Cr^{3+}$ with subordinate amounts of divalent cations whose average contents decrease in the sequence $Fe^{2+} \rightarrow Mg^{2+} \rightarrow Mn^{2+} \rightarrow Zn^{2+}$. The Ti^{4+} cations also incorporate into the $Y3$ position. Additional tetrahedral $T1$ positions contain $B^{3+} \rightarrow Al^{3+}$. Positions with triangular coordination can be populated by B^{3+}. Chloride anions replace OH^- groups only at the OH10 position, while F^- can replace hydroxyl ion at OH11 and OH10 positions.

Modular character of the crystal structures of VSMs determines their physical and crystal chemical properties. The presence of cationic rods $Y1$–$X4$–$X4$–$Y1$ with incomplete occupancy of the $Y1$ and $X4$ positions is an important stabilizing factor, which compensates stresses arising from the unequal distribution of cations in channel positions by cooperative crystal chemical adaptation. Another factor explaining a high stability of the VSMs structures is the presence of complex substitution schemes. For example, the incorporation of large cations in X positions is compensated by Al replacing Si in tetrahedra and by Mg replacing Al in octahedra.

Owing the population of the $Y1$ polyhedron by different cations (in the sequence $Al^{3+} \rightarrow Fe^{3+} \rightarrow Mn^{3+} \rightarrow Cu^{2+} \rightarrow Mg^{2+} \rightarrow Fe^{2+} \rightarrow Mn^{2+}$), its configuration widely changes and the polyhedral volume variations can exceed 30%.

The limit and mechanism of incorporation of the $[H_4O_4]^{4-}$ clusters into VSM structure is was studied in detail for the Si-deficient vesuvianite epitaxial crusts grows on wiluite. Among the eight studied Si-deficient VSMs, the maximal degree of the $[H_4O_4]^{4-} \leftrightarrow [SiO_4]^{4-}$ substitution (up to 40%) was observed in the sorosilicate $Z1$ and $Z2$ positions (Fig. 3.67), mainly in the first one. As a result of this substitution, the mean $< Z1,2\text{–O} >$ bond lengths can reach 1.710 Å, which significantly exceeds the lengths of similar bonds (1.62–1.64 Å) in the tetrahedral occupied completely by Si atoms. In the positions of the nesosilicate groups ($Z3$), a significant increase in the $\langle Si\text{–}O \rangle$ bond length does not occur; therefore, the maximum theoretical limit for the substitution of the hydrogarnet type in the structure of the VSMs is limited by the number of sorosilicate positions. As compared to the solid-state ^1H NMR spectrum of magnesiovesuvianite, the spectrum of Si-deficient vesuvianite has an additional peak of 3.38 ppm (Fig. 3.68), which corresponds to the appearance of an additional hydrogen position belonging to the $[H_4O_4]^{4-}$ cluster. The calculation of the O···O distance in the O–H···O hydrogen bond corresponding to the additional ^1H NMR band gives the value of 2.96 Å, which is close to the analogous bonds in synthetic katoite,

$Ca_3Al_2(SiO_4)_{3-x}(OH)_{4x}$ (x = 1.5–3). In contrast to the overwhelming majority of VSMs without Si deficiency, the samples of "hydrovesuvianite" showed an abnormally high content of H_2O (4.30 wt%) and two-stage mass loss. The first stage of mass loss (2.08 wt%) in the range 300–600 °C corresponds to the decomposition of the $[H_4O_4]^{4-}$ groups, the second stage and the corresponding endothermic peak with the minimum at a temperature of 1014 °C (2.22 wt%) corresponding to the loss of OH^- groups at the O10 and O11 sites are followed by the melting of the mineral.

For low-symmetry VSMs modifications (space groups *P*4/*n* and *P*4*nc*) the phenomenon of cooperative crystal-chemical adaptation takes place. The reduction of symmetry is associated with the ordering of cations in the chains of the group of channel positions *Y*1-*X*4-*X*4-*Y*1. Due to the different populations of square antiprisms *X*4(A,B) located in the channels of the structure, their volume varies widely from 22.9 to 28.7 Å^3. A change in the volume of the polyhedra causes stresses in the structure and, as a result, a change in the polyhedral volume of the *Y*3(A,B) octahedra The increase in the polyhedral volume of the pair of positions *X*4A + *Y*3A is accompanied by the decrease of the polyhedral volume of the pair of positions *X*4B + *Y*3B (Table 3.11).

The VSMs crystallized in the *P*4/*n* space group are characterized by the formation of merohedral twins. Twinning occurs on the two-fold axis along the [110] direction; the corresponding twinning matrix is (010 100 00-1). Refining the structure using twinning matrix reduced the value of *R*-factor from R_{1mean} = 0.053 (excluding twinning) to R_{1mean} = 0.03. The formation of merohedral twins is characteristic of the most highly ordered samples, the degree of order of which can be estimated by the formula $K_O = |\delta(X4A) - \delta(X4B)| \cdot 100\%$, where δ(*X*4A) and δ(*X*4B) are the occupancy values of the corresponding positions. The distribution of the values of the ordering coefficient clearly indicates the presence of merohedral twins in samples with $K_O > 68\%$ (Fig. 3.71b).

With decreasing in symmetry, the role of weak hydrogen bonds in VSMs increases, which causes a regular change in the IR spectra in the region of stretching vibrations of the O–H bonds. The absence of bands in the region of 3600–3670 cm^{-1} in the IR spectra of a high-symmetric vesuvianite (space group *P*4/*nnc*) was previously noted. The absence of even weak hydrogen bonds in the *P*4/*nnc* modification was determined by neutron diffraction: the angle ∠(DHA) for the O11–H1···O7 hydrogen bond is within 105–109°, i.e. less than the minimum angle of 120° for the formation weak hydrogen bond. The strength of the O11–H1···O7 hydrogen bond in low-symmetry vesuvianite increases with the angle ∠O11–H1···O7, reaching 148° in the *P*4*nc* structure of the vesuvianite from the Karmankul'skii cordon, South Urals and 142° in the structure of alumovesuvianite. For the aforementioned low-symmetry VSM, bands in the region of 3600–3670 cm^{-1} are clearly observed. This regularity is permanent, and it can be used to determine whether the VSM belongs to high-symmetry (space group *P*4/*nnc*) or low-symmetry modifications (space group *P*4/*n* and *P*4*nc*) using IR spectroscopy.

The use of solid-state NMR methods made it possible to determine the quantitative ratios of elements with different coordination numbers (CN): aluminum (CN = 4, 5, 6), boron (CN = 3, 4), sodium (CN = 8) and lanthanum (CN = 9). This confirms the critical importance of this method for the identification of VSMs in the absence of detailed structural data and for the development of the current VSMs nomenclature.

The study of the VSMs crystal chemistry by a complex of methods made it possible to confirm many conclusions of previous authors, as well as describe the mechanisms of the incorporation of Na^{+}, $S_2^{\cdot-}$, REE^{3+}, $H_4O_4^{4-}$ into the vesuvianite structure and their role in the structure for the first time. Determining the order of incorporation of cations and anions into various structural positions as well as the proposed method for calculating iron state according to the formula $Fe^{2+}/Fe^{3+} = 5.1816 \times \exp(-1.0156 \times Mg(apfu)$ provided the basis for the VSMs and rules for the calculation of the VSMs formulae having only data on the chemical composition of the mineral.

Different in color, crystal shape and chemical composition generations of VSMs were studied in the Phlogopite pit of the Kovdor alkaline massif. Primary apple-green vesuvianite (space group *P*4*nc*) is replaced by a *REE*-bearing brown vesuvianite prismatic habit (space group *P*4/*n*) and then with late crusts of orange manaevite (Ce) (space group *P*4/*nnc*) growing on the latter. A gradual increase increase in the *REE* content is accompanied by an increase of the unit cell parameters and an increase in their overall symmetry in the sequence *P*4*nc* → *P*4/*n* → *P*4/*nnc*. This transition is apparently associated with the incorporation of late carbonatite melts, which caused both the rare-earth specialization of the late VSMs. The crystallization temperature of VSMs in skarnoids is 370 °C, which is consistent with magnetite-ilmenite geothermometer data. And the intrusion of carbonatite melts first caused an increase in the crystallization temperature of VSMs in the range (300–470 °C), and then a decrease to 280 °C during the formation of manaevite (Ce).

Acknowledgements The study was carried out at the Department of Crystallography of St. Petersburg State University with financial support by state contract No. 16.518.11.7096, Russian Science Foundation grant 14-17-0071, Russian President grant for young candidates of Sciences MK-3296.2015.5, internal SPSU grant 3.38.136.2014, as well as scholarships of the Government of the Russian Federation (2015–2016). The part work related to garnet-vesuvianite relationships supported by MK-6240.2021.1.5 grant. A part of this work related to IR spectroscopy and scientific and technical text editing was carried out in accordance with the state task, state registration number 124013100858-3. Preparation of the parts related diffraction geothermometer were carried out in accordance with the state task, registration number FMEZ-2025-0046. At the different stages of the work the author was helped by A. A. Agakhanov, E. Yu. Avdontseva, S. M. Aksenov, M. Annenburg, G. F. Anastasenko, A. A. Antonov, C. A. Ballivián Justiniano, A. V. Bazai, G. V. Barkhudarova, E. V. Belogub, V. V. Bortnikov, S.N. Britvin, O. G. Bubnova, A. P. Chernyateva, U. Halenius, E. V. Galuskin, I. O. Galuskina, M.E. Generalov, O.S. Grunskiy, V.V. Gurzhy, P. Dzierżanowski, Yu. V. Erokhin, A. A. Zolotarev Jr., G. Yu. Ivanyuk, V. Yu. Karpenko, S. V. Krivovichev, A. S. Mazur, M. M. Moiseev, D. V. Spiridonova, G. L. Starova, M. V. Stifeeva, I. V. Pekov, N. V. Platonova, V. V. Shilovskikh, M. A. Vetrova, N. S. Vlasenko, and V. N. Yakovenchuk, which support is very appreciated by author. In addition, the author thanks the library of St. Petersburg state University for the opportunity of free access to relevant literature and emphasizes that this work would not have been possible without the understanding and support of all employees of the Department of Crystallography and "X-Ray Diffraction Research Centre" of St. Petersburg state University. I would like especially thanks my family: mother Olga and father Leonid, my wife Viia and children Aglaia, Ostap, Lisaveta and Rodion Panikorovskii.

Appendix

See Tables A.1, A.2, A.3, A.4, A.5, A.6, A.7, A.8, A.9, A.10, A.11, A12, A.13, A.14, A.15, A.16, A.17, A.18, A.19, A.20, A.21, A.22, A.23, A.24, A.25, A.26, A.27, A.28, A.29, A.30, A.31, A.32, A.33, A.34, A.35, A.36, A.37, A.38, A.39, A.40, A.41, A.42, A.43, A.44, A.45, A.46, A.47, A.48, A.49, A.50, A.51, A.52, A.53, A.54, A.55 and A.56.

T. L. Panikorovskii and N. V. Chukanov, *Vesuvianite-Supergroup Minerals*, Springer Mineralogy, https://doi.org/10.1007/978-3-032-06970-2

Table A.1 Structural complexity of VSM for *P*4/*nnc*, *P*4/*n* and *P*4*nc* modifications

Sample	Number of atoms (v)	I_G (bit/atom)	$I_{G,Total}$ (bit/cell)
Alumovesuvianite	274	5.244	1436.861
banat_nnc	284	4.389	1246.528
crust_on_wil_nnc	256	4.188	1072.000
Lager_1999_p4nnc	276	4.311	1189.953
mo_ves_ree_2_0m_n	270	5.255	1418.740
razsheplenie(wil_bereg)_nnc	298	4.427	1319.312
ves_ahm(2)(28975)-n	256	5.141	1316.000
ves_ahm(2)(28975)-nnc	258	4.221	1088.897
Ves_ak_chag_p4n	260	5.192	1349.816
Ves_ak_chag_p4nc	274	5.200	1424.861
Ves_ak_chag-without_H	256	5.094	1304.000
ves_alain_zanzo-thel(13988)-n	256	5.141	1316.000
ves_alain_zanzo-thel(13988)-nnc	258	4.221	1088.897
ves_an_bazh(cr)2 (2)-nnc	258	4.221	1088.897
ves_an_bazh(cr)2-n	254	5.115	1299.126
ves_asbestos(70026)_n	258	5.166	1332.897
ves_asbestos(70026)-nc	256	5.094	1304.000
ves_at_koy_xcal_nnc	256	4.188	1072.000
ves_azegous-2(67495)_n	256	5.141	1316.000
ves_azegous-2(67495)_nnc	256	4.188	1072.000
ves_barb(23700)_n	256	5.141	1316.000
ves_barb(23700)_nc	258	5.120	1320.897
ves_barb_new(23700)_n	256	5.141	1316.000
ves_barb_new(23700)_nnc	258	4.221	1088.897
ves_bazh(79479)_n	256	5.141	1316.000
ves_bazh(79479)_nc	254	5.067	1287.126
ves_bazh(79479)_nnc	258	4.221	1088.897
ves_bazh(antonov)_n	256	5.141	1316.000
ves_bazh(antonov)-nc	266	5.161	1372.705
ves_bazh(antonov)-nnc	258	4.221	1088.897
ves_bazh_new(79478)_n	256	5.141	1316.000
ves_bazh_new(79478)_nc	254	5.067	1287.126
ves_bazh_new(79478)_nnc	258	4.221	1088.897
ves_bellecombe(79621)_n	254	5.115	1299.126
ves_bellecombe(79621)_nc	272	5.176	1407.790
ves_bellecombe(79621)_nnc	258	4.221	1088.897
ves_blag(7_1 spuck)_nnc	268	4.320	1157.712

(continued)

Table A.1 (continued)

Sample	Number of atoms (v)	I_G (bit/atom)	$I_{G,Total}$ (bit/cell)
Ves_blag_n	266	5.206	1384.705
ves_blag_nnc	268	4.320	1157.712
ves_blag2_nnc	266	4.288	1140.705
ves_br_ptk_0m_nnc	256	4.188	1072.000
ves_collo_di_palom_nnc	282	4.359	1229.353
ves_crust(3_wiluites)_nnc	260	4.253	1105.816
ves_crust_on_wil_nnc	292	4.395	1283.429
ves_cyp(nor_fersmana)2_nnc	264	4.257	1123.720
Ves_cyp__mg_nc	270	5.210	1406.740
ves_cyp_mg_n	258	5.166	1332.897
ves_cyp_mn_n	254	5.115	1299.126
ves_erem_n	256	5.141	1316.000
ves_erem_nc	256	5.094	1304.000
ves_erem_nnc	258	4.221	1088.897
ves_F(fluorvesuvianite)_n	256	5.141	1316.000
ves_F(fluorvesuvianite)_nc	258	5.120	1320.897
ves_F(fluorvesuvianite)_nnc	256	4.188	1072.000
ves_Falza(56-3067)_nnc	266	4.288	1140.705
ves_frag_el_ma(70024)_nnc	256	4.188	1072.000
ves_frugard(13973)_nnc	256	4.188	1072.000
ves_gazlov(31902)_n	256	5.141	1316.000
ves_gazlov(31902)_nnc	256	4.188	1072.000
ves_georgetown(51073)_n	254	5.115	1299.126
ves_georgetown(51073)_nc	272	5.176	1407.790
ves_georgetown(51073)_nnc	258	4.221	1088.897
ves_gl(eger)14687_nnc	256	4.188	1072.000
ves_gl(svetovody)_n	254	5.115	1299.126
ves_gl(svetovody)_nnc	258	4.221	1088.897
ves_gl(vapenna jesenik)_nnc	256	4.188	1072.000
ves_gl_birhin(yellow)_n	266	5.206	1384.705
ves_gl_birhin(yellow)_nc	264	5.135	1355.720
ves_gl_birhin_new_n	266	5.206	1384.705
ves_gl_birhin_new_nnc	268	4.320	1157.712
ves_gl_doveren_new_nc	266	5.161	1372.705
ves_gl_irkutsk-dark_nnc	266	4.288	1140.705
ves_gl_landfall_nnc	284	4.389	1246.528
ves_gl_sludyanka_nnc	264	4.257	1123.720

(continued)

Table A.1 (continued)

Sample	Number of atoms (v)	I_G (bit/atom)	$I_{G,Total}$ (bit/cell)
ves_gl_stara_kamenica_n	256	5.141	1316.000
ves_gl_will_auto_zone-yellow_n	280	5.258	1472.199
ves_gl_will_auto_zone-yellow_nc	272	5.176	1407.790
ves_gl_will_auto_zone-yellow_nnc	268	4.320	1157.712
ves_gl_yellow(iordanov)_nc	272	5.176	1407.790
ves_gl_yellow(iordanov)_nnc	258	4.221	1088.897
ves_gtl(husha_gol)_nnc	260	4.253	1105.816
ves_gtl(husha_gol_n)	254	5.115	1299.126
ves_guan_sea(58654)_nnc	256	4.188	1072.000
ves_gukun(16908)_nnc	282	4.359	1229.353
ves_gulshad(43180)_nnc	268	4.320	1157.712
Ves_haslan(15978)_nnc	256	4.188	1072.000
ves_hessen(51056)_nnc	256	4.188	1072.000
ves_igul(43974)_n	254	5.115	1299.126
ves_Jandrik(29112)_nnc	264	4.257	1123.720
ves_jef(big-lil)_n	256	5.141	1316.000
ves_kar(21155)_n	274	5.244	1436.861
ves_kar_0m_sx_nc	272	5.176	1407.790
ves_ked(41304)_n	264	5.181	1367.720
ves_ked(41304)_nnc	268	4.320	1157.712
ves_kedabek(41323_nnc	258	4.221	1088.897
ves_keyvy(707)_nnc	256	4.188	1072.000
ves_keyvy_nnc	256	4.188	1072.000
ves_keyvy-976_nnc	256	4.188	1072.000
ves_khewza(65273)_n	256	5.141	1316.000
ves_kimito (13985)_nnc	264	4.257	1123.720
ves_kluchev(27952)_nc	272	5.176	1407.790
ves_korev(1557)_n	256	5.141	1316.000
ves_kov_ree-last2_nnc	268	4.320	1157.712
ves_kov-new(yakovencuk)_nc	272	5.176	1407.790
ves_pk_kov_REE_n	264	5.181	1367.720
ves_pk_kov_REE_new_n	264	5.181	1367.720
Ves_pk_kov_REE-last1_nnc	268	4.320	1157.712
ves_razshepleny_nnc	276	4.326	1193.953
ves_templeton(31413)_nnc	264	4.257	1123.720
ves_wil2(xcal)_n	282	5.281	1489.353
ves_will_auto_225-19594_n	280	5.258	1472.199

(continued)

Table A.1 (continued)

Sample	Number of atoms (v)	I_G (bit/atom)	$I_{G,Total}$ (bit/cell)
ves_will_auto_GV-7_nnc	268	4.320	1157.712
ves_will_auto_nnc	284	4.389	1246.528
ves_will_GV-8_nnc	268	4.320	1157.712
ves-wil_mother2(Mo-SN)_nnc	284	4.389	1246.528
white_crust_nnc	292	4.395	1283.429
will_auto-225-19594_n	272	5.220	1419.790
wiluite-1_nnc	284	4.389	1246.528

Table A.2 Studied samples, their occurrences and availability of the data on their chemical composition and crystal structures

Locality	Museum/collection number	Composition	Structure	Source	Remarks
Kristiansand, Vest-Agder (Norway)	13975	+	+	Fersman Museum	
Titovskoe locality, Tas-Khayakhtakh, Yakutia (Russia)	68750	+	+	Fersman Museum	
Savvateevo, East Transbaikalia (Russia)	57348	+	+	Fersman Museum	
Østerbø, Vest-Agder, Aurland (Norway)	67940	+	+	Fersman Museum	
Kemiö Island (Finland)	29112	+	+	Fersman Museum	
Gukunvara (Finland)	16908	+	+	Fersman Museum	
Myrseter area, Drammen, Buskerud (Norway)	60211	+	+	Fersman Museum	
Frugård, Mäntsälä, Etelä-Suomen Lääni, (Finland)	13973	+	+	Fersman Museum	
Bensheim, Hesse (Germany)	51056	+	+	Fersman Museum	
Vesuvio, Somma-Vesuvius Complex, Naples Province (Italy)	5228	+	+	Fersman Museum	
Maderanertal valley, Uri, (Shwitzerland)	27726	+	+	Fersman Museum	
Monte Rosa massif, Valais (Shwitzerland)	31407	+	+	Fersman Museum	
Maderanertal valley, Uri, (Shwitzerland)	27727			Fersman Museum	

(continued)

Table A.2 (continued)

Locality	Museum/collection number	Composition	Structure	Source	Remarks
Bellecombe, Châtillon (Italy)	79621	+	+	Fersman Museum	
Vesuvio, Somma-Vesuvius Complex, Naples Province (Italy)	27837	+	+	Fersman Museum	
Ala di Stura, Lanzo Valleys, Piedmont (Italy)	13988	+	+	Fersman Museum	
Collo di Palombaia, Elba Island, Tuscany (Italy)	28920	+	+	Fersman Museum	
Varaita Valley, Piedmont (Italy)	13963	+	+	Fersman Museum	
Vesuvio, Somma-Vesuvius Complex, Naples Province (Italy)	27844	+	+	Fersman Museum	
Eger, Karlovy Vary Region (Czech Republic)	15978	+	+	Fersman Museum	
Canzoccoli Quarry, Predazzo, Fiemme Valley (Italy)	27728	+	+	Fersman Museum	
Monzoni, Predazzo (Italy)	7359	+	+	Fersman Museum	
Loyane (Macedonia)	60294	+	+	Fersman Museum	Magnesiovesuvianite (holotype)
Loyane (Macedonia)	59783	+	+	Fersman Museum	
Jeffrey mine, Asbestos (Canada)	70026	+	+	Fersman Museum	
Templeton, Québec (Canada)	31413	+	+	Fersman Museum	
Obira mine, Oyta prefecture, (Japan)	54541	+	+	Fersman Museum	
Huo Xiang, Huangshi Prefecture, Hubei Province (China)	58654		+	Fersman Museum	

(continued)

Table A.2 (continued)

Locality	Museum/collection number	Composition	Structure	Source	Remarks
Frag el Ma, Marrakech-Safi Region (Marocco)	70024	+	+	Fersman Museum	
Rila Mt., Urdine Ezera, Kyustendil (Bulgaria)	77291	+	+	Fersman Museum	
Banat Mts., Caraş-Severin (Romania)	51109	+		Fersman Museum	
Kosmet, Kosovo (Serbia)	55211	+	+	Fersman Museum	
Azegour Mine, Souss-Massa Region (Marocco)	67495	+	+	Fersman Museum	
Kluchevskii mine, Monetnaya Dacha, Midlle Urals (Russia)	55291	+	+	Fersman Museum	
Siegel lode, Georgetown, California (USA)	51073	+	+	Fersman Museum	
Banat Mts., Caraş-Severin (Romania)	51108	+	+	Fersman Museum	
Loyane (Macedonia)	60295	+	+	Fersman Museum	
Gökum limestone quarry, Uppland (Sweden)	13967	+	+	Fersman Museum	
Fassa Valley, Trentino-Alto Adige (Italy)	51110	+		Fersman Museum	
Li-Uenthal (?), Tirol (Switzerland)	51085	+	+	Fersman Museum	
Hazlov, Karlovy Vary Region (Czech Republic)	61902	+	+	Fersman Museum	

(continued)

Table A.2 (continued)

Locality	Museum/collection number	Composition	Structure	Source	Remarks
Kemiö Island (Finland)	13985	+	+	Fersman Museum	
Oravița Skarns, Banat Mts, Caraș-Severin (Romania)	5136	+	+	Fersman Museum	
Kluchevskii mine, Monetnaya Dacha, Midlle Urals (Russia)	27952	+	+	Fersman Museum	
Ulug-Tanzek, Sangilen, Tuva (Russia)		+	+	Karpenko V. Yu	
Vesuvio, Somma-Vesuvius Complex, Naples Province (Italy)	51062	+	+	Fersman Museum	
Polyakovskii mine, South Urals (Russia)	26478	+	+	Fersman Museum	
Mt. Rila, Urdine Ezera, Kyustendil (Bulgaria)	77291	+	+	Fersman Museum	
Kristiansand, Vest-Agder (Norway)	14001	+	+	Fersman Museum	
Lopan gorge, Transcaucasia (South Osetia)	33623	+	+	Fersman Museum	
Kedabek, Magara, (Azerbaijan)	41323	+	+	Fersman Museum	
Lupikko, Karelia (Russia)	54594		+	Fersman Museum	
Sakharjok massif, Keivy, Kola Peninsula (Russia)	50578	+	+	Fersman Museum	
Nizhne-Isetskaya dacha, Middle Urals (Russia)	33530	+	+	Fersman Museum	

(continued)

Table A.2 (continued)

Locality	Museum/collection number	Composition	Structure	Source	Remarks
Asbestovkaya Kop', Govorukhinskiy razres, Korevo, Middle Urals (Russia)	23675	+	+	Fersman Museum	
Asbestovkaya Kop', Govorukhinskiy razres, Korevo, Middle Urals (Russia)	1557	+	+	Fersman Museum	
Kedabek, Magara, (Azerbaijan)	41304	+	+	Fersman Museum	
N. Kommercheskii mine, Blagodat' Mt., Middle Urals (Russia)	50624		+	Fersman Museum	
Barbotova Yama, Shishimskie Mts, South Urals (Russia)	23700	+	+	Fersman Museum	
Leykhtenbergitovaya kop', Shishimskiye Mts. South Urals (Russia)	9518	+	+	Fersman Museum	
Akhmatovskaya kop', South Urals (Russia)	27720	+	+	Fersman Museum	
Lyangar, Nurata (Uzbekistan)	40409	+	+	Fersman Museum	
Igikul', Ilmenskii park, South Urals (Russia)	43974	+	+	Fersman Museum	
Bazhenovskoe deposit, Middle Urals (Russia)	79478	+	+	Fersman Museum	
Sierra de la Cruz, Coahuila, Coahuila (Mexico)	81432	+		Fersman Museum	Reshoot, small mosaic crystalls

(continued)

Table A.2 (continued)

Locality	Museum/collection number	Composition	Structure	Source	Remarks
Hindubagh, Balochistan (Pakistan)	65272	+	+	Fersman Museum	
Rila Mt., Urdine Ezera, Kyustendil (Bulgaria)	73599	+	+	Fersman Museum	
Monetnaya Dacha, Midlle Urals (Russia)	17344	+	+	Fersman Museum	
Khewra, Punjab Region, Jhelum District (Pakistan)	65273	+	+	Fersman Museum	
Kristiansand, Vest-Agder (Norway)	15377	+	+	Fersman Museum	
Dahu Unur, Tuva (Russia)	62632	+		Fersman Museum	Metamict
Dahu Unur, Tuva (Russia)	64248	+		Fersman Museum	Metamict
Dahu Unur, Tuva (Russia)	61060	+		Fersman Museum	Metamict
Monetnaya Dacha, Midlle Urals (Russia)	17342	+	+	Fersman Museum	
Dahu Unur, Tuva (Russia)	61501			Fersman Museum	Metamict
N. Kommercheskii mine, Blagodat' Mt., Middle Urals (Russia)	50625	+	+	Fersman Museum	
Akhmatovskaya kop', South Urals (Russia)	28975	+	+	Fersman Museum	
Eremeevkaya kop', Nyazemskie Mts., South Urals, (Russia)	18983	+	+	Fersman Museum	

(continued)

Table A.2 (continued)

Locality	Museum/collection number	Composition	Structure	Source	Remarks
Berezovkaya dacha, Middle urals (Russia)	14258	+		Fersman Museum	Garnet
Gul'shad, Northern Balkhash (Kazakhstan)	43180	+	+	Fersman Museum	
Ak-Chagyl, Northern Balkhash (Kazakhstan)	38476	+	+	Fersman Museum	
Asbestovkaya Kop', Govorukhinskiy razres, Korevo, Middle Urals (Russia)	23677	+	+	Fersman Museum	
Bazhenovskoe deposit, Middle Urals (Russia)	80010			Fersman Museum	Diopside
Bazhenovskoe deposit, Middle Urals (Russia)	79479	+	+	Fersman Museum	
Nyazemskie Mts., South Urals (Russia)	67964	+	+	Fersman Museum	
Nikolae-Maximilyanovkaya kop', Nyazemskie Mts. South Urals (Russia)	67560	+	+	Fersman Museum	
Polyakovkiy mine, South Urals (Russia)	32801	+	+	Fersman Museum	
Karmankul'skiy cordon, South Urals (Russia)	21155	+	+	Fersman Museum	
Nyazemskie Mts. South Urals (Russia)	11805	+	+	Fersman Museum	

(continued)

Table A.2 (continued)

Locality	Museum/collection number	Composition	Structure	Source	Remarks
Sakharjok massif, Keivy, Kola Peninsula (Russia)	976	+		Bel'kov Museum	
Sakharjok massif, Keivy, Kola Peninsula (Russia)	707	+	+	Bel'kov Museum	
Alchuri, Shigar Valley (Pakistan)		+	+	Chukanov N. V	
Wessels mine, Kalahari (RSA)		+	+	Pekov I. V	Cyprine (Mn-zone)
Wessels mine, Kalahari (RSA)		+	+	Pekov I. V	Cyprine, holotype
Lupikko, Karelia (Russia)			+	Britvin S. N	
Lupikko, Karelia (Russia)	189/12409	+	+	SPSU Museum	
Lupikko, Karelia (Russia)		+	+	Yakovenchuk V. N	Fluorvesuvianite (holotype)
Bazhenovskoe deposit, Middle Urals (Russia)		+	+	Agakhanov A. A	
Mica Pit, Kovdor massif, Kola Peninsula (Russia)		+	+	Agakhanov A. A	
Hirata outcrop, Kushiro area, Hiroshima prefecture (Japan)		+		Pekov I. V	Pseudomorphs after vesuvianite
Mica Pit, Kovdor massif, Kola Peninsula (Russia)		+	+	Yakovenchuk V. N	
Wiluy River, Yakutia (Russia)		+	+	SPSU Museum	
Koytezek, Pamir (Tajikistan)			+	Agakhanov A. A	
Wiluy River, Yakutia (Russia)		+	+	Barkhudarova G. V	
Jeffrey mine, Asbestos (Canada)		+	+	Pekov I. V	
Lupikko, Karelia (Russia)		+	+	Pekov I. V	

(continued)

Table A.2 (continued)

Locality	Museum/collection number	Composition	Structure	Source	Remarks
Bazhenovskoe deposit, Middle Urals (Russia)		+	+	Pekov I. V	
Bazhenovskoe deposit, Middle Urals (Russia)		+	+	Antonov A. A	
Bazhenovskoe deposit, Middle Urals (Russia)		+	+	Antonov A. A	
Monetnaya Dacha, Midlle Urals (Russia)	186/12298	+	+	SPSU Museum	
Akhmatovskaya kop', South Urals (Russia)	198/14298	+	+	SPSU Museum	
Kristiansand, Vest-Agder (Norway)		+	+	SPSU Museum	
Øvstebø, Kleppan, Telemark (Norway)	171/11184	+	+	SPSU Museum	
Wiluy River, Yakutia (Russia)	202/14107	+	+	SPSU Museum	
Karmankul'skiy cordon, South Urals (Russia)	178/11282	+	+	SPSU Museum	
Shishimskaya kop', Shishimskie Mts., South Urals (Russia)	180/12292	+	+	SPSU Museum	
Ak-Chagyl, Northern Balkhash (Kazakhstan)	196/14101	+	+	SPSU Museum	
Jeffrey mine, Asbestos (Canada)		+	+	Chukanov N. V	Alumovesuvianite, holotype
Piliola (Finland)			+	Belogub E. V	

(continued)

Table A.2 (continued)

Locality	Museum/collection number	Composition	Structure	Source	Remarks
Kombat Mine, Grootfontein district, Otjozondjupa region, (Namibia)		+	+	Chukanov N. V	Milanriederite (holotype)
Lekhoylinsky, Voykar-Synyinsky massif, Polar Urals		+	+	Erokhin Yu. V	
Yulia deposit, Khakassia (Russia)	21370	+	+	Fersman Museum	
Monzoni, Predazzo (Italy)	56–3057	+	+	SPSU Museum	
Mica Pit, Kovdor massif, Kola Peninsula (Russia)		+	+	Moiseev M. M	Manaevite-Ce (holotype)
Mica Pit, Kovdor massif, Kola Peninsula (Russia)		+	+	Yakovenchuk V. N	(*REE*)
Mica Pit, Kovdor massif, Kola Peninsula (Russia)		+	+	Yakovenchuk V. N	(*REE*)
Mica Pit, Kovdor massif, Kola Peninsula (Russia)		+	+	Yakovenchuk V. N	(*REE*)
Mica Pit, Kovdor massif, Kola Peninsula (Russia)		+	+	Yakovenchuk V. N	(*REE*)
Mica Pit, Kovdor massif, Kola Peninsula (Russia)		+	+	Yakovenchuk V. N	(*REE*)
Wiluy River, Yakutia (Russia)		+	+	Galuskin E. V	
Landfall, Drammen, Buskerud (Norway)		+	+	Galuskin E. V	
Tas–Khayakhtakh, Yakutia (Russia)		+	+	Galuskin E. V	Light conductors

(continued)

Table A.2 (continued)

Locality	Museum/collection number	Composition	Structure	Source	Remarks
Bazhenovskoe deposit, Middle Urals (Russia)		+	+	Galuskin E. V	
Wiluy River, Yakutia (Russia)		+	+	Galuskin E. V	
Birhin massif, South Syberia (Russia)		+	+	Galuskin E. V	
Naslawice, Dolnośląskie (Poland)	wniz/m/10/2b	+		Silesian Univ. Museum	
Wiluy River, Yakutia (Russia)		+		Galuskin E. V	(crust)
Husha-Gol, East Sayan (Russia)		+	+	Galuskina I. O	
Bazhenovskoe deposit, Middle Urals (Russia)		+	+	Galuskin E. V	
Yoko-Dovyren, Northern Baikal (Russia)		+	+	Galuskin E. V	
Naslawice, Dolnośląskie (Poland)		+		Galuskin E. V	
Yoko-Dovyren, Northern Baikal (Russia)			+	Galuskin E. V	
Ilchir, Sayan (Russia)		+		Galuskina I. O	
Naslawice, Dolnośląskie (Poland)		+		Galuskin E. V	
Husha-Gol, East Sayan (Russia)		+	+	Galuskina I. O	
Stara Kamienica, Sudety (Poland)		+		Szełęg E	

(continued)

Table A.2 (continued)

Locality	Museum/collection number	Composition	Structure	Source	Remarks
N'Chwaning III, Kuruman, Northern Cape (RSA)		+	+	Galuskin E. V	
Wiluy River, Yakutia (Russia)		+	+	Galuskin E. V	
Naslawice, Dolnośląskie (Poland)		+	+	Galuskin E. V	
Wiluy River, Yakutia (Russia)		+	+	Galuskin E. V	Si-deficient crust
Wiluy River, Yakutia (Russia)				Galuskin E. V	Si-deficient crust
Jordanów, Dolnośląskie (Poland)	IV-4389	+		Silesian Univ. Museum	
Vapenna, Jesenik, Sudety (Czech Republic)	14688	+	+	Silesian Univ. Museum	
Eger, Karlovy Vary Region (Czech Republic)	14687	+	+	Silesian Univ. Museum	
Naslawice, Dolnośląskie (Poland)		+		Galuskin E. V	
Slyudyanka, Baikal (Russia)		+	+	Reznitskii L. Z	Light
Slyudyanka, Baikal (Russia)		+	+	Reznitskii L. Z	Dark

Table A.3 List of localities, locality types, mineral associations and short description and sizes of the VSMs samples studied in this work

Locality	Paragenesis	Samples number	Mineral assemladge	Description	Crystlal size (mm)
Jeffrey mine, Canada	Rodingite	3	Diopside, grossular, prehnite	Translucent, greenish, pink well shaped prismatic crystals	4 × 4 × 8
Tuydo, Macedonia	Rodingite	2	Diopside, grossular, prehnite, magnesiochromite	Translucent needle-like crystals or spherulites	0.1 × 0.1 × 9
Naslawice, Poland	Rodingite	2	Diopside, grossular, clinozoisite, prehnite, pumpellyite-Mg, tremolite, hydrogrossular, chlorite, actinolite	Aggregates of prismatic pink crystals	2 × 2 × 3.5
Kovdor, Russia	Skarnoid	4	Pargasite, biotite, nepheline, andradite	Green fine-grained mass	2 × 2
Kovdor, Russia	Hydrotermally altered skarnoid	10	Calcite, grossular, pyroaurite-3R, scolecite, pectolite, diopside, sphalerite	Apple green and brown crystals with hatching on the prisms	9 × 9 × 16
Akhmatovskaya Kop', Russia	Rodingite	4	Grossulyar, diopside, klinochlor, titanite, ludwigite	Olive-green with brownish titanic zones of prism crystals	12 × 12 × 17
Ak-Chagyl, Kazakhstan	Skarn	1	Epidote, wollastonite, andradite, actinolite, pargasite, malachite	Large (5–10 cm) grayish-green spherulites coated with malachite	4 × 4 × 30
Alain Zanzo Thel, Italy	Rodingite	1	Antigorite, clinochlore	Brown prismatic crystals	4 × 4 × 8
Azegous, Marocco		3			
Banat, Romania	Skarn	8	Wollastonite, calcite, grossular	Well-shaped large green crystals of pyramidal habit (1–2 cm)	15 × 15 × 25
Barbotova Yama, Russia	Rodingite	2	Diopside, grossular, prehnite	Short prismatic green crystals with grossular inclusions	5 × 5 × 7

(continued)

Table A.3 (continued)

Locality	Paragenesis	Samples number	Mineral assemladge	Description	Crystlal size (mm)
d'Aosta Valley, Italy	Contact zone of greenschist and limestone	2	Chlorite, grossular	Green well faceted prismatic crystals	4 × 4 × 8
Koytezek, Tajikistan	Skarn	1	Calcite, wollastonite	Large brown crystals of wiluite-like prismatic habit (up to 2–3 cm)	14 × 14 × 25
Bazhenovskoe, Russia	Rodingite	8	Calcite, prehnite, grossular, clinochlorine, hydrotalcite, diopside	Well-shaped elongated prismatic green crystals	6 × 6 × 22
Blagodat', Russia	Skarn	2	Grossular-andradite, wollastonite	Coarse-grained brown intergrowths with garnet	5 × 5
Pitkaranta, Russia	Skarn	2	Calcite	Brown short-prismatic sectorial crystals	5 × 5 × 17
Pitkaranta, Russia	Hydrotermally altered skarn	1	Diopside, clinochlore, sphalerite	Needle radial-radiant aggregates of transparent crystals	0.4 × 0.4 × 12
Collo di Palombio, Italy	Hornfel	1	Grossular, amphibole, plagioclase, scapolite	Large marsh-green grain in rock	12 × 12 × 15
Kleppan, Norway	Skarn	3	Piedmontite, scheelite, calcite, zoisite, albite, tremolite	Coarse-grained blue mass in plagioclase	5 × 5 × 7
Wessels, RSA	Low-temperature hydrothermal veins	2	Henritermierite, apatite, calcite, andradite	Large intergrowths of prismatic red crystals	12 × 12 × 16
Dahu Unur, Russia	Skarn	6	Nepheline, biotite, andradite, hastingsite, feldspar, calcite	Giant prismatic black crystals in nepheline	40 × 40 × 60

(continued)

Table A.3 (continued)

Locality	Paragenesis	Samples number	Mineral assemladge	Description	Crystlal size (mm)
Hazlov, Czech Republic	Skarn	4	Epidote, diopside, calcite, grossular, tremolite, wollastonite	Large short-prismatic green crystals	14 × 14 × 17
Georgeown, USA	Rodingite	1	Grossular, diopside, clinochlore, prehnite, opal	Large intergrowths of tabular crystals of bright green color	5 × 7 × 12
Birhin, Russia	Altered skarn	1	Wollastonite, fluorapatite, larnite, olivine, galuskinite, grossular, andradite	Greenish fine-grained mass	1 × 1 × 1.3
Yoko-Dovyren, Russia	Altered skarn	1	Diopside, wollastonite, dovyrenite, calcite, brucite	Fine-grained yellowish-green mass	2 × 2 × 3
Husha-Gol, Russia	Rodingite	2			
Sludyanka, Russia	Altered skarn	2	Diopside, scapolite, calcite, phlogopite, apatite	Large brown grains	15 × 15 × 25
Landfall, Sweden	Skarn	1	Calcite	Fractured dipyramidal crystals of green color	28 × 28 × 30
Tas-Khayakhtakh, Russia	Altered skarn	1	Fluorapatite, prehnite, calcite, quartz	Filamentary transparent radial aggregates	
Gukunvara, Finland	Skarn	1	Chlorite, diopside, andradite, calcite	Prismatic brown crystals in a chlorite "shirt"	7 × 7 × 12
Gulshad, Kazakhstan	Skarn	1	Epidote, calcite, andradite	Radiated marsh color aggregates	5 × 5 × 26
Hessen, Germany	Skarn	1	Wollastonite, diopside, calcite, scapolite	Dark green prismatic crystals (up to 2 cm) in calcite	14 × 14 × 30
Igukul', Russia	Rodingite	1	Grossular, diopside, clinochlore	Druse of fine-grained green vesuvian on grossularite	3 × 3 × 5
Ilchir, Russia		1			

(continued)

Table A.3 (continued)

Locality	Paragenesis	Samples number	Mineral assemladge	Description	Crystlal size (mm)
Iordanow, Poland	Rodingite	2	Grossular, diopside, clinochlore	Well-shapedd pinkish prismatic crystals	3 × 3 × 5
Kimito, Finland	Skarn	2	Calcite, grossular, diopside	Dark green prismatic crystals in calcite	5 × 5 × 8
Karmankul'skii cordon, Russia	Rodingite	6	Grossular, magnetite, clinochlor	Druse of multifaceted greenish-yellow crystals	4 × 4 × 7
Kedabek, Azerbaijan	Altered skarn	2	Calcite, grossular, diopside	Giant crystalline vesuvianite rock	50 × 50 × 75
Sakharyok, Russia	Fenitized gabbro	3	Nepheline, diopside, apatite, albite, phlogopite	Short-prismatic brownish yellow poikiloblasts	12 × 12 × 15
Khewra, Pakistan	Rodingite	1	Prehnite, calcite, chlorite	Druse of isometric green crystals	5 × 5 × 6
Kluchevskii mine, Russia	Rodingite	3	Uvarovite, manasseite, clinochlore, chromite, calcite	Prismatic bright green crystals with pronounced hatching on prisms	5 × 5 × 14
Korevo, Russia	Rodingite	1	Prehnite, grossular, calcite	Well faceted isometric pyramidal crystals	5 × 5 × 7
Kosmet, Serbia	Rodingite	2	Grossular, prehnite	Fragile prehnite-vesuvianite rock with transparent prismatic crystals	3 × 3 × 5
Kristiansand, Norway	Hornfel	2	Quartz, calcite, biotite	Large dark brown prismatic crystals	40 × 40 × 70
Leykhtenbergitovaya kop', Russia	Rodingite	1	Clinohlore, grossular	Green fractured crystals in "leykhetnbergite"	5 × 5 × 8

(continued)

Table A.3 (continued)

Locality	Paragenesis	Samples number	Mineral assemladge	Description	Crystlal size (mm)
Lopanskoe, S. Osetia	Rodingite	1	Grossular, prehnite	Massive vesuvianite with druses of well-shaped green crystals along the cracks	4 × 4 × 7
Lyangar, Uzbekistan	Skarn	1	Garnet, wollastonite, diopside, calcite	Green grains in calcite-vesuvianite rock	4 × 4 × 6
Maderaner, Switzerland	Green schists	1	Clinochlore, calcite, grossular	Druses of greenish-brown short-prismatic crystals	7 × 7 × 9
Moetnaya dacha, Russia	Rodingite	4	Chromite, uvarovite, prehnite, clinohlore	Flattened bright green crystals with hatching	4 × 4 × 8
Monte-Rosa, Switzerland	Green schists	1	Clinozoisite, almandine, clinochlore, calcite, titanite	Druses of multifaceted isometric brown crystals	6 × 6 × 9
Italy	Volcanic skarn	3	Epidote, clinozoisite, grossular, andradite	Short prismatic crystals of green – greenish-brown color in voids	9 × 9 × 16
N. Istetskaya dacha, Russia	Rodingite	1	Prehnite, grossular, chlorite	Druses of small prismatic crystals	3 × 3 × 5
N'channing III, RSA	Hydrotermally altered Mn-ores	1	Henritermierite, apatite, calcite, andradite	Aggregates of differently-oriented prismatic crystals with hatching on the prisms	5 × 5 × 12
Nikolae-Maximilyanovkaya kop', Russia	Rodingite	1	Epidote, diopside, klinochlor, magnetite, ilmenite, spinel, titanite, perovskite, grossular, scapolite	Druses of green prismatic crystals over cracks	3 × 3 × 5
Nyazemskie Mts., Russia	Rodingite	1	Clinochlore, andradite	Large prismatic crystals of brown in chlorite	9 × 9 × 15

(continued)

Table A.3 (continued)

Locality	Paragenesis	Samples number	Mineral assemladge	Description	Crystlal size (mm)
Oravica, Romania	Skarn	1	Calcite, wollastonite, diopside, epidote, muscovite	Large fissured green crystals	$35 \times 35 \times 60$
Ostebo, Norway	Skarn	1	Feldspar, zoisite, albite, tremolite, quartz	Coarse-grained mass of blue vesuvianite in plagioclase	$5 \times 5 \times 7$
Obira, Japan	Skarn	1	Tremolite, calcite, axinite	Black elongated prismatic crystals	$7 \times 7 \times 25$
Piedmont, Italy	Rodingite	1	Antigorite, actinolite	Bunches of elongated prismatic orange crystals	$4 \times 4 \times 10$
Polyakovskiy mine, Russia	Rodingite	2	Prehnite, clinohlore	Multifaced gem-quality green crystals on prehnite	$6 \times 6 \times 9$
Rila gora, Bulgaria	Skarn	2	Wollastonite, grossular, diopside, scapolite	Columnar swamp green	
Sierra de Cruces, Mexico	Skarn	1	Grossular	Giant poikiloblasts of marsh color	$22 \times 22 \times 36$
Savvateevo, Russia	Skarn	1	Actinolite, calcite, scapolite	Thick fibrous aggregates with calcite	$12 \times 12 \times 70$
Stara Kamienica, Poland	Altered skarn	1	Prehnite, clinochlore, calcite	Fractured prismatic tan crystals	$8 \times 8 \times 15$
Templtone, Canada	Skarn	2	Albite, calcite, lizardite	Zonal wiluite-like crystals	$15 \times 15 \times 25$
Alchuri, Pakistan	Alpine-cleft-type	1	Diopside, epidote, clinozoisite	Brushes of brown lamellar crystals	$14 \times 14 \times 26$
Canzocolli, Italy	Contact areole of monzonite in limestone		Calcite, grossular, epidote, wollastonite, tremolite, gehlenite, monticellite, dolomite	Large prismatic greenish brown crystals	$13 \times 13 \times 24$

(continued)

Table A.3 (continued)

Locality	Paragenesis	Samples number	Mineral assemladge	Description	Crystlal size (mm)
Manzoni, Italy	Contact areole of monzonite in limestone		Calcite, antigorite, epidote, wollastonite, tremolite, gehlenite, monticellite, dolomite	Large yellow dipyramidal poikiloblasts	22 × 22 × 26
Titovskoe, Russia	Skarn	1	Calcite, phlogopite, forsterite, andradite, monticellite, datolite	Large green fractured calcite mass	13 × 15 × 30
Ulug-Tanzek, Russia	Altere gabbro	1		Fractured brownish yellow crystals	9 × 9 × 16
Yulia mine, Russia	Skarn	1	Calcite, actinolite, andradite, diopside	Large short prismatic fractured zonal crystals of brown color	24 × 24 × 36
Gökum, Sweden	Skarn	1	Calcite, wollastonite, diopside, epidote, chlorite	Brown wiluite-like crystals	8 × 8 × 14
Vapenna Jessenik, Poland	Skarn	1	Diopside, calcite, grossular, epidote	Finegrained brown crystals	3 × 3 × 5
Wiluy, Russia	Altered gabbro	7	Gehlenite, hydrogrossular, calcite, apatite, diopside	Large prismatic green poikilocrystals	20 × 20 × 32
Wiluy, Russia	Low-temperature hydrothermal vein	10	Hydrogrossular, wiluite, prehnite, montmorillonite	Thin crusts and split crystals in Achtaragdite	0.5 × 10 × 18
Zi Uental, Switzerland	Rodingite	1	Clinohlore, grossular	Growths of brown short-prismatic crystals in chlorite	6 × 6 × 10
Shishimskaya kop', Russia	Rodingite	1	Clinochlore, prehnite, calcite, magnetite	Druses of large green prism crystals on cracks	13 × 13 × 22
Kazakhstan	Skarn	1	Pargasite, calcite	Radiated aggregates of pale green color	4 × 4 × 23
Norway	Skarn	1			
Guan See, China	Skarn	1		Larg aggregates of black columnar vesuvianite	35 × 35 × 60

Table A.4 Symmetry and wavenumbers of bands in the IR spectra of VSMs

Sample	Sp. Gr	ν(O–H)	ν(B–O)	ν(Si–O), δ(*M*···O–H)	δ(SiO_4), δ(Si_2O_7), ν(*M*–O)
67940	*P4/nnc*	3580sh, 3567, 3449w, 3205, 3170sh	1165sh, 1075sh	1024, 985, 911, 875sh, 799	693w, 637, 605sh, 578, 492, 448, 416, 378
15978	*P4/nnc*	3645sh, 3631w, 3590sh, 3564, 3462, 3196, 3120sh	1160sh, 1075sh	1022, 984, 905, 875sh, 799	636, 600sh, 578, 560w, 490, 445, 415, 376
23700	*P4/n*	3672w, 3628, 3563w, 3505, 3454, 3252, 3156	1160sh, 1076	1019, 973, 915, 904, 865sh, 807, 765sh	714w, 655sh, 627, 604, 576, 488, 434, 412, 388, 380sh
28975	*P4/n*	3674, 3629, 3575, 3488, 3424, 3300w, 3157	1165sh, 1075sh	1017, 969, 916, 901, 807, 785sh, 760sh	685sh, 645sh, 618, 601, 571, 525, 484, 430, 415sh, 377
31413	*P4/nnc*	3625sh, 3565, 3443, 3220	1170sh, 1075sh	1018, 980, 910, 875sh, 805sh	630sh, 600sh, 572, 488, 443, 420sh, 380sh
61902	*P4/nnc*	3640sh, 3586, 3563, 3448, 3203	1165sh, 1075sh	1022, 985, 904, 880sh, 798	713w, 637, 600sh, 578, 490, 447, 414, 374
65273	*P4/n*	3672w, 3632, 3521, 3423, 3150	1423w, 1160sh, 1075sh	1019, 965, 916, 884, 861, 799	709w, 660sh, 635sh, 604, 572, 520sh, 484, 436, 409, 375
70024	*P4/nnc*	3634w, 3585sh, 3562, 3457, 3194	1165sh, 1075sh	1022, 986, 905, 875sh, 800	713w, 636, 605sh, 579, 558, 491, 447, 415, 398, 375
73599	*P4/nnc*	3630sh, 3585sh, 3566, 3449, 3199	1165sh, 1075sh	1022, 986, 910, 880sh, 800	695sh, 638, 600sh, 578, 561, 491, 447, 415, 375
976	*P4/nnc*	3640w, 3564, 3453, 3217	1165sh, 1077	1019, 985, 908, 875sh, 804	634, 576, 554w, 488, 444, 413, 395, 380sh

Table A.5 Powder diffraction data for vesuvianite from Vesuvius (sample 27844 FM). The eight most intense lines are shown in bold

d_{meas} Å	*hkl*	I_{meas}	d_{meas} Å	*hkl*	I_{meas}
10.83	110	4	2.081	623	5
5.85	002	4	2.060	543	4
4.673	202	4	2.038	730	9
4.000	222	6	2.022	642	4
3.857	400	2	2.008	731	2
3.463	**322**	**15**	1.9941	633	8
3.234	402	4	1.9602	651	6
3.0615	313	5	1.9281	116	2

(continued)

Table A.5 (continued)

d_{meas} Å	*hkl*	I_{meas}	d_{meas} Å	*hkl*	I_{meas}
3.038	510	8	1.9171	713	2
2.995	501	9	1.9029	206	4
2.939	**511**	**18**	1.8823	216	7
2.897	323	5	1.8647	515	2
2.744	**440**	**100**	1.7960	831	1
2.659	530	3	1.7614	714	9
2.586	**531**	**58**	1.7158	910	2
2.522	314	6	**1.6768**	**734**	**16.85**
2.492	611	1	**1.6603**	**436**	**22.28**
2.454	**620**	**45**	**1.6215**	**526**	**29.76**
2.433	324	7	1.5853	824	2
2.367	541	9	1.5712	327	5
2.345	404	5	1.5580	616	7
2.320	105	8	1.5397	735	2
2.293	334	5	1.5318	626	1
2.269	631	2	1.5235	942	4
2.193	710	5	1.5089	1021	2
2.177	701	5	1.5028	1002	1
2.158	711	3	1.4979	636	8
2.141	305	2	1.4753	1022	1
2.121	315	11			

Table A.6 Powder diffraction data for vesuvianite from Bazhenovskoe deposit (A. A. Agakhanov's sample). The eight most intense lines are shown in bold

d_{meas} Å	*hkl*	I_{meas}	d_{meas} Å	*hkl*	I_{meas}
10.95	110	4	2.298	334	1
5.89	002	5	2.274	631	3
4.699	202	3	2.200	533	3
4.020	222	8	2.163	711	2
3.877	400	5	2.132	315	8
3.472	**322**	**12**	2.121	514	11
3.246	402	6	2.085	623	4
3.073	313	4	2.061	543	2
3.044	510	11	2.041	730	9
3.004	501	7	2.025	642	4

(continued)

Table A.6 (continued)

d_{meas} Å	*hkl*	I_{meas}	d_{meas} Å	*hkl*	I_{meas}
2.955	**004**	**39**	2.010	444	5
2.905	323	2	1.9984	633	7
2.850	114	1	1.9592	651	6
2.751	**432**	**100**	1.9182	713	2
2.668	530	5	1.9113	206	5
2.603	**423**	**20**	1.8829	624	4
2.594	**522**	**60**	1.7646	714	5
2.533	314	3	1.7179	841	2
2.494	611	1	1.6798	734	7
2.457	**620**	**33**	1.6652	921	10
2.441	503	3	**1.6280**	**526**	**13**
2.373	541	6	**1.6230**	**804**	**15**
2.354	404	3	1.5715	913	7
2.326	523	4	1.5581	616	4

Table A.7 Powder diffraction data for vesuvianite from Kovdor (A. A. Agakhanov's sample). The eight most intense lines are shown in bold

d_{meas} Å	*hkl*	I_{meas}	d_{meas} Å	*hkl*	I_{meas}
10.97	110	5	2.241	125	2
5.92	002	3	2.212	533	3
4.026	222	11	2.152	305	9
3.642	330	0.9	2.132	135	12
3.480	240	8	2.124	461	8
3.224	402	19	2.070	543	5
3.083	133	4	2.044	730	5
3.028	510	12	2.014	731	4
3.013	431	9	2.001	633	10
2.992	**501**	**21**	1.9653	651	4
2.955	**004**	**36**	1.9372	614	1
2.942	**511**	**40**	1.9235	713	2
2.755	**342**	**100**	1.9054	206	1
2.668	530	3	1.8862	126	2
2.601	**243**	**60**	1.8322	182	2
2.560	531	11	1.7671	714	9
2.512	611	2	1.7207	190	2
2.461	**620**	**24**	**1.6677**	**734**	**21**

(continued)

Table A.7 (continued)

d_{meas} Å	*hkl*	I_{meas}	d_{meas} Å	*hkl*	I_{meas}
2.444	503	6	1.6309	346	14
2.379	451	5	**1.6264**	**526**	**21**
2.355	404	5	1.5850	824	4
2.330	414	6	1.5747	327	5
2.302	334	6	1.5618	616	7
2.279	631	3			

Table A.8 Powder diffraction data for vesuvianite from Koytezek pass (A. A. Agakhanov's sample). The eight most intense lines are shown in bold

d_{meas} Å	*hkl*	I_{meas}	d_{meas} Å	*hkl*	I_{meas}
10.82	110	4	2.156	711	6
5.47	220	2	2.134	305	2
4.678	202	2	2.122	315	8
4.004	222	9	**2.115**	**641**	**12**
3.880	400	2	2.079	623	4
3.465	322	8	2.060	325	4
3.235	402	5	2.035	730	6
3.060	313	3	2.019	642	3
3.033	510	9	2.005	731	7
2.998	501	5	1.9908	415	9
2.942	**511**	**28**	1.9574	651	8
2.894	323	3	1.9251	703	3
2.741	**204**	**100**	1.9149	713	2
2.660	530	3	1.9009	206	2
2.584	**531**	**54**	1.8806	216	4
2.520	314	3	1.7909	831	1
2.490	611	3	1.7585	714	9
2.452	**620**	**23**	1.7114	910	4
2.432	503	6	1.6751	734	5
2.364	541	9	**1.6582**	**436**	**17**
2.342	404	4	**1.6216**	**526**	**15**
2.315	105	5	**1.6180**	**931**	**23**
2.289	334	3	1.5670	770	9
2.268	631	6	1.5557	616	6
2.192	710	7			

Table A.9 Powder diffraction data for vesuvianite from Kluchevskii mine (sample 27962 FM). The eight most intense lines are shown in bold

d_{meas} Å	*hkl*	I_{meas}	d_{meas} Å	*hkl*	I_{meas}
11.03	110	5	2.210	533	7
5.89	002	5	2.196	710	9
4.698	202	12	2.182	701	2
4.026	321	7	2.162	711	2
3.474	**322**	**20**	2.131	315	15
3.251	402	5	2.121	641	12
3.078	313	3	2.066	543	4
3.050	510	7	2.043	730	8
3.005	501	10	2.026	642	5
2.952	**511**	**40**	2.013	731	7
2.908	323	4	1.9993	633	9
2.814	521	0.7	1.9643	651	6
2.750	**440**	**100**	1.9350	604	2
2.670	530	3	1.8846	652	6
2.604	**531**	**25**	1.7648	714	13
2.594	**600**	**72**	1.7170	910	3
2.533	314	5	1.6804	734	16
2.499	611	3	**1.6658**	**436**	**23**
2.459	**620**	**27**	1.6290	526	11
2.376	541	5	**1.6247**	**804**	**23**
2.354	404	4	1.5715	327	9
2.328	523	8	1.5611	616	8
2.301	334	3	1.5253	942	2
2.273	631	5	1.5008	1002	5
2.237	215	1	1.4785	1022	3

Table A.10 Powder diffraction data for Si-deficient B-bearing vesuvianite from Wiluy (E. ssssssV. Galuskin's sample). The eight most intense lines are shown in bold

d_{meas} Å	*hkl*	I_{meas}	d_{meas} Å	*hkl*	I_{meas}
5.88	002	4	2.198	710	9
5.48	220	3	2.177	701	2
4.68	202	2	2.160	711	3
4.015	222	4	2.126	315	4
3.882	400	3	2.120	641	11
3.472	**322**	**23**	2.0596	543	2
3.242	402	3	**2.040**	**730**	**12**
3.066	313	2	2.024	642	2
3.046	**510**	**16**	2.011	731	3
3.004	501	9	1.9963	415	4
2.945	**511**	**29**	1.9618	006	5
2.902	323	2	1.9037	206	3
2.747	**204**	**100**	1.8844	624	7
2.663	530	7	1.7966	831	2
2.591	**531**	**35**	1.7628	535	4
2.528	314	2	1.7170	841	2
2.497	611	1	1.6775	734	3
2.457	**620**	**66**	1.6608	436	3
2.437	503	4	**1.6214**	**526**	**14**
2.377	541	2	1.5710	913	6
2.347	404	3	1.5570	616	3
2.320	523	3	1.5248	546	3
2.294	334	1	1.5108	427	2
2.273	631	3	1.3777	656	0.4
2.207	533	2			

Table A.11 Unit cell parameters, symmetry, occupancy of the positions of square pyramid *Y*1A,B and values of *R-factor* for structurally studied vesuvianite samples

Sample name	Parameter *a* (Å)	Parameter *c* (Å)	Number of violating reflections for the space group *P*4/*nnc*	Number of violating reflections for the space group *P*4/*n*	Space group	Boron content (*apfu*)	Occupan-cy of *Y*1A (calcula-ted as Fe)	Occupan-cy of *Y*1B (calcula-ted as Fe)	R_1-factor
ves_pk_kov_REE-last1	15.8591(4)	11.9686(5)	19	0	*P*4/*n*	2.70	0.196(5)	0.344(5)	0.0349
ves_pk_kov_REE_new	15.7092(2)	11.8940(3)	194	19	*P*4/*n*	2.13	0.377(11)	0.498(11)	0.0578
ves_ahm2(28975)	15.5572(4)	11.8425(4)	315	1	*P*4/*n*		0.862(4)	0.05(5)	0.0266
ves_ak_chag(38476)	15.5775(4)	11.7998(5)	76	28	*P*4/*n*		0.37(1)	0.38(1)	0.0508
ves_alain_ zanzo-thel(13988)	15.5780(3)	11.8409(4)	248	2	*P*4/*n*		0.578(14)	0.297(12)	0.0285
ves_aosta_valley(aks)	15.5450(6)	11.8193(9)	9		*P*4/*nnc*				0.0333
ves_asbestos(70026)	15.4728(2)	11.7864(2)	68	68	*P*4*nc*		0.37(2)	0.27(2)	0.0243
ves_at_koy(xcal)	15.5602(6)	11.8153(8)	29		*P*4/*nnc*				0.0302
ves_azegous-2(67495)	15.5744(2)	11.8203(2)	21	4	*P*4/*n*		0.489(5)	0.467(5)	0.0294
ves_barb_new(23700)	15.6224(7)	11.8381(8)	608	0	*P*4/*n*		0.846(3)	0.053(3)	0.0271
ves_bazh(79478)	15.5648(4)	11.8493(7)	65	34	*P*4/*n*		0.456(14)	0.537(14)	0.0794
ves_bazh_new(79478)	15.5877(2)	11.8461(2)	68	47	*P*4*nc*		0.722(7)	0.188(7)	0.0258
ves_bazh(79479)	15.5569(2)	11.8240(3)	269	47	*P*4/*n*		0.380(11)	0.349(10)	0.0653
ves_bellecombe(79621)	15.6024(4)	11.8844(5)	577	27	*P*4/*n*		0.128(5)	0.793(4)	0.0267
ves_cyp	15.4889(5)	11.7730(4)	0	0	*P*4/*nnc*				0.0402
ves_cyp(15377)	15.4840(2)	11.7552(2)	34		*P*4/*nnc*	0.98			0.0234
ves_gl(eger)14687	15.556(1)	11.7330(2)	4		*P*4/*nnc*				0.0347

(continued)

Table A.11 (continued)

Sample name	Parameter *a* (Å)	Parameter *c* (Å)	Number of violating reflections for the space group *P*4/*nnc*	Number of violating reflections for the space group *P*4/*n*	Space group	Boron content (*apfu*)	Occupan-cy of *Y*1A (calcula-ted as Fe)	Occupan-cy of *Y*1B (calcula-ted as Fe)	R_1-factor
ves_erem(18983)	15.6188(3)	11.8910(4)	2296	18	*P*4/*n*		0.012(3)	0.865(3)	0.0157
ves_f(fluorvesuvianite)	15.5480(4)	11.7939(5)	120	5	*P*4/*n*		0.439(7)	0.482(7)	0.0381
ves_56-3057	15.6062(2)	11.8036(2)	119	163	*P*4*nc*	1.16	0.654(7)	0.108	0.0243
ves_Falza(56–3067)	15.5390(2)	11.7702(2)	42		*P*4/*nnc*				0.0544
ves_frag_el_ma(70024)	15.5228(5)	11.7787(9)	1		*P*4/*nnc*				0.0517
ves_frugard(13973)	15.5423(3)	11.7854(4)	6		*P*4/*nnc*				0.0272
ves_gazlov(61902)	15.5595(4)	11.8269(6)	32	12	*P*4/*nnc*				0.019
ves_georgetown(51073)	15.5533(2)	11.8309(3)	202	17	*P*4/*n*		0.226(9)	0.426(9)	0.0578
ves_gl(eger)14687	15.5557(1)	11.7330(2)	4		*P*4/*nnc*				0.0347
ves_gl_husha_gol	15.5683(3)	11.8500(5)	162	5	*P*4/*n*		0.221(10)	0.673(12)	0.0681
ves_gl_svetovody	15.4976(2)	11.7925(3)	216	8	*P*4/*n*		0.282(10)	0.358(11)	0.063
Ves_haslan(15978)	15.5431(3)	11.7921(4)	2		*P*4/*nnc*				0.0221
ves_hessen(51056)	15.5455(2)	11.7883(2)	16	6	*P*4/*nnc*				0.0238
ves_gl_husha_gol	15.5683(3)	11.8500(5)	162	5	*P*4/*n*		0.221(10)	0.673(12)	0.0681
ves_igul(43974)	15.5393(4)	11.8159(5)	210	4	*P*4/*n*		0.727(7)	0.176(7)	0.0516
ves_gl_ yellow(iordanov)	15.5876(2)	11.8465(2)	84	37	*P*4*nc*		0.760(6)	0.104(7)	0.0243
ves_kar	15.6218(8)	11.8775(6)	56	56	*P*4*nc*		0.789(5)	0.085(5)	0.0341

(continued)

Table A.11 (continued)

Sample name	Parameter *a* (Å)	Parameter *c* (Å)	Number of violating reflections for the space group *P*4/*nnc*	Number of violating reflections for the space group *P*4/*n*	Space group	Boron content (*apfu*)	Occupan-cy of *Y*1A (calcula-ted as Fe)	Occupan-cy of *Y*1B (calcula-ted as Fe)	R_1-factor
ves_kar(21155)	15.5243(3)	11.8293(5)	80	10	*P*4/*n*		0.491(7)	0.391(6)	0.0414
ves_keyvy(50578)	15.5168(2)	11.7700(4)	0	0	*P*4/*nnc*				0.0339
ves_keyvy(707)	15.5141(4)	11.7828(5)	22		*P*4/*nnc*				0.0268
ves_keyvy-976	15.4801(5)	11.7556(6)	128	45	*P*4/*nnc*				0.0245
ves_khewza(65273)	15.5711(4)	11.8423(5)	132	10	*P*4/*n*		0.416(9)	0.495(9)	0.0549
ves_kimito (13985)	15.6086(7)	11.8792(8)	5		*P*4/*nnc*				0.0275
ves_md(xcalfull)	15.5685(1)	11.8365(1)	1542	278	*P*4/*n*		0.389(10)	0.440(10)	0.0577
ves_kluchev(27952)	15.5508(2)	11.8331(3)	93	86	*P*4*nc*		0.753(6)	0.032(5)	0.0257
ves_korev(1557)	15.5580(3)	11.8298(3)	238	5	*P*4/*n*		0.777(4)	0.123(4)	0.032
ves_korev(23677)	15.5260(3)	11.8092(3)	1099	4	*P*4/*n*		0.797(10)	0.116(9)	0.0573
ves_korev(23675)	15.5790(3)	11.8524(3)	240	2	*P*4/*n*		0.070(5)	0.839(5)	0.0363
ves_kosmet(55211)	15.5554(3)	11.8269(3)	506	166	*P*4*nc*		0.715(9)	0.178(10)	0.0311
ves_ kov-new(yakovencuk)	15.6051(2)	11.8595(3)	274	79	*P*4*nc*		0.126(6)	0.723(6)	0.0242
ves_kov_ree-last2	15.8213(2)	11.9209(2)	110	0	*P*4/*n*	2.45	0.391(9)	0.172(9)	0.0442
ves_ kov-new(yakovencuk)	15.6051(2)	11.8595(3)	274	79	*P*4*nc*		0.126(6)	0.723(6)	0.0242

(continued)

Table A.11 (continued)

Sample name	Parameter *a* (Å)	Parameter *c* (Å)	Number of violating reflections for the space group *P*4/*nnc*	Number of violating reflections for the space group *P*4/*n*	Space group	Boron content (*apfu*)	Occupan-cy of *Y*1A (calcula-ted as Fe)	Occupan-cy of *Y*1B (calcula-ted as Fe)	R_1-factor
ves_kristiansand(13975)_skol_po_100)	15.5145(2)	11.7836(2)	303	111	*P*4/*nnc*				0.0216
ves_kristiansand(14001)	15.5223(1)	11.7671(2)	20		*P*4/*nnc*				0.0252
ves_leychtenberg(9578)	15.7190(18)	11.866(2)	13		*P*4/*nnc*	1.12			0.0488
ves_lopanskiy(33623)	15.5632(2)	11.8221(2)	237	5	*P*4/*n*		0.653(5)	0.256(4)	0.0364
ves_lyangar(40409)	15.5376(3)	11.7880(5)	7		*P*4/*nnc*				0.0295
ves_maderaner(27726)	15.5769(1)	11.8560(1)	1465	7	*P*4/*n*		0.681(12)	0.221(11)	0.0692
ves_monentaya(17342)	15.5775(3)	11.8300(4)	56	3	*P*4/*n*		0.492(6)	0.308(6)	0.0494
ves_md(17344)	15.5657(3)	11.8259(4)	51	25	*P*4/*n*		0.396(10)	0.437(10)	0.0577
ves_md(55291)	15.5582(4)	11.8536(5)	291	8	*P*4/*n*		0.440(16)	0.200(14)	0.0818
ves_monte_rosa(31407)	15.5281(6)	11.8330(8)	193	0	*P*4/*n*		0.326(12)	0.572(13)	0.0729
ves_n_isetskaya_new(33530)	15.5509(5)	11.8307(5)	618	11	*P*4/*n*		0.036(7)	0.867(8)	0.0429
	15.5930(3)	11.8611(4)	137	3	*P*4/*n*		0.238(8)	0.636(9)	0.0599
ves_n_isetskaya_new(33530)	15.5509(5)	11.8307(5)	618	11	*P*4/*n*		0.036(7)	0.867(8)	0.0429
ves_nyaz(11805)	15.5573(4)	11.8312(6)	173	0	*P*4/*n*		0.124(9)	0.778(9)	0.0655

(continued)

Table A.11 (continued)

Sample name	Parameter *a* (Å)	Parameter *c* (Å)	Number of violating reflections for the space group *P*4/*nnc*	Number of violating reflections for the space group *P*4/*n*	Space group	Boron content (*apfu*)	Occupan-cy of *Y*1A (calcula-ted as Fe)	Occupan-cy of *Y*1B (calcula-ted as Fe)	R_1-factor
ves_ nyazemskie(2)(67964)	15.5883(2)	11.8622(3)	215	1	*P*4/*n*		0.881(4)	0	0.0312
ves_ostebo(67940)	15.5323(3)	11.7901(3)	20		*P*4/*nnc*				0.0248
ves_oyta(54541)_new	15.5977(5)	11.7523(5)	54	27	*P*4/*nnc*	1.29			0.0267
ves_piemont(13963)	15.5596(2)	11.8354(3)	752	9	*P*4/*n*		0.307(15)	0.528(17)	0.0872
ves_pk_kov_REE	15.7051(3)	11.9015(3)	304	5	*P*4/*n*	2.05	0.213(7)	0.640(7)	0.037
ves_pol(26478)	15.5541(4)	11.8297(4)	250	160	*P*4*nc*		0.771(9)	0.082(6	0.0256
ves_pol(32801)	15.5303(4)	11.8312(4)	117	29	*P*4*nc*		0.779(11)	0.131(12)	0.036
ves_ptk	15.642(3)	11.802(2)	0	0	*P*4/*nnc*	2.48			0.037
ves_ptk(xcal)	15.6386(2)	11.7929(2)	12		*P*4/*nnc*	2.3			0.0348
ves_rila (77291)	15.5139(2)	11.7782(2)	29	3	*P*4/*nnc*				0.0298
ves_rila(2)(73599)	15.5073(3)	11.7815(4)	3		*P*4/*nnc*				0.0293
ves_rila-2(77291)	15.5311(2)	11.7934(3)	25		*P*4/*nnc*				0.027
ves_razshepleny	15.6128(6)	11.8017(6)	0	0	*P*4/*nnc*	2.39			0.065
ves_savvateevo(57348)	15.5751(5)	11.8255(7)	2	0	*P*4/*nnc*				0.0365
ves_gl_stara_kamenica	15.5302(2)	11.8114(2)	502	17	*P*4/*n*		0.759(7)	0.086(8)	0.0277
ves_gl_stara_ kamenieca_new	15.5409(2)	11.8176(2)	629	8	*P*4/*n*		0.764(4)	0.025(3)	0.021

(continued)

Table A.11 (continued)

Sample name	Parameter *a* (Å)	Parameter *c* (Å)	Number of violating reflections for the space group *P*4/*nnc*	Number of violating reflections for the space group *P*4/*n*	Space group	Boron content (*apfu*)	Occupan-cy of *Y*1A (calcula-ted as Fe)	Occupan-cy of *Y*1B (calcula-ted as Fe)	R_1-factor
ves_templeton(31413)	15.7118(4)	11.9075(4)	3	0	*P*4/*nnc*	2.24			0.0281
ves_ti_chuk	15.5703(4)	11.8217(5)	29		*P*4/*nnc*				0.0276
ves_tirol(27728)	15.6165(2)	11.8310(3)	98	35	*P*4/*nnc*	3.35			0.0278
ves_ tirol-methacryst(7359)	15.5915(3)	11.7878(3)	18		*P*4/*nnc*	2.24			0.0449
ves_ulug-tanz	15.5568(3)	11.7908(4)	10		*P*4/*nnc*				0.0318
ves_upland_ jokum(13967)	15.5652(4)	11.8157(6)	84	5	*P*4/*nnc*				0.0282
Ves_gl_vapenna Jesenik	15.5334(4)	11.7786(5)	11	4	*P*4/*nnc*				0.0227
ves_vezuvium_27844_ new	15.5720(3)	11.8158(5)	79		*P*4/*nnc*				0.0273
ves_vezuvium(51062)	15.5459(3)	11.7988(4)	15	3	*P*4/*nnc*	1.34			0.0306
ves_vezuvium(5228)	15.5308(2)	11.7798(3)	21	10	*P*4/*nnc*	1.18			0.0208
ves_yak_kov_brown	15.6657(3)	11.9295(4)	352	70	*P*4/*n*	1.60	0.466(7)	0.441(7)	0.0351
ves_zi_uenthal(51085)	15.5919(3)	11.8598(4)	400	0	*P*4/*n*		0.065(4)	0.846(4)	0.0304
ves_jef(big-lil)	15.5380(2)	11.8175(3)	538	14	*P*4/*n*		0.446(11)	0.290(11)	0.0614
ves-jef(lil)(Xcal)	15.5570(3)	11.8215(3)	260	4	*P*4/*n*		0.184(8)	0.568(8)	0.0449
ves_an_bazh(cr)(xcal)	15.5301(3)	11.8259(4)	145	4	*P*4/*n*		0.276(6)	0.523(6)	0.0438

(continued)

Table A.11 (continued)

Sample name	Parameter *a* (Å)	Parameter *c* (Å)	Number of violating reflections for the space group *P*4/*nnc*	Number of violating reflections for the space group *P*4/*n*	Space group	Boron content (*apfu*)	Occupan-cy of *Y*1A (calcula-ted as Fe)	Occupan-cy of *Y*1B (calcula-ted as Fe)	R_1-factor
ves_an_bazh(ves-garnet)(Xcal)	15.5527(2)	11.8452(3)	1538	13	*P*4/*n*		0.645(13)	0.253(12)	0.0742
ves-cyprine	15.4889(5)	11.7730(4)	0	0	*P*4/*nnc*				0.0394
Ves_kz	15.5698(6)	11.7942(4)	0		*P*4/*nnc*				0.0327
ves_nor	15.5674(2)	11.843(2)	0	0	*P*4/*nnc*				0.0337
ves_shk	15.547(2)	11.833(2)	52	0	*P*4/*n*		0.459(5)	0.411(5)	0.052
ves_guan_sea(58654)	15.5314(3)	11.7624(4)	3		*P*4/*nnc*				0.0257
ves_myrsaeter(60211)	15.5945(4)	11.8299(5)	22	12	*P*4/*n*		0.458(8)	0.375(8)	0.0555
ves_piilola	15.5004(1)	11.7720(2)	439	80	*P*4/*nnc*				0.0268
ves_pk_wes	15.4843(3)	11.7753(4)	201	26	*P*4/*n*		0.561(8)	0.503(8)	0.0571
ves_tirol_ziuenthal(51085)	15.5849(8)	11.8748(8)	648	0	*P*4/*n*		0.869(4)	0	0.0257
ves_tuydo(59783)New	15.4942(12)	11.809(1)	36	1	*P*4/*n*		0.259(12)	0.134(10)	0.0697
ves_zi_uenthal_new(51085)	15.5996(2)	11.8663(2)	547	1	*P*4/*n*		0.128(5)	0.799(4)	0.023
will_auto-225-19594	15.7486(5)	11.828(1)	24	2	*P*4/*n*	1.80	0.369(8)	0.472(8)	0.0518
Razsheplenie(wil_bereg)	15.6901(6)	11.8413(6)	9		*P*4/*nnc*	0.40			0.0638
ves_pk_lp	15.5955(4)	11.8588(3)	167	11	*P*4/*nnc*	0.80			0.0365

(continued)

Table A.11 (continued)

Sample name	Parameter *a* (Å)	Parameter *c* (Å)	Number of violating reflections for the space group *P*4/*nnc*	Number of violating reflections for the space group *P*4/*n*	Space group	Boron content (*apfu*)	Occupan-cy of *Y*1A (calcula-ted as Fe)	Occupan-cy of *Y*1B (calcula-ted as Fe)	R_1-factor
ves_will_auto_ 225-19594	15.7059(1)	11.8151(2)	721	66	*P*4/*n*	1.88	0.354(7)	0.498(7)	0.0408
ves_white_crust on wil	15.5841(4)	11.7682(6)	1		*P*4/*nnc*	1.00			0.0436
ves_ monte-somma(27844)	15.5752(12)	11.8208(6)	1		*P*4/*nnc*	2.00			0.0292
ves_kedabek(41323)	15.5630(6)	11.8322(8)	119	8	*P*4/*n*	2.34	0.334(8)	0.361(8)	0.0559
ves_Jandrik(29112)	15.5618(5)	11.8202(6)	3		*P*4/*nnc*	1.60			0.0269
ves_gukun(16908)	15.6246(5)	11.7891(8)	2		*P*4/*nnc*	1.81			0.0263
ves_gl_will_auto_ zone-yellow	15.6545(3)	11.8007(4)	33	10	*P*4*nc*	1.50	0.560(9)	0.217(10)	0.0327
ves_gl_birhin_new	15.5610(2)	11.7871(2)	69	7	*P*4/*n*	1.56	0.417(5)	0.397(5)	0.0326
ves_gl_ doveren(dark-green)_ new	15.6605(3)	11.8552(3)	34	30	*P*4*nc*	1.50	0.612(9)	0.190(13)	0.0282
ves_br_ptk	15.552(2)	11.772(2)	17		*P*4/*nnc*	1.00			0.035
ves_will_auto_GV-7	15.6449(3)	11.7694(5)	19	16	*P*4*nc*	2.23	0.610(10)	0.138(10)	0.0399
ves_templtone_rim	15.5397(4)	11.7825(4)	29		*P*4/*nnc*				0.0267

* The boron content was calculated from structural data, based on the sum of the population of T1 + T2 positions and using atomic boron scattering curves (i.e., the partial population of these positions by Al^{3+} atoms was not taken into account)

Table A.12 Crystallographic data and refinement parameters for crystal structure of vesuvianite from Vesuvius (sample 27844 FM)

Crystallographic parameters	
Symmetry	Tetragonal
Space group	*P4/nnc*
Unit cell parameters *a*, *c* (Å)	15.5720(3), 11.8158(5)
Unit cell volume (Å^3)	2865.16(16)
Z	2
Calculated density (g/cm^3)	3.40
Absorption coefficient (mm^{-1})	2.839
Crystal size (mm)	0.21 × 0.15 × 0.14
Data collection parameters	
2θ range (°)	5.23–55.00
h, *k*, *l* range	−20 → 19, −15 → 20, −15 → 14
Number reflections	14428
Unique reflections (R_{int})	1653 (0.044)
Unique reflections c $F > 4\sigma F$	1448
Refinement parameters	
R_1 [$F > 4\sigma(F)$], wR_2 [$F > 4\sigma(F)$]	0.0267, 0.0717
R_1, wR_2 all data	0.0275, 0.0722
S	1.184
ρ_{max}, ρ_{min}, e Å^{-3}	0.54, −1.05
Refinment method	Full matrix least-squares on F^2

Table A.13 Atom fractional coordinates, displacement parameters (Å^2) and site occupancy for crystal structure of vesuvianite from Vesuvius (sample 27844 FM)

Site	Occupancy	*x*/*a*	*y*/*b*	*z*/*c*	U_{ani}	Multiplicity
*X*1	Ca	¾	¼	¼	0.0060(2)	4*c*
*X*2	Ca	0.68906(3)	0.04449(3)	0.12069(3)	0.0052(1)	16*k*
*X*3	Ca	0.68033(3)	0.89939(3)	−0.11177(4)	0.0126(1)	16*k*
*X*4	Ca	¾	¾	0.14474(16)	0.0077(3)	4*e*
*Z*1	Si	¾	¼	0	0.0048(2)	4*d*
*Z*2	Si	0.54083(4)	0.18088(4)	0.12847(5)	0.0048(2)	16*k*
*Z*3	Si	0.84929(4)	0.91656(4)	0.13547(5)	0.0056(2)	16*k*
*Y*1	$Fe_{0.36}Mg_{0.14}$	¾	¾	0.0484(2)	0.0187(7)	4*e*
*Y*2	Al	½	0	½	0.0044(2)	8*f*
*Y*3	$Al_{0.90}Fe_{0.10}$	0.61247(4)	0.12099(4)	0.37386(4)	0.0054(2)	16*k*
O1	O	0.7794(1)	0.1726(1)	0.0856(1)	0.0071(3)	16*k*
O2	O	0.61756(9)	0.1601(1)	0.2205(1)	0.0070(3)	16*k*

(continued)

Table A.13 (continued)

Site	Occupancy	*x*/*a*	*y*/*b*	*z*/*c*	U_{ani}	Multiplicity
O3	O	0.5477(1)	0.2220(1)	0.4247(1)	0.0069(3)	16*k*
O4	O	0.6060(1)	0.0614(1)	0.5299(1)	0.0066(3)	16*k*
O5	O	0.8294(1)	0.0141(1)	0.1791(1)	0.0086(3)	16*k*
O6	O	0.7724(1)	−0.1197(1)	0.0582(1)	0.0110(3)	16*k*
O7	O	0.4442(1)	0.1731(1)	0.1789(1)	0.0099(3)	16*k*
O8	O	0.59105(9)	0.93943(9)	0.0671(1)	0.0064(3)	16*k*
O9	O	0.8548(1)	−0.1452(1)	¼	0.0082(4)	8*h*
O10	O	¾	¾	−0.1346(3)	0.0179(8)	4*e*
O11	$O_{0.75}F_{0.25}$	0.5039(1)	0.0616(1)	0.3643(1)	0.0078(3)	16*k*
*T*1	$Al_{0.11}$	0.4447(4)	0.0553(4)	¼	0.010(3)	8*h*
*T*2	$B_{0.25}$	¾	¾	¾	0.002(5)	2*a*

Table A.14 Selected bond lengths (Å) in the crystal structure of vesuvianite from Vesuvius (sample 27844 FM)

Bond	Bond length (Å)	Bond	Bond length (Å)	Bond	Bond length (Å)
*Z*1–O1 4×	1.6387(15)	*X*2–O6	2.9606(17)	*Y*1–O6 4×	2.0621(17)
		*X*2–O2	2.4231(15)	*Y*1–O10	2.161(5)
*Z*2–O7	1.6226(16)	*X*2–O5	2.4300(16)	<*Y*1–O>	2.082
*Z*2–O2	1.6471(16)	*X*2–O4	2.4532(15)		
*Z*2–O3	1.6411(16)	*X*2–O1	2.4755(16)	*Y*2–O8 2×	1.8785(15)
*Z*2–O4	1.6786(16)	<*X*2–O>	2.4724	*Y*2–O11 2×	1.8696(15)
<*Z*2–O>	1.6473			*Y*2–O4 2×	1.9401(15)
		*X*3–O7	2.5155(17)	<*Y*2–O>	1.8961
*Z*3–O6	1.6085(17	*X*3–O3	2.4368(16)		
*Z*3–O8	1.6239(15)	*X*3–O11	2.4791(16)	*Y*3–O1	1.9310(16)
*Z*3–O5	1.6333(16)	*X*3–O6	2.9728(17)	*Y*3–O11	1.9265(16)
*Z*3–O9	1.6626(10)	*X*3–O6	2.4850(17)	*Y*3–O2	1.9141(16)
<*Z*3–O>	1.6321	*X*3–O7	2.5723(17)	*Y*3–O3	1.9632(16)
		*X*3–O10	2.5809(6)	*Y*3–O5	1.9953(17
*X*1–O1 4x	2.3319(15)	*X*3–O7	2.3802(16)	*Y*3–O4	2.0669(16)
*X*1–O2 4x	2.5167(15)	*X*3–O8	2.6055(16)	<*Y*3–O>	1.9662
<*X*1–O>	2.424	<*X*3–O>	2.5586		
				*T*1–O11 2×	1.637(4)
*X*2–O8	2.3252(15)	*X*4–O6 4×	2.2987(18)	*T*1–O7 2×	2.017(7)
*X*2–O5	2.3402(16	*X*4–O9 4×	2.621(2)	<*T*1–O>	1.827
*X*2–O3	2.3719(16)	<*X*4–O>	2.460		

Table A.15 Crystallographic data and refinement parameters for crystal structure of vesuvianite from Akhmatovskaya pit (sample 28975 FM)

Crystallographic parameters	
Symmetry	Tetragonal
Space group	$P4/n$
Unit cell parameters a, c (Å)	15.5572(4), 11.8425(4)
Unit cell volume (Å^3)	2866.22(16)
Z	2
Calculated density (g/cm^3)	3.402
Absorption coefficient (mm^{-1})	2.839
Crystal size (mm)	0.24 × 0.19 × 0.17
Data collection parameters	
2θ range (°)	5.05–55.00
h, k, l range	$-9 \rightarrow 20$, $-18 \rightarrow 16$, $-15 \rightarrow 7$
Number reflections	6787
Unique reflections (R_{int})	3201 (0.023)
Unique reflections c $F > 4\sigma F$	3110
Refinement parameters	
R_1 [$F > 4\sigma(F)$], wR_2 [$F > 4\sigma(F)$],	0.0270, 0.0638
R_1, wR_2 all data	0.0336, 0.0670
S	1.105
ρ_{max}, ρ_{min}, e Å^{-3}	0.60, –0.43
Refinement method	Full matrix least-squares on F^2

Table A.16 Atom fractional coordinates, displacement parameters (Å^2) and site occupancy for vesuvianite crystal structure from Akhmatovskaya pit (sample 28975 FM)

Site	Occupancy	x/a	y/b	z/c	U_{ani}	Multiplicity
Z1A	Si	¼	¾	0	0.0060(3)	2*a*
Z1B	Si	¾	¼	½	0.0061(3)	2*b*
Z2A	Si	0.67993(5)	0.45841(5)	0.12595(6)	0.0053(2)	8*g*
Z2B	Si	0.46142(5)	0.68145(5)	0.36912(6)	0.0056(2)	8*g*
Z3A	Si	0.34948(5)	0.42066(5)	0.13648(6)	0.0059(2)	8*g*
Z3B	Si	0.41106(5)	0.34853(5)	0.36518(6)	0.0060(2)	8*g*
*X*1	Ca	¾	¼	0.25007(6)	0.0087(2)	4*f*
*X*2A	Ca	0.54235(4)	0.30997(4)	0.12162(4)	0.0073(1)	8*g*
*X*2B	Ca	0.31195(4)	0.54596(4)	0.38081(4)	0.0075(1)	8*g*

(continued)

Table A.16 (continued)

Site	Occupancy	*x*/*a*	*y*/*b*	*z*/*c*	U_{ani}	Multiplicity
*X*3A	Ca	0.60393(4)	0.68372(4)	0.10432(5)	0.0105(1)	8*g*
*X*3B	Ca	0.67673(4)	0.59723(4)	0.37569(4)	0.0098(1)	8*g*
*X*4A	$Ca_{0.02}$	¾	¾	0.563(5)	0.001	2*c*
*X*4B	$Ca_{0.98}$	¼	¼	0.1464(1)	0.0084(4)	2*c*
*Y*2A	Al	½	½	0	0.0052(2)	4*d*
*Y*2B	Al	½	½	½	0.0041(2)	4*e*
*Y*3A	$Al_{0.873}Fe_{0.127}$	0.39023(5)	0.62244(5)	0.12692(5)	0.0060(2)	8*g*
*Y*3B	$Al_{0.781}Fe_{0.219}$	0.61856(4)	0.38768(5)	0.37298(5)	0.0081(2)	8*g*
*Y*1A	$Fe_{0.85}$	¾	¾	0.53063(7)	0.0049(3)	2*c*
*Y*1B	$Fe_{0.02}$	¼	¼	0.075(3)	0.003(6)	2*c*
O1A	O	0.2822(1)	0.6736(1)	0.0856(2)	0.0076(4)	8*g*
O1B	O	0.6719(1)	0.2764(1)	0.4138(1)	0.0094(4)	8*g*
O2A	O	0.6587(1)	0.3817(1)	0.2167(1)	0.0080(4)	8*g*
O2B	O	0.3848(1)	0.6619(1)	0.2769(1)	0.0075(4)	8*g*
O3A	O	0.4509(1)	0.7226(1)	0.0742(1)	0.0073(4)	8*g*
O3B	O	0.7226(1)	0.4530(1)	0.4238(1)	0.0076(4)	8*g*
O4A	O	0.6057(1)	0.4373(1)	0.0272(1)	0.0069(4)	8*g*
O4B	O	0.4401(1)	0.6064(1)	0.4658(2)	0.0086(4)	8*g*
O5A	O	0.3321(1)	0.5189(1)	0.1801(1)	0.0077(4)	8*g*
O5B	O	0.5083(1)	0.3259(1)	0.3221(2)	0.0089(4)	8*g*
O6A	O	0.2715(1)	0.3867(1)	0.0615(2)	0.0105(4)	8*g*
O6B	O	0.3726(1)	0.2722(1)	0.4423(1)	0.0102(4)	8*g*
O7A	O	0.6716(1)	0.5551(1)	0.1737(1)	0.0095(4)	8*g*
O7B	O	0.5573(1)	0.6766(1)	0.3173(1)	0.0103(4)	8*g*
O8A	O	0.4398(1)	0.4105(1)	0.0700(1)	0.0080(4)	8*g*
O8B	O	0.4069(1)	0.4383(1)	0.4358(1)	0.0083(4)	8*g*
O9	O	0.3510(1)	0.3583(1)	0.2498(1)	0.0090(4)	8*g*
O10A	O	¾	¾	0.3782(3)	0.0127(9)	2*c*
O10B	O	¾	¾	0.1422(3)	0.0110(9)	2*c*
O11A	O	0.4965(1)	0.5645(1)	0.1335(2)	0.0084(4)	8*g*
O11B	O	0.5595(1)	0.4955(1)	0.3636(2)	0.0083(4)	8*g*
H11A	H	0.529(3)	0.558(4)	0.197(3)	0.11(2)	8*g*
H11B	H	0.520(3)	0.495(3)	0.305(3)	0.10(2)	8*g*
H10B	H	¾	¾	0.221(5)	0.015(17)	2*c*

Table A.17 Selected bond length (Å) in the crystal structure of vesuvianite c from Akhmatovskaya pit (sample 28975 FM)

Bond	Bond length (Å)	Bond	Bond length (Å)	Bond	Bond length (Å)
Z1A–O1A 4×	1.6410(18)	X2A–O6A	2.824(2)	X4B–O6A 4×	2.3763(19)
		X2A–O8A	2.3154(19)	X4B–O9 4×	2.6104(19)
Z1B–O1B	1.6390(18)	<X2A–O>	2.4650	<X4B–O>	2.4933
Z2A–O2A	1.6393(19)	X2B–O1B	2.4249(19)	Y1A–O6B 4×	1.9644(19)
Z2A–O3A	1.640(2)	X2B–O2B	2.4594(19)	Y1A–O10A	1.805(4)
Z2A–O4A	1.6760(19)	X2B–O3B	2.3750(18)	<Y1A–O>	1.9325
Z2A–O7A	1.613(2)	X2B–O4B	2.4238(19)		
<Z2A–O>	1.6421	X2B–O5B	2.3295(19)	Y1B–O6A 4×	2.159(3)
		X2B–O5A	2.4343(18)	Y1B–10B	2.571(3)
Z2B–O2B	1.6456(19)	X2B–O6B	3.085(2)	<Y1B–O>	2.2414
Z2B–O3B	1.6329(19)	X2B–O8B	2.3259(19)		
Z2B–O4B	1.6680(19)	<X2B–O>	2.4822	Y2A–O4A 2×	1.9395(18)
Z2B–O7B	1.615(2)			Y2A–O8A 2×	1.8708(18)
<Z2B–O>	1.6404	X3A–O3A	2.4828(19)	Y2A–O11A 2×	1.8731(2)
		X3A–O6A	2.968(2)	<Y2A–O>	1.8946
Z3A–O5A	1.636(2)	X3A–O6A	2.3955(19)		
Z3A–O6A	1.594(2)	X3A–O7A	2.513(2)	Y2B–O4B 2×	1.9428(19)
Z3A–O8A	1.6192(19)	X3A–O7A	2.406(2)	Y2B–O8B 2×	1.8951(18)
Z3A–O9	1.6564(18)	X3A–O7B	2.6270(19)	Y2B–O11B 2×	1.8633(18)
<Z3A–O>	1.6264	X3A–O8A	2.6232(19)	<Y2B–O>	1.9004
		X3A–O10B	2.5355(9)		
Z3B–O5B	1.635(2)	X3A–O11A	2.520(2)	Y3A–O1A	1.922(2)
Z3B–O6B	1.614(2)	<X3A–O>	2.5634	Y3A–O2B	1.8812(18)
Z3B–O8B	1.630(2)			Y3A–O3A	1.925(2)
Z3B–O9	1.6611(19)	X3B–O3B	2.4230(19)	Y3A–O4A	2.0489(19)
<Z3B–O>	1.6350	X3B–O6B	3.060(2)	Y3A–O5A	1.952(2)
		X3B–O6B	2.6587(19)	Y3A–O11A	1.885(2)
X1–O1A × 2	2.3359(19)	X3B–O7B	2.464(2)	<Y3A–O>	1.9357
X1–O1B × 2	2.3252(19)	X3B–O7B	2.336(2)		

(continued)

Table A.17 (continued)

Bond	Bond length (Å)	Bond	Bond length (Å)	Bond	Bond length (Å)
*X*1–O2A × 2	2.5247(19)	*X*3B–O7A	2.4806(19)	*Y*3B–O1B	1.980(2)
*X*1–O2B × 2	2.5248(19)	*X*3B–O8B	2.6432(19)	*Y*3B–O2A	1.9557(18)
<*X*1–O>	2.4277	*X*3B–O10A	2.6360(6)	*Y*3B–O3B	2.003(2)
		*X*3B–O11B	2.419(2)	*Y*3B–O4B	2.1181(19)
*X*2A–O1A	2.5313(19)	<*X*3B–O>	2.5689	*Y*3B–O5B	2.057(2)
*X*2A–O2A	2.4057(19)			*Y*3B–O11B	1.916(2)
*X*2A–O3A	2.3755(18)	*X*4A–O6B 4×	1.939(3)	<*Y*3B–O>	2.0050
*X*2A–O4A	2.4785(19)	*X*4A–O9 4×	3.195(2)		
*X*2A–O5A	2.3446(19)	<*X*4A–O>	2.567		
*X*2A–O5B	2.4449(19)				

Table A.18 Crystallographic data and refinement parameters for crystal structure of vesuvianite from Iordanow

Crystallographic parameters	
Symmetry	Tetragonal
Space group	*P*4*nc*
Unit cell parameters *a*, *c* (Å)	15.5876(2), 11.8465(2)
Unit cell volume (Å^3)	2878.38(9)
Z	2
Calculated density (g/cm^3)	3.349
Absorption coefficient (mm^{-1})	2.759
Crystal size (mm)	0.19 × 0.17 × 0.15
Data collection parameters	
2θ range (°)	5.33–55.00
h, *k*, *l* range	−15 → 17, −10 → 20, −15 → 9
Number reflections	6959
Unique reflections (R_{int})	2524 (0.024)
Unique reflections c $F > 4\sigma F$	2414
Refinement parameters	
R_1 [$F > 4\sigma(F)$], wR_2 [$F > 4\sigma(F)$],	0.0268, 0.0724
R_1, wR_2 all data	0.0285, 0.0733
S	1.118
ρ_{max}, ρ_{min}, e Å^{-3}	0.69, −0.52
Refinment method	0.11(2)
Crystallographic parameters	Full matrix least-squares on F^2

Table A.19 Atom fractional coordinates, displacement parameters (Å^2) and site occupancy for vesuvianite crystal structure from Iordanow

Site	Occupancy	*x*/*a*	*y*/*b*	*z*/*c*	U_{ani}	Multiplicity
*X*1	Ca	½	0	0.7167(2)	0.0079(3)	4*b*
*X*2A	Ca	0.56154(8)	0.79364(9)	0.5859(1)	0.0075(3)	8*c*
*X*2B	Ca	0.29490(9)	0.93992(9)	0.3450(1)	0.0078(3)	8*c*
*X*3A	Ca	0.56580(9)	0.6475(1)	0.3610(1)	0.0146(4)	8*c*
*X*3B	Ca	0.42911(9)	0.65121(9)	0.0846(1)	0.0124(3)	8*c*
*X*4A	$Ca_{0.87}$	½	½	0.6124(3)	0.009(1)	2*a*
*X*4B	$Ca_{0.13}$	½	½	0.814(2)	0.014(7)	2*a*
*Z*1	Si	½	0	0.4660(3)	0.0065(3)	4*b*
*Z*2A	Si	0.4310(1)	0.7912(1)	0.3379(2)	0.0056(4)	8*c*
*Z*2B	Si	0.5696(1)	0.7910(1)	0.0947(2)	0.0066(4)	8*c*
*Z*3A	Si	0.5991(1)	0.6629(1)	0.8297(2)	0.0065(4)	8*c*
*Z*3B	Si	0.6700(1)	0.5993(1)	0.6008(2)	0.0063(4)	8*c*
*Y*1A	$Fe_{0.08}Al_{0.05}$	½	½	0.514(2)	0.021(8)	2*a*
*Y*1B	$Fe_{0.75}Al_{0.12}$	½	½	0.9292(2)	0.0099(7)	2*a*
*Y*2	Al	0.7518(2)	0.7509(1)	0.4647(2)	0.0062(3)	8*c*
*Y*3A	$Al_{0.936(7)}Fe_{0.064(7)}$	0.6384(1)	0.8697(1)	0.8387(2)	0.0089(6)	8*c*
*Y*3B	$Al_{0.974(7)}Fe_{0.026(7)}$	0.3729(1)	0.8632(1)	0.0910(2)	0.0054(6)	8*c*
O1A	O	0.4693(3)	0.9226(3)	0.5519(4)	0.0105(10)	8*c*
O1B	O	0.5292(3)	0.9229(3)	0.8807(4)	0.0070(9)	8*c*
O2A	O	0.6321(3)	0.9092(3)	0.6849(4)	0.0100(10)	8*c*
O2B	O	0.4108(3)	0.8665(3)	0.2441(4)	0.0091(10)	8*c*
O3A	O	0.5279(3)	0.7974(3)	0.3908(4)	0.0079(10)	8*c*
O3B	O	0.4724(3)	0.7991(3)	0.0420(3)	0.0088(10)	8*c*
O4A	O	0.3567(3)	0.8143(3)	0.4353(4)	0.0079(9)	8*c*
O4B	O	0.6446(3)	0.8094(3)	0.9960(4)	0.0086(10)	8*c*
O5A	O	0.7675(3)	0.5813(3)	0.6441(4)	0.0096(10)	8*c*
O5B	O	0.5774(3)	0.7608(3)	0.7867(4)	0.0106(10)	8*c*
O6A	O	0.6339(3)	0.5222(3)	0.5254(4)	0.0122(11)	8*c*
O6B	O	0.5217(3)	0.6258(3)	0.9056(4)	0.0140(11)	8*c*
O7A	O	0.3047(3)	0.5752(3)	0.1469(4)	0.0089(9)	8*c*
O7B	O	0.4200(3)	0.6936(3)	0.2901(4)	0.0123(10)	8*c*
O8A	O	0.8432(3)	0.8115(3)	0.4002(4)	0.0083(9)	8*c*
O8B	O	0.6610(3)	0.6899(3)	0.5334(4)	0.0086(10)	8*c*
O9	O	0.6072(3)	0.6033(3)	0.7144(4)	0.0107(7)	8*c*
O10A	O	½	½	0.0917(9)	0.014(2)	2*a*
O10B	F	½	½	0.3239(7)	0.0141(17)	2*a*

(continued)

Table A.19 (continued)

Site	Occupancy	*x/a*	*y/b*	*z/c*	U_{ani}	Multiplicity
O11A	O	0.3137(3)	0.7546(2)	0.0993(4)	0.0076(9)	8*c*
O11B	O	0.6918(3)	0.7533(3)	0.3267(4)	0.0118(10)	8*c*

Table A.20 Selected bond lengths (Å) in the crystal structure of vesuvianite from Iordanow

Bond	Bond length (Å)	Bond	Bond length (Å)	Bond	Bond length (Å)
*Z*1–O1A 2×	1.649(5)	*X*2A–O5B	2.445(5)	<*X*4A–O>	2.485
*Z*1–O1B 2×	1.635(5)	*X*2A–O6A	2.902(5)		
<*Z*1–O>	1.642	*X*2A–O8B	2.325(5)	*X*4B–O6B 4×	2.264(13)
		<*X*2A–O>	2.475	*X*4B–O9 4×	2.606(12)
*Z*2A–O2B	1.647(5)			<*X*4B–O>	2.435
*Z*2A–O3A	1.638(5)	*X*2B–O1B	2.469(4)		
*Z*2A–O4A	1.674(5)	*X*2B–O2B	2.450(5)	*Y*1A–O6A 4×	2.120(5)
*Z*2A–O7B	1.632(5)	*X*2B–O3B	2.389(4)	*Y*1A–O10B 4×	2.25(3)
<*Z*2A–O>	1.648	*X*2B–O4A	2.430(4)	<*Y*1A–O>	2.146
		*X*2B–O5B	2.313(5)		
*Z*2B–O2A	1.640(5)	*X*2B–O5A	2.440(5)	*Y*1B–O6B 4×	2.010(5)
*Z*2B–O3B	1.644(5)	*X*2B–O6B	3.014(5)	*Y*1B–10A	1.925(12)
*Z*2B–O4B	1.678(5)	*X*2B–O8A	2.335(5)	<*Y*1B–O>	1.993
*Z*2B–O7A	1.616(4)	<*X*2B–O>	2.480		
<*Z*2B–O>	1.645			*Y*2–O4A	1.970(5)
		*X*3A–O3A	2.436(4)	*Y*2–O4B	1.924(5)
*Z*3A–O5A	1.636(2)	*X*3A–O6A	2.392(5)	*Y*2–O8A	1.873(5)
*Z*3A–O6B	1.613(5)	*X*3A–O6A	2.954(5)	*Y*2–O8B	1.890(5)
*Z*3A–O8A	1.628(5)	*X*3A–O7A	2.648(5)	*Y*2–O11A	1.866(5)
*Z*3A–O9	1.656(5)	*X*3A–O7B	2.528(5)	*Y*2–O11B	1.884(5)
<*Z*3A–O>	1.633	*X*3A–O7B	2.404(4)	<*Y*2–O>	1.901
		*X*3A–O8B	2.610(5)		
*Z*3B–O5A	1.642(5)	*X*3A–O10B	2.555(2)	*Y*3A–O1B	1.958(5)
*Z*3B–O6A	1.628(5)	*X*3A–O11B	2.598(5)	*Y*3A–O2A	1.925(5)
*Z*3B–O8B	1.629(5)	<*X*3A–O>	2.569	*Y*3A–O3A	1.983(5)
*Z*3B–O9	1.666(5)			*Y*3A–O4B	2.089(5)
<*Z*3B–O>	1.641	*X*3B–O3B	2.454(5)	*Y*3A–O5B	2.041(5)
		*X*3B–O6B	2.595(4)	*Y*3A–O11B	1.946(5)

(continued)

Table A.20 (continued)

Bond	Bond length (Å)	Bond	Bond length (Å)	Bond	Bond length (Å)
*X*1–O1A 2×	2.344(5)	*X*3B–O6B	3.051(5)	<*Y*3A–O>	1.990
*X*1–O1B 2×	2.330(5)	*X*3B–O7B	2.527(5)		
*X*1–O2A 2×	2.527(5)	*X*3B–O7A	2.390(4)	*Y*3B–O1A	1.885(5)
*X*1–O2B 2×	2.523(5)	*X*3B–O7A	2.491(5)	*Y*3B–O2B	1.908(5)
<*X*1–O>	2.431	*X*3B–O8A	2.628(5)	*Y*3B–O3B	1.934(5)
		*X*3B–O10A	2.6046(14)	*Y*3B–O4A	2.061(5)
*X*2A–O1A	2.505(5)	*X*3B–O11A	2.422(4)	*Y*3B–O5A	1.960(5)
*X*2A–O2A	2.414(5)	<*X*3B–O>	2.574	*Y*3B–O11A	1.931(4)
*X*2A–O3A	2.372(5)			<*Y*3B–O>	1.947
*X*2A–O4B	2.472(4)	*X*4A–O6A 4×	2.353(5)		
*X*2A–O5A	2.366(5)	*X*4A–O9 4×	2.616(4)		

Table A.21 Powder diffraction data for wiluite from Yakutia (G. V. Barkhudarova's sample). The eight most intense lines are shown in bold

d_{meas} Å	*hkl*	I_{meas}	d_{meas} Å	*hkl*	I_{meas}
11.01	110	4	2.138	641	10
4.013	222	4	2.113	721	10
3.920	400	4	2.076	712	7
3.502	420	11	2.062	524	14
3.256	402	5	2.028	731	9
3.074	510	9	2.006	633	12
3.058	313	8	1.9821	651	5
3.029	501	9	1.9436	425	2
3.006	422	9	1.9294	713	4
2.978	511	6	1.9041	820	5
2.921	**004**	**27**	1.8894	624	5
2.771	**440**	**66**	1.8764	821	3
2.764	**432**	**100**	1.8625	515	1
2.689	214	6	1.7681	714	14
2.606	**600**	**63**	1.7452	644	2
2.583	**224**	**21**	1.7356	841	4
2.517	611	7	1.6851	921	13
2.479	**620**	**46**	1.6566	107	13

(continued)

Table A.21 (continued)

d_{meas} Å	*hkl*	I_{meas}	d_{meas} Å	*hkl*	I_{meas}
2.443	532	10	**1.6360**	**931**	**54**
2.388	602	6	1.6304	804	9
2.343	404	4	**1.6213**	**814**	**21**
2.316	414	7	1.5943	824	3
2.293	631	8	1.5873	770	13
2.218	710	8	1.5653	664	4
2.203	701	16	1.5564	327	7
2.180	711	10	1.5492	1011	7

Table A.22 Crystallographic data and refinement parameters for the crystal structure of Wiluite from Yakutia

Crystallographic parameters	
Symmetry	Tetragonal
Space group	*P4/nnc*
Unit cell parameters *a*, *c* (Å)	15.7027(3), 11.7008(3)
Unit cell volume (Å^3)	2885.12(12)
Z	2
Calculated density (g/cm^3)	3.412
Absorption coefficient (mm^{-1})	2.730
Crystal size (mm)	0.21 × 0.15 × 0.11
Data collection parameters	
2θ range (°)	6.77–55.00
h, *k*, *l* range	$-11 \rightarrow 20$, $-20 \rightarrow 13$, $-15 \rightarrow 11$
Number reflections	8686
Unique reflections (R_{int})	1643 (0.044)
Unique reflections c $F > 4\sigma F$	1533
Refinement parameters	
R_1 [$F > 4\sigma(F)$], wR_2 [$F > 4\sigma(F)$],	0.0220, 0.0586
R_1, wR_2 all data	0.0240, 0.0600
S	1.091
ρ_{max}, ρ_{min}, e Å^{-3}	0.31, –0.58
Refinment method	Full matrix least-squares on F^2

Table A.23 Fractional atom coordinates, displacement parameters (Å^2) and site occupancy for wiluite crystal structure

Site	Occupancy	*x*/*a*	*y*/*b*	*z*/*c*	U_{ani}	Multiplicity
*X*1	Ca	¾	¼	¼	0.0064(1)	4*c*
*X*2	Ca	0.69058(2)	0.04532(2)	0.12003(3)	0.0072(1)	16*k*
*X*3	Ca	0.67936(3)	–0.10140(3)	–0.10357(4)	0.0140(1)	16*k*
*X*4	$Ca_{0.5}$	¾	¾	0.1430(1)	0.0070(3)	4*e*
*Z*1	Si	¾	¼	0	0.0062(2)	4*d*
*Z*2	Si	0.54358(4)	0.17851(3)	0.12776(4)	0.0095(1)	16*k*
*Z*3	Si	0.65021(3)	−0.08393(3)	0.36489(4)	0.0066(1)	16*k*
*Y*1	$Mg_{0.28}Fe_{0.22}$	¾	¾	0.0535(2)	0.0127(6)	4*e*
*Y*2	Al	½	0	½	0.0061(2)	8*f*
*Y*3	$Al_{0.91}Fe_{0.09}$	0.61162(3)	0.12009(3)	0.37214(4)	0.0071(2)	16*k*
O1	O	0.78022(8)	0.17262(8)	0.0844(1)	0.0093(3)	16*k*
O2	O	0.62126(9)	0.16061(8)	0.2171(1)	0.0111(3)	16*k*
O3	O	0.54227(9)	0.22479(9)	0.4217(1)	0.0118(3)	16*k*
O4	O	0.60441(8)	0.06182(8)	0.5284(1)	0.0092(3)	16*k*
O5	O	0.82837(8)	0.01183(8)	0.1797(1)	0.0098(3)	16*k*
O6	O	0.62133(9)	–0.22357(8)	0.0532(1)	0.0116(3)	16*k*
O7A	$O_{0.50}$	0.4450(5)	0.1745(3)	0.1814(6)	0.012(1)	16*k*
O7B	$O_{0.50}$	0.4556(5)	0.1463(3)	0.1924(7)	0.016(1)	16*k*
O8	O	0.59229(8)	−0.06024(8)	0.0664(1)	0.0078(3)	16*k*
O9	O	0.64627(8)	−0.14627(8)	¼	0.0094(4)	8 *h*
O10	$O_{0.64(1)}$	¾	¾	–0.1301(4)	0.025(2)	4*e*
O11	O	0.44210(9)	–0.00206(9)	0.1457(1)	0.0143(3)	16*k*
O12	$O_{0.19(1)}$	0.6770(6)	–0.2265(5)	–0.2045(7)	0.031(3)	16*k*
*T*1	$B_{0.51}Al_{0.18}$	0.4465(2)	0.0535(2)	¼	0.018(1)	8*h*
*T*2	B	¾	¾	¾	0.015(1)	2*a*

Table A.24 Selected bond lengths (Å) in the crystal structure of Wiluite from Yautia

Bond	Bond length (Å)	Bond	Bond length (Å)	Bond	Bond length (Å)
*Z*1–O1 4x	1.6364(13)	*X*2–O5	2.4201(13)	<*Y*1–O>	2.0797
		*X*2–O4	2.4697(13)		
*Z*2–O7(A,B)	1.672(7)	*X*2–O1	2.4800(13)	*Y*2–O8 2×	1.8971(12)
*Z*2–O2	1.6309(14)	<*X*2–O>	2.4850	*Y*2–O11 2×	1.9321(14)
*Z*2–O3	1.6249(15)			*Y*2–O4 2×	1.9340(13)
*Z*2–O4	1.6699(14)	*X*3–O7A	2.441(7)	<*Y*2–O>	1.9211
<*Z*2–O>	1.6472	*X*3–O3	2.3845(14)		
		*X*3–O11	2.5534(15)	*Y*3–O1	1.9553(14)
*Z*3–O6	1.6093(14)	*X*3–O7A	2.574(5)	*Y*3–O11	1.9892(15)
*Z*3–O8	1.6308(13)	*X*3–O6	2.4056(13)	*Y*3–O2	1.9283(14)
*Z*3–O5	1.6268(14)	*X*3–O7A	2.620(7)	*Y*3–O3	2.0554(15)
*Z*3–O9	1.6641(8)	*X*3–O10	2.6021(7)	*Y*3–O5	2.0360(14)
<*Z*3–O>	1.6328	*X*3–O7B	2.464(8)	*Y*3–O4	2.0472(13)
		*X*3–O8	2.4986(13)	<*Y*3–O>	2.0019
*X*1–O1 4×	2.3355(13)	*X*3–O7B	2.601(8)		
*X*1–O2 4×	2.4910(13)	<*X*3–O>	2.5144	*T*1–O11 2×	1.502(2)
<*X*1–O>	2.4133			*T*1–O7B 2×	1.612(5)
		*X*4–O6 4x	2.3145(15)	<*T*1–O>	1.557
*X*2–O8	2.3501(13)	*X*4–O9 4x	2.6218(18)		
*X*2–O5	2.3336(13)	<*X*4–O>	2.4685	*T*2–O10 2×	1.403(5)
*X*2–O3	2.3829(13)			*T*1–O12	1.317(9)
*X*2–O6	3.0457(14)	*Y*1–O6 4x	2.0626(14)	<*T*2–O>	1.374
*X*2–O2	2.3984(14)	*Y*1–O10	2.148(5)		

Table A.25 Powder diffraction data for cyprine from Wessels. The eight most intense lines are shown in bold

d_{meas} Å	*hkl*	I_{meas}	d_{meas} Å	*hkl*	I_{meas}
9.40	101	7	**2.324**	**144**	**10**
5.89	**002**	**12**	2.275	631	2
5.50	220	2	2.208	533	2
5.192	112	2	2.201	710	5
4.916	310	3	2.186	701	1
4.704	202	5	2.162	711	2
4.497	212	3	2.141	504	3
4.019	222	9	2.122	154	5
3.886	400	3	2.086	623	3
3.777	312	2	2.064	254	3
3.667	330	0.6	2.042	730	7
3.482	240	10	2.026	642	2
3.244	402	7	2.014	731	3
3.073	313	3	1.9976	363	4
3.048	510	8	1.9635	651	4
3.007	**431**	**12**	1.9205	713	3
2.950	**004**	**47**	1.8885	624	10
2.906	323	5	1.7975	282	2
2.752	**432**	**100**	1.7634	714	8
2.666	530	4	1.7184	910	4
2.594	**522**	**76**	**1.6224**	**672**	**28**
2.530	314	5	1.5714	770	5
2.498	611	1	1.5593	166	9
2.459	**620**	**35**	1.5233	655	2
2.441	503	7	1.4983	844	5
2.376	602	4	1.4753	517	4
2.351	404	7			

Table A.26 Crystallographic data and refinement parameters for crystal structure of cyprine from the Wessels mine

Crystallographic parameters		
Zone	Mg (1)	Mn (2)
Symmetry	Tetragonal	
Space group	*P*4/*n*	
Unit cell parameters *a*, *c* (Å)	15.5173(4), 11.8230(5)	15.5699(2), 11.8042(2)
Unit cell volume ($Å^3$)	2846.8(2)	2861.6(1)
Z	2	
Calculated density (g/cm^3)	3.44	3.49
Absorption coefficient (mm^{-1})	3.125	3.504
Crystal size (mm)	0.15 × 0.14 × 0.12	0.07 × 0.07 × 0.05
Data collection parameters		
2θ range (°)	5.06–57.84	5.06 – 53.00
h, *k*, *l* range	−20 → 10, −20 → 18, −15 → 15	−18 → 15, −19 → 13, −13 → 14
Number reflections	14877	6688
Unique reflections (R_{int})	3254 (0.028)	2869 (0.037)
Unique reflections c $F > 4\sigma F$	2848	2071
Refinement parameters		
R_1 [$F > 4\sigma(F)$], wR_2 [$F > 4\sigma(F)$]	0.034, 0.077	0.035, 0.070
R_1, wR_2 all data	0.042, 0.080	0.051, 0.077
S	1.265	0.917
ρ_{max}, ρ_{min}, e $Å^{-3}$	0.43, –0.64	1.49, –1.42
Refinment method	Full matrix least-squares on F^2	

Table A.27 Atom fractional coordinates, displacement parameters (Å^2) and site occupancy for Mg-zone of cyprine crystal structure from the Wessels mine, RSA

Site	Occupancy	*x*/*a*	*y*/*b*	*z*/*c*	U_{ani}	Multiplicity
*Z*1A	$Si_{0.942(11)}$	¼	¾	0	0.0046(7)	2*a*
*Z*1B	$Si_{0.969(11)}$	¾	¼	½	0.0067(7)	2*b*
*Z*2A	Si	0.68089(6)	0.45926(6)	0.12898(8)	0.0075(2)	8*g*
*Z*2B	Si	0.45930(6)	0.68104(6)	0.37077(8)	0.0074(2)	8*g*
*Z*3A	Si	0.34872(6)	0.41624(6)	0.13498(8)	0.0083(2)	8*g*
*Z*3B	Si	0.41597(6)	0.34874(6)	0.36497(8)	0.0082(2)	8*g*
*X*1	Ca	¾	¼	0.25008(9)	0.0084(2)	4*f*
*X*2A	Ca	0.54365(4)	0.31110(4)	0.12061(6)	0.0083(2)	8*g*
*X*2B	Ca	0.31112(4)	0.54380(4)	0.37940(6)	0.0083(2)	8*g*
*X*3A	Ca	0.60100(5)	0.68042(5)	0.11168(7)	0.0134(2)	8*g*
*X*3B	Ca	0.68009(5)	0.60087(5)	0.38776(7)	0.0134(2)	8*g*
*X*4A	$Ca_{0.475(3)}$	¾	¾	0.3479(3)	0.0096(9)	2*c*
*X*4B	$Ca_{0.525(3)}$	¼	¼	0.1519(2)	0.0119(8)	2*c*
*Y*1A	$Cu_{0.525(3)}$	¾	¾	−0.5551(2)	0.0123(6)	2*c*
*Y*1B	$Cu_{0.475(3)}$	¼	¼	0.0555(2)	0.0101(7)	2*c*
*Y*2A	Al	½	½	0	0.0052(3)	4*d*
*Y*2B	Al	½	½	½	0.0047(3)	4*e*
*Y*3A	$Al_{0.825}Mn_{0.175}$	0.38891(5)	0.61991(5)	0.12585(7)	0.0072(3)	8*g*
*Y*3B	$Al_{0.803}Mn_{0.197}$	0.61974(5)	0.38891(5)	0.37396(7)	0.0075(3)	8*g*
O1A	O	0.28064(15)	0.67262(15)	0.0858(2)	0.0083(5)	8*g*
O1B	O	0.67257(15)	0.28057(15)	0.4145(2)	0.0088(5)	8*g*
O2A	O	0.65991(15)	0.38220(15)	0.2210(2)	0.0104(5)	8*g*
O2B	O	0.38234(15)	0.65996(15)	0.2788(2)	0.0096(5)	8*g*
O3A	O	0.45105(15)	0.72188(15)	0.0754(2)	0.0093(5)	8*g*
O3B	O	0.72199(15)	0.45100(15)	0.4246(2)	0.0090(5)	8*g*
O4A	O	0.60683(15)	0.43767(15)	0.0303(2)	0.0099(5)	8*g*
O4B	O	0.43773(15)	0.60683(15)	0.4696(2)	0.0097(5)	8*g*
O5A	O	0.32925(15)	0.51429(16)	0.1782(2)	0.0117(5)	8*g*
O5B	O	0.51397(16)	0.32904(15)	0.3218(2)	0.0112(5)	8*g*
O6A	O	0.27089(16)	0.38036(17)	0.0591(2)	0.0142(5)	8*g*
O6B	O	0.37996(17)	0.27087(16)	0.4414(2)	0.0141(5)	8*g*
O7A	O	0.67360(15)	0.55614(15)	0.1781(2)	0.0118(5)	8*g*
O7B	O	0.55612(15)	0.67387(16)	0.3218(2)	0.0118(5)	8*g*
O8A	O	0.43898(15)	0.40857(15)	0.0667(2)	0.0092(5)	8*g*
O8B	O	0.40845(15)	0.43869(15)	0.4340(2)	0.0095(5)	8*g*
O9	O	0.35485(15)	0.35480(15)	0.2498(2)	0.0112(5)	8*g*

(continued)

Table A.27 (continued)

Site	Occupancy	*x*/*a*	*y*/*b*	*z*/*c*	U_{ani}	Multiplicity
O10A	O	¾	¾	0.3641(5)	0.0146(11)	2*c*
O10B	O	¾	¾	0.1360(5)	0.0135(11)	2*c*
O11A	O	0.49504(16)	0.56201(16)	0.1359(2)	0.0099(5)	8*g*
O11B	O	0.56196(16)	0.49479(16)	0.3641(2)	0.0097(5)	8*g*
H11A	H	0.508(4)	0.526(4)	0.186(5)	0.07(2)	8*g*
H11B	H	0.533(4)	0.507(4)	0.305(4)	0.050	8*g*
H10B	H	¾	¾	0.220(7)	0.01(2)	2*c*

Table A.28 Fractional atom coordinates, displacement parameters (Å^2) and site occupancy for crystal structure of cyprine Mn-zone from the Wessels mine, RSA

Site	Occupancy	*x*/*a*	*y*/*b*	*z*/*c*	U_{ani}	Multiplicity
*Z*1A	$Si_{0.946(10)}$	¼	¾	0	0.0039(9)	2*a*
*Z*1B	$Si_{0.965(10)}$	¾	¼	½	0.0047(9)	2*b*
*Z*2A	Si	0.68159(10)	0.45943(10)	0.13062(10)	0.0105(3)	8*g*
*Z*2B	Si	0.45848(9)	0.68077(10)	0.37213(10)	0.0086(3)	8*g*
*Z*3A	Si	0.34811(9)	0.41400(10)	0.13464(10)	0.0083(3)	8*g*
*Z*3B	Si	0.41745(10)	0.34884(10)	0.36434(10)	0.0082(3)	8*g*
*X*1	Ca	¾	¼	0.25024(10)	0.0072(3)	4*f*
*X*2A	Ca	0.54471(7)	0.31183(7)	0.12073(7)	0.0075(2)	8*g*
*X*2B	Ca	0.31081(7)	0.54320(7)	0.37878(7)	0.0075(2)	8*g*
*X*3A	Ca	0.59908(7)	0.67778(7)	0.11509(8)	0.0130(3)	8*g*
*X*3B	Ca	0.68077(7)	0.60093(7)	0.39007(8)	0.0125(3)	8*g*
*X*4A	$Ca_{0.663(5)}$	¾	¾	0.3528(2)	0.0093(9)	2*c*
*X*4B	$Ca_{0.337(5)}$	¼	¼	0.1460(5)	0.0081(17)	2*c*
*Y*1A	$Cu_{0.337(5)}$	¾	¾	−0.5547(3)	0.0062(14)	2*c*
*Y*1B	$Cu_{0.663(5)}$	¼	¼	0.05286(16)	0.0098(7)	2*c*
*Y*2A	$Al_{0.736}Mn_{0.264}$	½	½	0	0.0060(5)	4*d*
*Y*2B	$Al_{0.866}Mn_{0.134}$	½	½	½	0.0069(6)	4*e*
*Y*3A	$Al_{0.528}Mn_{0.472}$	0.38879(7)	0.61802(7)	0.12502(7)	0.0067(3)	8*g*
*Y*3B	$Al_{0.752}Mn_{0.248}$	0.62064(8)	0.38878(8)	0.37464(8)	0.0058(4)	8*g*
O1A	O	0.2797(2)	0.6725(2)	0.0858(2)	0.0069(8)	8*g*
O1B	O	0.6731(2)	0.2812(2)	0.4147(2)	0.0073(8)	8*g*
O2A	O	0.6609(2)	0.3824(2)	0.2230(2)	0.0095(8)	8*g*
O2B	O	0.3815(2)	0.6599(2)	0.2803(2)	0.0082(8)	8*g*
O3A	O	0.4516(2)	0.7219(2)	0.0749(2)	0.0077(8)	8*g*
O3B	O	0.7218(2)	0.4507(2)	0.4251(2)	0.0078(8)	8*g*
O4A	O	0.6078(2)	0.4372(2)	0.0320(2)	0.0084(8)	8*g*

(continued)

Table A.28 (continued)

Site	Occupancy	*x*/*a*	*y*/*b*	*z*/*c*	U_{ani}	Multiplicity
O4B	O	0.4371(2)	0.6071(2)	0.4714(2)	0.0085(8)	8*g*
O5A	O	0.3274(2)	0.5109(2)	0.1783(2)	0.0110(8)	8*g*
O5B	O	0.5151(2)	0.3294(2)	0.3212(2)	0.0102(8)	8*g*
O6A	O	0.2717(2)	0.3770(2)	0.0564(2)	0.0120(8)	8*g*
O6B	O	0.3824(2)	0.2709(2)	0.4406(2)	0.0132(8)	8*g*
O7A	O	0.6742(2)	0.5560(2)	0.1803(2)	0.0151(9)	8*g*
O7B	O	0.5552(2)	0.6729(2)	0.3239(2)	0.0119(8)	8*g*
O8A	O	0.4384(2)	0.4075(2)	0.0654(2)	0.0076(8)	8*g*
O8B	O	0.4087(2)	0.4392(2)	0.4322(2)	0.0073(8)	8*g*
O9	O	0.3554(2)	0.3531(2)	0.2504(2)	0.0087(8)	8*g*
O10A	O	¾	¾	0.3634(5)	0.0147(17)	2*c*
O10B	O	¾	¾	0.1398(5)	0.0172(17)	2*c*
O11A	O	0.4939(2)	0.5616(2)	0.1380(2)	0.0084(8)	8*g*
O11B	O	0.5624(2)	0.4951(2)	0.3630(2)	0.0084(8)	8*g*

Table A.29 Selected bond length (Å) in the crystal structure of Mg-zone of cyprine from the Wessels mine, RSA

Bond	Bond length (Å)	Bond	Bond length (Å)	Bond	Bond length (Å)
*Z*1A–O1A 4×	1.642(2)	*X*2A–O6A	2.927(3)	*X*4B–O6A 4×	2.325(3)
		*X*2A–O8A	2.309(2)	*X*4 × B–O9 4×	2.576(3)
*Z*1B–O1B	1.640(2)	<*X*2A–O>	2.467	<*X*4B–O>	2.451
*Z*2A–O2A	1.649(2)	*X*2B–O1B	2.488(2)	*Y*1A–O6B 4×	2.043(3)
*Z*2A–O3A	1.641(2)	*X*2B–O2B	2.426(2)	*Y*1A–O10A	2.259(6)
*Z*2A–O4A	1.672(3)	*X*2B–O3B	2.374(3)	<*Y*1A–O>	2.086
*Z*2A–O7A	1.616(2)	*X*2B–O4B	2.440(2)		
<*Z*2A–O>	1.645	*X*2B–O5B	2.325(2)	*Y*1B–O6A 4×	2.049(3)
		*X*2B–O5A	2.439(3)	*Y*1B–10B	2.264(6)
*Z*2B–O2B	1.648(2)	*X*2B–O6B	2.936(3)	<*Y*1B–O>	2.094
*Z*2B–O3B	1.639(2)	*X*2B–O8B	2.315(2)		
*Z*2B–O4B	1.675(3)	<*X*2B–O>	2.468	*Y*2A–O4A 2×	1.952(2)
*Z*2B–O7B	1.614(3)			*Y*2A–O8A 2×	1.879(2)

(continued)

Table A.29 (continued)

Bond	Bond length (Å)	Bond	Bond length (Å)	Bond	Bond length (Å)
<*Z*2B–O>	1.644	*X*3A–O3A	2.452(2)	*Y*2A–O11A 2×	1.875(2)
		*X*3A–O6A	2.476(3)	<*Y*2A–O>	1.902
*Z*3A–O5A	1.633(3)	*X*3A–O6A	2.987(3)		
*Z*3A–O6A	1.604(3)	*X*3A–O7A	2.367(2)	*Y*2B–O4B 2×	1.952(2)
*Z*3A–O8A	1.621(2)	*X*3A–O7A	2.496(2)	*Y*2B–O8B 2×	1.879(2)
*Z*3A–O9	1.661(3)	*X*3A–O7B	2.582(3)	*Y*2B–O11B 2×	1.875(2)
<*Z*3A–O>	1.630	*X*3A–O8A	2.596(3)	<*Y*2B–O>	1.902
		*X*3A–O10B	2.567(1)		
*Z*3B–O5B	1.633(3)	*X*3A–O11A	2.482(3)	*Y*3A–O1A	1.928(2)
*Z*3B–O6B	1.609(3)	<*X*3A–O>	2.556	*Y*3A–O2B	1.915(2)
*Z*3B–O8B	1.621(2)			*Y*3A–O3A	1.946(2)
*Z*3B–O9	1.662(3)	*X*3B–O3B	2.453(2)	*Y*3A–O4A	2.052(3)
<*Z*3B–O>	1.631	*X*3B–O6B	2.480(3)	*Y*3A–O5A	1.981(3)
		*X*3B–O6B	2.985(3)	*Y*3A–O11A	1.880(3)
*X*1–O1A × 2	2.333(3)	*X*3B–O7B	2.365(3)	<*Y*3A–O>	1.950
*X*1–O1B × 2	2.334(3)	*X*3B–O7B	2.495(3)		
*X*1–O2A × 2	2.506(2)	*X*3B–O7A	2.576(3)	*Y*3B–O1B	1.931(2)
*X*1–O2B × 2	2.507(2)	*X*3B–O8B	2.590(3)	*Y*3B–O2A	1.915(3)
<*X*1–O>	2.420	*X*3B–O10A	2.571(1)	*Y*3B–O3B	1.951(2)
		*X*3B–O11B	2.480(3)	*Y*3B–O4B	2.054(3)
*X*2A–O1A	2.490(2)	<*X*3B–O>	2.555	*Y*3B–O5B	1.984(3)
*X*2A–O2A	2.425(2)			*Y*3B–O11B	2.054(3)
*X*2A–O3A	2.375(2)	*X*4A–O6B × 4	2.324(3)	<*Y*3B–O>	1.952
*X*2A–O4A	2.441(2)	*X*4A–O9 × 4	2.576(3)		
*X*2A–O5A	2.327(2)	<*X*4A–O>	2.450		
*X*2A–O5B	2.439(3)				

Table A.30 Selected bond lengths (Å) in the crystal structure of Mn-zone of cyprine crystal structure from the Wessels mine, RSA

Bond	Bond length (Å)	Bond	Bond length (Å)	Bond	Bond length (Å)
*Z*1A–O1A 4×	1.642(3)	*X*2A–O6A	3.010(4)	*X*4B–O6A 4×	2.275(4)
		*X*2A–O8A	2.319(4)	*X*4B–O9 4×	2.602(4)
*Z*1B–O1B	1.640(3)	<*X*2A–O>	2.474	<*X*4B–O>	2.439
*Z*2A–O2A	1.652(3)	*X*2B–O1B	2.514(3)	*Y*1A–O6B 4×	2.091(4)
*Z*2A–O3A	1.645(4)	*X*2B–O2B	2.419(3)	*Y*1A–O10A	2.260(7)
*Z*2A–O4A	1.672(3)	*X*2B–O3B	2.372(3)	<*Y*1A–O>	2.125
*Z*2A–O7A	1.619(4)	*X*2B–O4B	2.459(3)		
<*Z*2A–O>	1.647	*X*2B–O5B	2.326(4)	*Y*1B–O6A 4×	2.010(4)
		*X*2B–O5A	2.434(3)	*Y*1B–10B	2.274(7)
*Z*2B–O2B	1.650(3)	*X*2B–O6B	2.898(4)	<*Y*1B–O>	2.063
*Z*2B–O3B	1.643(4)	*X*2B–O8B	2.312(4)		
*Z*2B–O4B	1.674(3)	<*X*2B–O>	2.467	*Y*2A–O4A 2×	1.980(3)
*Z*2B–O7B	1.615(4)			*Y*2A–O8A 2×	1.896(3)
<*Z*2B–O>	1.646	*X*3A–O3A	2.443(4)	*Y*2A–O11A 2×	1.893(3)
		*X*3A–O6A	2.525(3)	<*Y*2A–O>	1.923
*Z*3A–O5A	1.625(4)	*X*3A–O6A	2.982(3)		
*Z*3A–O6A	1.611(4)	*X*3A–O7A	2.356(4)	*Y*2B–O4B 2×	1.964(3)
*Z*3A–O8A	1.630(4)	*X*3A–O7A	2.521(4)	*Y*2B–O8B 2×	1.887(3)
*Z*3A–O9	1.667(3)	*X*3A–O7B	2.558(3)	*Y*2B–O11B 2×	1.888(3)
<*Z*3A–O>	1.633	*X*3A–O8A	2.577(3)	<*Y*2B–O>	1.913
		*X*3A–O10B	2.621(1)		
*Z*3B–O5B	1.632(4)	*X*3A–O11A	2.454(4)	*Y*3A–O1A	1.954(3)
*Z*3B–O6B	1.605(4)	<*X*3A–O>	2.560	*Y*3A–O2B	1.948(3)
*Z*3B–O8B	1.624(4)			*Y*3A–O3A	1.980(4)
*Z*3B–O9	1.657(3)	*X*3B–O3B	2.460(4)	*Y*3A–O4A	2.044(3)
<*Z*3B–O>	1.630	*X*3B–O6B	2.456(3)	*Y*3A–O5A	2.024(4)
		*X*3B–O6B	2.456(3)	*Y*3A–O11A	1.864(4)
*X*1–O1A × 2	2.331(3)	*X*3B–O7B	2.384(4)	<*Y*3A–O>	1.969
*X*1–O1B × 2	2.332(3)	*X*3B–O7B	2.512(4)		

(continued)

Table A.30 (continued)

Bond	Bond length (Å)	Bond	Bond length (Å)	Bond	Bond length (Å)
*X*1–O2A × 2	2.506(3)	*X*3B–O7A	2.574(3)	*Y*3B–O1B	1.921(4)
*X*1–O2B × 2	2.508(3)	*X*3B–O8B	2.595(3)	*Y*3B–O2A	1.900(3)
<*X*1–O>	2.419	*X*3B–O10A	2.578(1)	*Y*3B–O3B	1.943(4)
		*X*3B–O11B	2.493(4)	*Y*3B–O4B	2.027(3)
*X*2A–O1A	2.482(3)	<*X*3B–O>	2.501	*Y*3B–O5B	1.990(4)
*X*2A–O2A	2.435(3)			*Y*3B–O11B	1.893(4)
*X*2A–O3A	2.369(3)	*X*4A–O6B 4×	2.336(4)	<*Y*3B–O>	1.946
*X*2A–O4A	2.423(3)	*X*4A–O9 4×	2.592(3)		
*X*2A–O5A	2.330(4)	<*X*4A–O>	2.464		
*X*2A–O5B	2.426(3)				

Table A.31 Powder diffraction data for alumovesuvianite from Asbestos. The eight most intense lines are shown in bold

d_{meas} Å	*hkl*	I_{meas}	d_{meas} Å	*hkl*	I_{meas}
11.07	211	12	2.073	623	5
5.98	220	5	2.065	543	3
5.52	112	3	2.043	712	2
5.23	202	1	2.029	730	5
4.71	231	6	2.015	642	4
4.05	400	10	2.000	731	5
3.903	312	3	1.9632	363	12
3.795	330	1	1.9279	651	6
3.680	322	1	1.9219	372	5
3.488	402	18	1.9062	713	6
3.260	313	4	1.8882	811	15
3.082	510	5	1.8729	652	1
3.058	431	6	1.8001	515	3
3.016	004	16	**1.7658**	**831**	**20**
2.969	323	**22**	1.7178	714	14
2.912	**432**	6	1.6813	910	10
2.761	530	**100**	**1.6672**	**734**	**20**
2.662	**224**	16	1.6479	436	7
2.612	**600**	**61**	**1.6247**	**912**	**21**
2.593	**314**	**25**	1.5870	931	7
2.539	620	5	1.5718	725	9

(continued)

Table A.31 (continued)

d_{meas} Å	*hkl*	I_{meas}	d_{meas} Å	*hkl*	I_{meas}
2.459	541	15	1.5573	770	6
2.384	404	5	1.5435	664	1
2.354	523	8	1.5381	754	2
2.334	334	14	1.5251	626	2
2.308	631	4	1.5115	1020	2
2.279	215	3	1.5030	950	4
2.245	170	1	1.4853	1002	4
2.201	701	9	1.4540	914	1
2.186	711	2	1.4460	228	2
2.165	504	7	1.4221	1032	3
2.146	315	3	1.3882	308	4
2.138	433	8	1.3750	774	5
2.090	712	2	**1.3443**	**880**	**22**

Table A.32 Crystallographic data and refinement parameters for crystal structure of alumovesuvianite from Asbestos

Crystallographic parameters	
Symmetry	Tetragonal
Space group	*P4/n*
Unit cell parameters *a*, *c* (Å)	15.5103(2), 11.8096(2)
Unit cell volume (Å^3)	2841.05(9)
Z	2
Calculated density (g/cm^3)	3.370
Absorption coefficient (mm^{-1})	2.563
Crystal size (mm)	0.21 × 0.17 × 0.14
Data collection parameters	
2θ range (°)	5.30–55.00
h, *k*, *l* range	−20 → 20, −15 → 19, −15 → 15
Number reflections	12750
Unique reflections (R_{int})	3265 (0.025)
Unique reflections c $F > 4\sigma F$	3098
Refinement parameters	
R_1 [$F > 4\sigma(F)$], wR_2 [$F > 4\sigma(F)$],	0.0363, 0.0809
R_1, wR_2 all data	0.0386, 0.0816
S	1.323
ρ_{max}, ρ_{min}, *e* Å^{-3}	0.66, −1.25
Refinment method	Full matrix least-squares on F^2

Table A.33 Atom fractional coordinates, displacement parameters (Å^2) and site occupancy for alumovesuvianite crystal structure from Asbestos

Site	Occupancy	*x*/*a*	*y*/*b*	*z*/*c*	U_{ani}	Multiplicity
*Z*1A	Si	¼	¾	0	0.0057(4)	2*a*
*Z*1B	Si	¾	¼	½	0.0051(4)	2*b*
*Z*2A	Si	0.67964(6)	0.45785(6)	0.12717(8)	0.0041(2)	8*g*
*Z*2B	Si	0.46007(6)	0.68203(6)	0.36915(8)	0.0042(2)	8*g*
*Z*3A	Si	0.34938(6)	0.42027(6)	0.13655(8)	0.0043(2)	8*g*
*Z*3B	Si	0.41117(6)	0.34803(6)	0.36691(8)	0.0050(2)	8*g*
*X*1	Ca	¾	¼	0.24990(8)	0.0046(2)	4*f*
*X*2A	Ca	0.54225(4)	0.30981(4)	0.12159(6)	0.0045(2)	8*g*
*X*2B	Ca	0.31188(4)	0.54586(4)	0.38088(6)	0.0048(2)	8*g*
*X*3A	Ca	0.60407(5)	0.68557(5)	0.10418(6)	0.0073(2)	8*g*
*X*3B	Ca	0.67828(5)	0.59957(5)	0.37759(6)	0.0081(2)	8*g*
*X*4A	$Ca_{0.07}$	¾	¾	0.3493(17)	0.0062(3)	2*c*
*X*4B	$Ca_{0.93}$	¼	¼	0.15023(13)	0.0062(3)	2*c*
*Y*2A	Al	½	½	0	0.0038(3)	4*d*
*Y*2B	Al	½	½	½	0.0049(3)	4*e*
*Y*3A	Al	0.38897(7)	0.62265(7)	0.12661(9)	0.0045(2)	8*g*
*Y*3B	Al	0.61928(7)	0.38642(7)	0.37467(9)	0.0062(2)	8*g*
*Y*1A	$Al_{0.73}Mn_{0.20}$	¾	¾	0.52998(17)	0.0061(6)	2*c*
*Y*1B	$Fe_{0.07}$	¼	¼	0.0392(13)	0.0061(6)	2*c*
O1A	O	0.28235(16)	0.67278(16)	0.0856(2)	0.0052(5)	8*g*
O1B	O	0.67189(16)	0.27863(16)	0.4144(2)	0.0060(5)	8*g*
O2A	O	0.65748(16)	0.38149(16)	0.2194(2)	0.0063(5)	8*g*
O2B	O	0.38316(16)	0.66151(16)	0.2766(2)	0.0054(5)	8*g*
O3A	O	0.45044(15)	0.72259(16)	0.0754(2)	0.0046(5)	8*g*
O3B	O	0.72111(16)	0.45215(16)	0.4236(2)	0.0059(5)	8*g*
O4A	O	0.60579(15)	0.43623(16)	0.0269(2)	0.0053(5)	8*g*
O4B	O	0.43894(16)	0.60696(16)	0.4675(2)	0.0066(5)	8*g*
O5A	O	0.33226(16)	0.51859(16)	0.1800(2)	0.0067(5)	8*g*
O5B	O	0.50913(17)	0.32674(16)	0.3240(2)	0.0082(5)	8*g*
O6A	O	0.27047(16)	0.38608(17)	0.0620(2)	0.0079(5)	8*g*
O6B	O	0.37121(18)	0.27168(16)	0.4436(2)	0.0106(5)	8*g*
O7A	O	0.67021(16)	0.55562(16)	0.1737(2)	0.0064(5)	8*g*
O7B	O	0.55609(16)	0.67557(16)	0.3166(2)	0.0078(5)	8*g*
O8A	O	0.43956(16)	0.41000(16)	0.0687(2)	0.0063(5)	8*g*
O8B	O	0.40697(16)	0.43836(16)	0.4370(2)	0.0055(5)	8*g*
O9	O	0.35166(16)	0.35762(16)	0.2507(2)	0.0069(5)	8*g*

(continued)

Table A.33 (continued)

Site	Occupancy	*x*/*a*	*y*/*b*	*z*/*c*	U_{ani}	Multiplicity
O10A	O	¾	¾	0.3741(4)	0.0101(11)	2*c*
O10B	O	¾	¾	0.1423(4)	0.0066(10)	2*c*
O11A	O	0.49569(16)	0.56346(17)	0.1349(2)	0.0063(5)	8*g*
O11B	O	0.55924(17)	0.49466(16)	0.3629(2)	0.0077(5)	8*g*
H11A	H	0.549(4)	0.563(6)	0.182(7)	0.11(3)	8*g*
H11B	H	0.524(5)	0.495(5)	0.293(4)	0.09(3)	8*g*
H10B	H	¾	¾	0.2275(18)	0.02(3)	2*c*

Table A.34 Selected bond lengths (Å) in the crystal structure of alumovesuvianite from Asbestos

Bond	Bond length (Å)	Bond	Bond length (Å)	Bond	Bond length (Å)
Z1A–O1A 4×	1.645(2)	X2A–O6A	2.813(3)	X4B–O6A 4×	2.375(3)
		X2A–O8A	2,311(3)	X4B–O9 4×	2.585(3)
Z1B–O1B	1.639(2)	<X2A–O>	2.461	<X4B–O>	2.480
Z2A–O2A	1.645(3)	X2B–O1B	2.439(3)	Y1A–O6B 4×	1.935(3)
Z2A–O3A	1.639(3)	X2B–O2B	2.441(3)	Y1A–O10A	1.841(6)
Z2A–O4A	1.677(3)	X2B–O3B	2.366(3)	<Y1A–O>	1.916
Z2A–O7A	1.617(3)	X2B–O4B	2.414(3)		
<Z2A–O>	1.646	X2B–O5B	2.431(3)	Y1B–O6A 4×	2.151(4)
		X2B–O5A	2.323(3)	Y1B–10B	2.14(2)
Z2B–O2B	1.649(3)	X2B–O6B	3.093(3)	<Y1B–O>	2.149
Z2B–O3B	1.639(3)	X2B–O8B	2.323 (3)		
Z2B–O4B	1.677(3)	<X2B–O>	2.479	Y2A–O4A 2×	1.942(2)
Z2B–O7B	1.616(3)			Y2A–O8A 2×	1.867(2)
<Z2B–O>	1.645	X3A–O3A	2.474(2)	Y2A–O11A 2×	1.874(2)
		X3A–O6A	2.369(3)	<Y2A–O>	1.894
Z3A–O5A	1.631(3)	X3A–O6A	2.979(3)		
Z3A–O6A	1.598(3)	X3A–O7A	2.406(3)	Y2B–O4B 2×	1.948(2)
Z3A–O8A	1.620(3)	X3A–O7A	2.498(3)	Y2B–O8B 2×	1.884(2)
Z3A–O9	1.662(3)	X3A–O7B	2.621(3)	Y2B–O11B 2×	1.863(2)
<Z3A–O>	1.628	X3A–O8A	2.612(3)	<Y2B–O>	1.898
		X3A–O10B	2.515(1)		
Z3B–O5B	1.635(3)	X3A–O11A	2.558(3)	Y3A–O1A	1.890(3)
Z3B–O6B	1.615(3)	<X3A–O>	2.559	Y3A–O2B	1.873(3)
Z3B–O8B	1.629(3)			Y3A–O3A	1.918(3)
Z3B–O9	1.661(3)	X3B–O3B	2.442(3)	Y3A–O4A	2.032(3)
<Z3B–O>	1.635	X3B–O6B	2.601 (3)	Y3A–O5A	1.943(3)

(continued)

Table A.34 (continued)

Bond	Bond length (Å)	Bond	Bond length (Å)	Bond	Bond length (Å)
		*X*3B–O6B	3.006(3)	*Y*3A–O11A	1.895(3)
*X*1–O1A 2×	2.335(2)	*X*3B–O7B	2.472(3)	<*Y*3A–O>	1.925
*X*1–O1B 2×	2.333(3)	*X*3B–O7B	2.345(3)		
*X*1–O2A 2×	2.520(2)	*X*3B–O7A	2.506(3)	*Y*3B–O1B	1.919(3)
*X*1–O2B 2×	2.500(2)	*X*3B–O8B	2.624(2)	*Y*3B–O2A	1.929(3)
<*X*1–O>	2.422	*X*3B–O10A	2.585(1)	*Y*3B–O3B	1.967(3)
		*X*3B–O11B	2.467(3)	*Y*3B–O4B	2.073(3)
*X*2A–O1A	2.515(2)	<*X*3B–O>	2.561	*Y*3B–O5B	2.033(3)
*X*2A–O2A	2.401(3)			*Y*3B–O11B	1.925(3)
*X*2A–O3A	2.383(2)	*X*4A–O6B 4×	2.210(10)	<*Y*3B–O>	1.974
*X*2A–O4A	2.463(2)	*X*4A–O9 4×	2.575(9)		
*X*2A–O5A	2.338(2)	<*X*4A–O>	2.393		
*X*2A–O5B	2.460(3)				

Table A.35 Powder diffraction data for magnesiovesuvianite from Tuydo. The eight most intense lines are shown in bold

d_{meas} Å	*hkl*	I_{meas}	d_{meas} Å	*hkl*	I_{meas}
10.96	**110**	**23**	2.058	254	2
5.48	220	10	2.036	730	17
4.039	231	2	2.020	642	3
3.872	400	8	2.007	731	5
3.653	330	2	1.9930	363	3
3.464	**240**	**22**	1.9581	651	6
3.238	402	2	1.9251	614	3
3.038	**510**	**33**	1.9205	713	1
2.997	431	9	1.8840	624	12
2.943	004	9	1.7602	714	4
2.740	**432**	**100**	1.7120	910	9
2.658	530	10	1.6751	734	2
2.583	**522**	**21**	1.6600	436	4
2.525	314	2	**1.6165**	**672**	**25**
2.491	611	3	1.5670	770	16
2.365	**620**	**94**	1.5551	616	3
2.320	144	4	1.5184	655	11
2.265	631	4	1.4930	844	3

(continued)

Table A.35 (continued)

d_{meas} Å	*hkl*	I_{meas}	d_{meas} Å	*hkl*	I_{meas}
2.192	**710**	**19**	1.4739	517	2
2.177	701	4	1.3841	774	3
2.156	711	8	1.3711	1004	5
2.117	154	12	1.3500	1052	2
2.080	623	2	1.3409	954	4

Table A.36 Crystallographic data and refinement parameters for crystal structure of magnesiovesuvianite from Tuydo

Crystallographic parameters	
Symmetry	Tetragonal
Space group	*P*4/*n*
Unit cell parameters *a*, *c* (Å)	15.5362(2), 11.7960(3)
Unit cell volume (Å^3)	2847.26(11)
Z	2
Calculated density (g/cm^3)	3.334
Absorption coefficient (mm^{-1})	2.493
Crystal size (mm)	0.22 × 0.04 × 0.03
Data collection parameters	
2θ range (°)	5.07–55.00
h, *k*, *l* range	−7 → 20, −19 → 11, −15 → 7
Number reflections	6737
Unique reflections (R_{int})	3266 (0.014)
Unique reflections c $F > 4\sigma F$	3097
Refinement parameters	
R_1 [$F > 4\sigma(F)$], wR_2 [$F > 4\sigma(F)$]	0.0270, 0.0702
R_1, wR_2 all data	0.03292, 0.0712
S	1.24
ρ_{max}, ρ_{min}, *e* Å^{-3}	0.50, –0.48
Refinment method	Full matrix least-squares on F^2

Table A.37 Atom fractional coordinates, displacement parameters (Å^2) and site occupancy for magnesiovesuvianite crystal structure from Tuydo

Site	Occupancy	*x*/*a*	*y*/*b*	*z*/*c*	U_{ani}	Multiplicity
Z1A	$Si_{0.95}$	¼	¾	0	0.0053(8)	2*a*
Z1B	$Si_{0.88}$	¾	¼	½	0.0029(8)	2*b*
Z2A	Si	0.6818(1)	0.4594(1)	0.1297(1)	0.0051(3)	8*g*
Z2B	Si	0.4583(1)	0.6801(1)	0.3717(1)	0.0045(3)	8*g*

(continued)

Table A.37 (continued)

Site	Occupancy	*x*/*a*	*y*/*b*	*z*/*c*	U_{ani}	Multiplicity
*Z*3A	Si	0.34848(9)	0.41410(9)	0.1339(1)	0.0058(3)	8*g*
*Z*3B	Si	0.41879(9)	0.34919(9)	0.3642(1)	0.0048(3)	8*g*
*X*1	Ca	¾	¼	0.2506(1)	0.0053(2)	4*f*
*X*2A	Ca	0.54434(7)	0.31139(7)	0.11937(8)	0.0054(2)	8*g*
*X*2B	Ca	0.30987(7)	0.54202(7)	0.37878(8)	0.0052(2)	8*g*
*X*3A	Ca	0.60084(8)	0.67943(8)	0.11666(8)	0.0126(2)	8*g*
*X*3B	Ca	0.68415(7)	0.60320(8)	0.39280(7)	0.0096(2)	8*g*
*X*4A	$Ca_{0.726(5)}$	¾	¾	0.3487(2)	0.0083(7)	2*c*
*X*4B	$Ca_{0.274(5)}$	¼	¼	0.1519(6)	0.010(2)	2*c*
*Y*1A	$Mg_{0.194}Al_{0.08}$	¾	¾	−0.5434(9)	0.0100(9)	2*c*
*Y*1B	$Mg_{0.406}Al_{0.32}$	¼	¼	0.0326(3)	0.0027(5)	2*c*
*Y*2A	Al	½	½	0	0.0046(3)	4*d*
*Y*2B	Al	½	½	½	0.0048(3)	4*e*
*Y*3A	Al	0.3877(1)	0.6194(1)	0.1261(1)	0.0058(3)	8*g*
*Y*3B	Al	0.6219(1)	0.3882(1)	0.3730(1)	0.0054(3)	8*g*
O1A	O	0.2797(2)	0.6722(2)	0.0864(2)	0.0072(7)	8*g*
O1B	O	0.6713(2)	0.2818(2)	0.4137(2)	0.0068(7)	8*g*
O2A	O	0.6609(2)	0.3825(2)	0.2232(3)	0.0061(7)	8*g*
O2B	O	0.3824(2)	0.6583(2)	0.2793(3)	0.0058(7)	8*g*
O3A	O	0.4504(2)	0.7205(2)	0.0765(3)	0.0065(7)	8*g*
O3B	O	0.7223(2)	0.4505(2)	0.4238(3)	0.0048(7)	8*g*
O4A	O	0.6067(2)	0.4372(2)	0.0306(2)	0.0061(7)	8*g*
O4B	O	0.4371(2)	0.6065(2)	0.4722(2)	0.0064(7)	8*g*
O5A	O	0.3286(2)	0.5123(2)	0.1774(3)	0.0086(7)	8*g*
O5B	O	0.5172(2)	0.3318(2)	0.3199(3)	0.0068(7)	8*g*
O6A	O	0.2708(2)	0.3773(3)	0.0573(2)	0.0141(9)	8*g*
O6B	O	0.3855(2)	0.2702(2)	0.4390(2)	0.0093(7)	8*g*
O7A	O	0.6738(2)	0.5557(2)	0.1812(3)	0.0092(7)	8*g*
O7B	O	0.5557(2)	0.6721(2)	0.3243(3)	0.0080(7)	8*g*
O8A	O	0.4394(2)	0.4075(2)	0.0647(2)	0.0045(7)	8*g*
O8B	O	0.4102(2)	0.4388(2)	0.4328(2)	0.0067(7)	8*g*
O9	O	0.3575(3)	0.3523(3)	0.2496(2)	0.0063(4)	8*g*
O10A	O	¾	¾	0.3612(5)	0.011(1)	2*c*
O10B	O	¾	¾	0.1293(5)	0.012(2)	2*c*
O11A	O	0.4944(2)	0.5601(2)	0.1375(3)	0.0094(7)	8*g*
O11B	O	0.5628(2)	0.4960(2)	0.3639(2)	0.0046(7)	8*g*

Table A.38 Selected bond length (Å) in the crystal structure of magnesiovesuvianite from Tuydo

Bond	Bond length (Å)	Bond	Bond length (Å)	Bond	Bond length (Å)
*Z*1A–O1A 4×	1.647(3)	*X*2A–O6A	2.984(4)	*X*4B–O6A 4×	2.293(5)
		*X*2A–O8A	2.303(3)	*X*4B–O9 4×	2.578(4)
*Z*1B–O1B	1.666(3)	<*X*2A–O>	2.468	<*X*4B–O>	2.436
*Z*2A–O2A	1.658(4)	*X*2B–O1B	2.496(4)	*Y*1A–O6B 4×	2.138(4)
*Z*2A–O3A	1.649(4)	*X*2B–O2B	2.431(4)	*Y*1A–O10A	2.149(13)
*Z*2A–O4A	1.688(4)	*X*2B–O3B	2.385(3)	<*Y*1A–O>	2.140
*Z*2A–O7A	1.619(4)	*X*2B–O4B	2.474(4)		
<Z2A–O>	1.654	*X*2B–O5B	2.341(3)	*Y*1B–O6A 4×	2.025(4)
		*X*2B–O5A	2.437(3)	*Y*1B–10B	1.910(8)
*Z*2B–O2B	1.642(4)	*X*2B–O6B	2.822(4)	<*Y*1B–O>	2.002
*Z*2B–O3B	1.640(4)	*X*2B–O8B	2.326(3)		
*Z*2B–O4B	1.678(3)	<*X*2B–O>	2.464	*Y*2A–O4A 2×	1.957(4)
*Z*2B–O7B	1.620(4)			*Y*2A–O8A 2×	1.880(3)
<Z2B–O>	1.645	*X*3A–O3A	2.469(4)	*Y*2A–O11A 2×	1.874(3)
		*X*3A–O6A	2.518(3)	<*Y*2A–O>	1.904
*Z*3A–O5A	1.639(4)	*X*3A–O6A	2.993(3)		
*Z*3A–O6A	1.612(4)	*X*3A–O7A	2.358(3)	*Y*2B–O4B 2×	1.950(4)
*Z*3A–O8A	1.635(4)	*X*3A–O7A	2.505(4)	*Y*2B–O8B 2×	1.865(3)
*Z*3A–O9	1.674(4)	*X*3A–O7B	2.550(3)	*Y*2B–O11B 2×	1.880(3)
<Z3A–O>	1.640	*X*3A–O8A	2.607(3)	<*Y*2B–O>	1.898
		*X*3A–O10B	2.568(1)		
*Z*3B–O5B	1.639(3)	*X*3A–O11A	2.496(3)	*Y*3A–O1A	1.925(4)
*Z*3B–O6B	1.598(3)	<*X*3A–O>	2.563	*Y*3A–O2B	1.908(3)
*Z*3B–O8B	1.616(3)			*Y*3A–O3A	1.939(4)
*Z*3B–O9	1.654(4)	*X*3B–O3B	2.473(4)	*Y*3A–O4A	2.048(3)
<Z3B–O>	1.627	*X*3B–O6B	2.398(3)	*Y*3A–O5A	1.994(4)
		*X*3B–O6B	2.996(3)	*Y*3A–O11A	1.901(4)
*X*1–O1A × 2	2.329(3)	*X*3B–O7B	2.404(4)	<*Y*3A–O>	1.953
*X*1–O1B × 2	2.333(3)	*X*3B–O7B	2.487(4)		

(continued)

Table A.38 (continued)

Bond	Bond length (Å)	Bond	Bond length (Å)	Bond	Bond length (Å)
$X1$–O2A × 2	2.502(4)	$X3$B–O7A	2.608(3)	$Y3$B–O1B	1.885(4)
$X1$–O2B × 2	2.524(4)	$X3$B–O8B	2.609(3)	$Y3$B–O2A	1.870(3)
<$X1$–O>	2.422	$X3$B–O10A	2.527(2)	$Y3$B–O3B	1.930(4)
		$X3$B–O11B	2.538(4)	$Y3$B–O4B	2.044(3)
$X2$A–O1A	2.469(4)	<$X3$B–O>	2.560	$Y3$B–O5B	1.951(4)
$X2$A–O2A	2.449(4)			$Y3$B–O11B	1.912(4)
$X2$A–O3A	2.364(3)	$X4$A–O6B 4×	2.380(4)	<$Y3$B–O>	1.932
$X2$A–O4A	2.420(3)	$X4$A–O9 4×	2.585(3)		
$X2$A–O5A	2.334(3)	<$X4$A–O>	2.483		
$X2$A–O5B	2.423(3)				

Table A.39 Crystallographic data and refinement parameters for crystal structure of F-bearing vesuvianite from Lupikko

Crystallographic parameters	
Symmetry	Tetragonal
Space group	*P4/nnc*
Unit cell parameters *a*, *c* (Å)	15.5480(4), 11.7939(5)
Unit cell volume (Å^3)	2851.08(17)
Z	2
Calculated density (g/cm^3)	3.371
Absorption coefficient (mm^{-1})	2.771
Crystal size (mm)	0.21 × 0.04 × 0.04
Data collection parameters	
2θ range (°)	6.80–55.00
h, *k*, *l* range	$-17 \rightarrow 20$, $-15 \rightarrow 19$, $-13 \rightarrow 15$
Number reflections	8548
Unique reflections (R_{int})	1641 (0.0258)
Unique reflections c $F > 4\sigma F$	1567
Refinement parameters	
R_1 [$F > 4\sigma(F)$], wR_2 [$F > 4\sigma(F)$]	0.0287, 0.0804
R_1, wR_2 all data	0.0302, 0.0812
S	1.222
ρ_{max}, ρ_{min}, *e* Å^{-3}	0.60, –0.45
Refinment method	Full matrix least-squares on F^2

Table A.40 Atom fractional coordinates, displacement parameters (Å^2) and site occupancy for crystal structure of F-bearing vesuvianite from Lupikko

Site	Occupancy	*x*/*a*	*y*/*b*	*z*/*c*	U_{ani}	Multiplicity
*X*1	Ca	¾	¼	¼	0.0084(2)	4*c*
*X*2	Ca	0.68900(4)	0.04335(4)	0.12039(5)	0.0076(2)	16*k*
*X*3	Ca	0.67986(4)	−0.1006(1)	−0.1134(1)	0.0121(2)	16*k*
*X*4	$Ca_{0.50}$	¾	¾	0.1427(2)	0.0102(5)	4*e*
*Z*1	$Si_{0.874(8)}$	¾	¼	0	0.0052(6)	4*d*
*Z*2	Si	0.54079(5)	0.18123(5)	0.12872(6)	0.0064(2)	16*k*
*Z*3	Si	0.58291(5)	−0.15077(5)	0.13521(6)	0.0062(2)	16*k*
*Y*1	$Fe_{0.44}Mg_{0.06}$	¾	¾	0.0493(2)	0.0200(8)	4*e*
*Y*2	Al	½	0	½	0.0069(3)	8*f*
*Y*3	$Al_{0.96}Fe_{0.04}$	0.61242(5)	0.12059(5)	0.37338(6)	0.0071(3)	16*k*
O1	O	0.7806(1)	0.1716(1)	0.0872(2)	0.0095(4)	16*k*
O2	O	0.6172(1)	0.1593(1)	0.2217(1)	0.0085(4)	16*k*
O3	O	0.5500(1)	0.2211(1)	0.4237(2)	0.0079(4)	16*k*
O4	O	0.6067(1)	0.0626(1)	0.5290(1)	0.0082(4)	16*k*
O5	O	0.8303(1)	0.0153(1)	0.1788(2)	0.0096(4)	16*k*
O6	O	0.6178(1)	−0.22899(1)	0.0595(2)	0.0138(5)	16*k*
O7	O	0.4439(1)	0.1743(1)	0.1776(2)	0.0104(4)	16*k*
O8	O	0.5908(1)	−0.0608(1)	0.0663(2)	0.0080(4)	16*k*
O9	O	0.6446(1)	−0.1446(1)	¼	0.0098(6)	8*h*
O10	O	¾	¾	−0.1339(4)	0.0153(9)	4*e*
O11	$O_{0.60}F_{0.40}$	0.4376(1)	−0.0047(1)	0.13641(16)	0.0052(4)	16*k*

Table A.41 Selected bond lengths (Å) in the crystal structure of F-bearing vesuvianite from Lupikko

Bond	Bond length (Å)	Bond	Bond length (Å)	Bond	Bond length (Å)
*Z*1–O1 4×	1.665(2)	*X*2–O3	2.375(2)	*X*4–O9 4×	2.641(3)
		*X*2–O6	2.902(2)	<*X*4–O>	2.471
*Z*2–O7	1.617(2)	*X*2–O2	2.434(2)		
*Z*2–O2	1.652(2)	*X*2–O5	2.344(2)	*Y*1–O6 4×	2.084(2)
*Z*2–O3	1.647(2)	*X*2–O4	2.449(2)	*Y*1–O10	2.161(5)
*Z*2–O4	1.685(2)	*X*2–O1	2.481(2)	<*Y*1–O>	2.099
<*Z*2–O>	1.650	<*X*2–O>	2.4659		
				*Y*2–O8 2×	1.8701(19)

(continued)

Table A.41 (continued)

Bond	Bond length (Å)	Bond	Bond length (Å)	Bond	Bond length (Å)
*Z*3–O6	1.603(2)	*X*3–O7	2.488(2)	*Y*2–O11 2×	1.8801(19)
*Z*3–O8	1.623(2)	*X*3–O3	2.466(2)	*Y*2–O4 2×	1.954(2)
*Z*3–O5	1.637(2)	*X*3–O11	2.467(2)	<*Y*2–O>	1.9014
*Z*3–O9	1.662(13)	*X*3–O6	2.498(2)		
<*Z*3–O>	1.631	*X*3–O6	3.012(2)	*Y*3–O1	1.900(2)
		*X*3–O7	2.562(2)	*Y*3–O11	1.908(2)
*X*1–O1 4×	2.324(2)	*X*3–O10	2.5775(8)	*Y*3–O2	1.889(2)
*X*1–O2 4×	2.523(2)	*X*3–O7	2.364(2)	*Y*3–O3	1.932(2)
<*X*1–O>	2.424	*X*3–O8	2.607(2)	*Y*3–O5	1.963(2)
		<*X*3–O>	2.5602	*Y*3–O4	2.047(2)
*X*2–O8	2.315(2)			<*Y*3–O>	1.9398
*X*2–O5	2.427(2)	*X*4–O6 4×	2.301(2)		

Table A.42 Crystallographic data and refinement parameters for crystal structure of Mn-bearing vesuvianite from N'chwaning III

Crystallographic parameters	
Symmetry	Tetragonal
Space group	*P*4/*n*
Unit cell parameters *a*, *c* (Å)	15.5495(6), 11.7788(6)
Unit cell volume (Å^3)	2848.0(3)
Z	2
Calculated density (g/cm^3)	3.450
Absorption coefficient (mm^{-1})	3.239
Crystal size (mm)	0.17 × 0.15 × 0.12
Data collection parameters	
2θ range (°)	6.28–53.00
h, *k*, *l* range	$-19 \rightarrow 8$, $-14 \rightarrow 15$, $-13 \rightarrow 14$
Number reflections	6500
Unique reflections (R_{int})	2963 (0.021)
Unique reflections c $F > 4\sigma F$	2698
Refinement parameters	
R_1 [$F > 4\sigma(F)$], wR_2 [$F > 4\sigma(F)$]	0.0413, 0.0970
R_1, wR_2 all data	0.0466, 0.0993
S	1.288
ρ_{max}, ρ_{min}, e Å^{-3}	1.1, –0.88
Refinment method	Full matrix least-squares on F^2

Table A.43 Atom fractional coordinates, displacement parameters (Å^2) and site occupancy for crystal structure of Mn-bearing vesuvianite from N'chwaning III

Site	Occupancy	*x/a*	*y/b*	*z/c*	U_{ani}	Multiplicity
*Z*1A	Si	¼	¾	0	0.0039(5)	2*a*
*Z*1B	Si	¾	¼	½	0.0051(5)	2*b*
*Z*2A	Si	0.6810(1)	0.4592(1)	0.1294(1)	0.0046(3)	8*g*
*Z*2B	Si	0.4592(1)	0.6811(1)	0.3708(1)	0.0050(3)	8*g*
*Z*3A	Si	0.34874(8)	0.41601(8)	0.13499(10)	0.0054(3)	8*g*
*Z*3B	Si	0.41611(8)	0.34877(8)	0.36492(10)	0.0055(3)	8*g*
*X*1	Ca	¾	¼	0.2501(1)	0.0072(3)	4*f*
*X*2A	Ca	0.54416(6)	0.31103(6)	0.12070(8)	0.0050(2)	8*g*
*X*2B	Ca	0.31102(6)	0.54419(6)	0.37923(8)	0.0052(2)	8*g*
*X*3A	Ca	0.60082(6)	0.68001(6)	0.11170(8)	0.0098(2)	8*g*
*X*3B	Ca	0.68007(6)	0.60086(6)	0.38864(8)	0.0100(2)	8*g*
*X*4A	$Ca_{0.523(4)}$	¼	¼	0.3501(3)	0.0092(9)	2*c*
*X*4B	$Ca_{0.477(4)}$	¼	¼	0.1500(3)	0.0043(9)	2*c*
*Y*1A	$Cu_{0.477(4)}$	¼	¼	0.4451(2)	0.0088(7)	2*c*
*Y*1B	$Cu_{0.523(4)}$	¼	¼	0.0546(2)	0.0127(7)	2*c*
*Y*2A	$Al_{0.89}Mn_{0.11}$	½	½	0	0.0022(3)	4*d*
*Y*2B	Al	½	½	½	0.0022(3)	4*e*
*Y*3A	$Al_{0.77}Mn_{0.23}$	0.38878(7)	0.61976(7)	0.12567(9)	0.0040(4)	8*g*
*Y*3B	$Al_{0.78}Mn_{0.22}$	0.61983(7)	0.38879(7)	0.37437(9)	0.0039(4)	8*g*
O1A	O	0.2808(2)	0.6726(2)	0.0849(3)	0.0045(6)	8*g*
O1B	O	0.6724(2)	0.2811(2)	0.4146(3)	0.0053(6)	8*g*
O2A	O	0.6597(2)	0.3825(2)	0.2216(3)	0.0065(6)	8*g*
O2B	O	0.3826(2)	0.6601(2)	0.2783(3)	0.0078(7)	8*g*
O3A	O	0.4512(2)	0.7218(2)	0.0753(3)	0.0061(6)	8*g*
O3B	O	0.7218(2)	0.4514(2)	0.4245(3)	0.0057(6)	8*g*
O4A	O	0.6068(2)	0.4374(2)	0.0301(3)	0.0075(6)	8*g*
O4B	O	0.4375(2)	0.6071(2)	0.4700(3)	0.0065(6)	8*g*
O5A	O	0.3290(2)	0.5139(2)	0.1781(3)	0.0080(7)	8*g*
O5B	O	0.5142(2)	0.3289(2)	0.3219(3)	0.0079(7)	8*g*
O6A	O	0.2711(2)	0.3800(2)	0.0583(3)	0.0099(7)	8*g*
O6B	O	0.3801(2)	0.2714(2)	0.4417(3)	0.0106(7)	8*g*
O7A	O	0.6736(2)	0.5562(2)	0.1788(3)	0.0097(7)	8*g*
O7B	O	0.5560(2)	0.6732(2)	0.3215(3)	0.0098(7)	8*g*
O8A	O	0.4387(2)	0.4083(2)	0.0661(3)	0.0056(6)	8*g*
O8B	O	0.4082(2)	0.4391(2)	0.4335(3)	0.0061(6)	8*g*
O9	O	0.3546(2)	0.3544(2)	0.2495(3)	0.0081(6)	8*g*

(continued)

Table A.43 (continued)

Site	Occupancy	*x*/*a*	*y*/*b*	*z*/*c*	U_{ani}	Multiplicity
O10A	O	¼	¼	0.6366(6)	0.009(1)	2*c*
O10B	O	¾	¾	0.1354(6)	0.010(1)	2*c*
O11A	O	0.4945(2)	0.5619(2)	0.1369(3)	0.0058(6)	8*g*
O11B	O	0.5619(2)	0.4946(2)	0.3630(3)	0.0059(6)	8*g*

Table A.44 Selected bond lengths (Å) in the crystal structure of Mn-bearing vesuvianite from N'chwaning III

Bond	Bond length (Å)	Bond	Bond length (Å)	Bond	Bond length (Å)
*Z*1A–O1A 4×	1.637(3)	*X*2A–O6A	2.947(4)	*X*4B–O6A 4×	2.315(4)
		*X*2A–O8A	2.322(3)	*X*4B–O9 4×	2.580(4)
*Z*1B–O1B	1.643(3)	<*X*2A–O>	2.468	<*X*4B–O>	2.448
*Z*2A–O2A	1.647(3)	*X*2B–O1B	2.491(3)	*Y*1A–O6B 4×	2.050(3)
*Z*2A–O3A	1.644(3)	*X*2B–O2B	2.429(3)	*Y*1A–O10A	2.256(7)
*Z*2A–O4A	1.677(3)	*X*2B–O3B	2.369(3)	<*Y*1A–O>	2.091
*Z*2A–O7A	1.621(3)	*X*2B–O4B	2.443(3)		
<*Z*2A–O>	1.647	*X*2B–O5B	2.326(3)	*Y*1B–O6A 4×	2.048(3)
		*X*2B–O5A	2.432(3)	*Y*1B–10B	2.238(7)
*Z*2B–O2B	1.646(3)	*X*2B–O6B	2.949(4)	<*Y*1B–O>	2.086
*Z*2B–O3B	1.643(3)	*X*2B–O8B	2.316(3)		
*Z*2B–O4B	1.674(3)	<*X*2B–O>	2.469	*Y*2A–O4A 2×	1.957(3)
*Z*2B–O7B	1.618(3)			*Y*2A–O8A 2×	1.883(3)
<*Z*2B–O>	1.645	*X*3A–O3A	2.453(3)	*Y*2A–O11A 2×	1.880(3)
		*X*3A–O6A	2.470(3)	<*Y*2A–O>	1.907
*Z*3A–O5A	1.633(3)	*X*3A–O6A	2.975(3)		
*Z*3A–O6A	1.609(3)	*X*3A–O7A	2.507(3)	*Y*2B–O4B 2×	1.960(3)
*Z*3A–O8A	1.622(3)	*X*3A–O7A	2.369(3)	*Y*2B–O8B 2×	1.883(3)
*Z*3A–O9	1.657(4)	*X*3A–O7B	2.570(3)	*Y*2B–O11B 2×	1.880(3)
<*Z*3A–O>	1.630	*X*3A–O8A	2.578(3)	<*Y*2B–O>	1.908
		*X*3A–O10B	2.578(1)		

(continued)

Table A.44 (continued)

Bond	Bond length (Å)	Bond	Bond length (Å)	Bond	Bond length (Å)
*Z*3B–O5B	1.636(3)	*X*3A–O11A	2.489(3)	*Y*3A–O1A	1.929(3)
*Z*3B–O6B	1.606(3)	<*X*3A–O>	2.554	*Y*3A–O2B	1.906(3)
*Z*3B–O8B	1.625(3)			*Y*3A–O3A	1.953(3)
*Z*3B–O9	1.665(4)	*X*3B–O3B	2.450(3)	*Y*3A–O4A	2.040(3)
<*Z*3B–O>	1.633	*X*3B–O6B	2.470(3)	*Y*3A–O5A	1.989(3)
		*X*3B–O6B	2.969(4	*Y*3A–O11A	1.879(3)
*X*1–O1A 2×	2.338(3)	*X*3B–O7B	2.369(3)	<*Y*3A–O>	1.949
*X*1–O1B 2×	2.334(3)	*X*3B–O7B	2.513(3)		
*X*1–O2A 2×	2.516(3)	*X*3B–O7A	2.570(3)	*Y*3B–O1B	1.922(3)
*X*1–O2B 2×	2.514(3)	*X*3B–O8B	2.581(3	*Y*3B–O2A	1.906(3)
<*X*1–O>	2.426	*X*3B–O10A	2.579(1)	*Y*3B–O3B	1.951(3)
		*X*3B–O11B	2.490(3)	*Y*3B–O4B	2.039(3)
*X*2A–O1A	2.491(3)	<*X*3B–O>	2.554	*Y*3B–O5B	1.987(3)
*X*2A–O2A	2.423(3)			*Y*3B–O11B	1.881(3)
*X*2A–O3A	2.365(3)	*X*4A–O6B 4×	2.316(4)	<*Y*3B–O>	1.948
*X*2A–O4A	2.438(3)	*X*4A–O9 4×	2.586(4)		
*X*2A–O5A	2.329(3)	<*X*4A–O>	2.451		
*X*2A–O5B	2.431(3)				

Table A.45 Occupancy of key cationic sites in the crystal structure of hongheite, after (Xu et al. 2019)

Site	Population
*X*1-3	18Ca
*X*4	□1.6 + 0.4Ca
*Y*1	0.9 Fe^{2+} + 0.1 Mg
*Y*2	4.0Al
*Y*3	2.99Fe^{3+} + 2.51 Mg + 2.31Al + 0.18Mn + 0.01Cr
*Z*1-3	18Si
*T*1	1.44B + 0.2Al + 2.36□
*T*2	1B

Table A.46 Powder diffraction data for Manaevite-(Ce) from Kovdor (M. M. Moiseev's sampel). The eight most intense lines are shown in bold

d_{meas} Å	*hkl*	I_{meas}	d_{meas} Å	*hkl*	I_{meas}
5.98	**002**	**27**	2.414	541	5
4.63	**311**	**30**	2.382	404	6
4.124	321	10	2.357	105	6
3.946	400	11	2.259	215	5
3.883	103	7	2.155	315	11
3.534	322	10	2.076	730	11
3.289	**402**	**31**	2.032	444	8
3.113	150	11	1.9408	713	4
3.081	313	6	1.9108	216	5
3.046	431	15	1.8686	802	11
2.991	**511**	**100**	1.6870	842	8
2.797	**204**	**95**	**1.6494**	**217**	**25**
2.718	530	19	1.5805	327	4
2.636	**423**	**81**	1.5284	862	7
2.503	**620**	**47**	1.4598	835	20
2.437	513	4	1.3655	935	7

Table A.47 Crystallographic data and refinement parameters for crystal structure of manaevite (Ce) from Kovdor

Crystallographic parameters	
Symmetry	Tetragonal
Space group	*P4/nnc*
Unit cell parameters *a*, *c* (Å)	15.9247(13), 11.9661(10)
Unit cell volume (Å^3)	3034.6(6)
Z	2
Calculated density (g/cm^3)	3.503
Absorption coefficient (mm^{-1})	5.385
Crystal size (mm)	0.10 × 0.10 × 0.03
Data collection parameters	
2θ range (°)	4.25–55.00
h, *k*, *l* range	−20 → 10, −20 → 11, −11 → 15
Number reflections	13868
Unique reflections (R_{int})	1757 (0.029)

(continued)

Table A.47 (continued)

Crystallographic parameters	
Symmetry	Tetragonal
Unique reflections c $F > 4\sigma F$	1527
Refinement parameters	
R_1 [$F > 4\sigma(F)$], wR_2 [$F > 4\sigma(F)$]	0.0348, 0.1067
R_1, wR_2 all data	0.0424, 0.1128
S	1.157
ρ_{max}, ρ_{min}, e Å^{-3}	1.06, –0.93
Refinment method	Full matrix least-squares on F^2

Table A.48 Atom fractional coordinates, displacement parameters (Å^2) and site occupancy for Manaevite-(Ce) crystal structure from Kovdor

Site	Occupancy	*x*/*a*	*y*/*b*	*z*/*c*	U_{ani}	Multiplicity
*X*1	Ca	¾	¼	¼	0213(4)	4*c*
*X*2	Ca	0.68860(6)	0.04625(6)	0.12282(7)	0.0176(3)	16*k*
*X*3	Ca	0.67859(3)	–0.10001(3)	–0.10732(5)	0.0270(2)	16*k*
*X*4	$Ca_{0.50}$	¾	¾	0.1471(3)	0.0328(10)	4*e*
*Z*1	$Si_{0.60}$	¾	¼	0	0.0112(7)	4*d*
*Z*2	$Si_{0.75}$	0.5421(1)	0.1810(1)	0.1260(1)	0.0170(4)	16*k*
*Z*3	$Si_{0.75}Al_{0.25}$	0.6504(1)	–0.0845(1)	0.3626(1)	0.0207(3)	16*k*
*Y*1	$Mg_{0.50}$	¾	¾	0.0425(8)	0.0366(18)	4*e*
*Y*2	$Al_{0.79}Fe_{0.21}$	½	0	½	0.0170(5)	8*f*
*Y*3	$Mg_{0.68}Fe_{0.32}$	0.61175(7)	0.11922(7)	0.37266(9)	0.0169(4)	16*k*
O1	O	0.7784(2)	0.1700(2)	0.0888(3)	0.0300(8)	16*k*
O2	O	0.6183(2)	0.1617(2)	0.2200(3)	0.0301(8)	16*k*
O3	O	0.5465(2)	0.2215(2)	0.4276(3)	0.0247(8)	16*k*
O4	O	0.6061(2)	0.0627(2)	0.5294(3)	0.0236(7)	16*k*
O5	O	0.8286(2)	0.0109(2)	0.1804(3)	0.0257(8)	16*k*
O6	O	0.6207(3)	–0.2232(2)	0.0551(4)	0.0399(10)	16*k*
O7	O	0.4425(3)	0.1768(3)	0.1794(4)	0.0501(12)	16*k*
O8	O	0.5939(2)	–0.0615(2)	0.0704(3)	0.0231(8)	16*k*
O9	O	0.6477(2)	–0.1477(2)	¼	0.0271(11)	8*h*
O10	$O_{0.80}$	¾	¾	–0.1319(12)	0.077(4)	4*e*
O11	$O_{0.75}F_{0.25}$	0.4366(2)	0.0000(2)	0.1404(3)	0.0257(8)	16*k*
Ow	$(H_2O)_{0.2}$	0.6682(19)	–0.0903(19)	–0.128(2)	0.010(7)	16*k*
*T*1	$Al_{0.08}$	0.4352(15)	0.0648(15)	¼	0.034(7)	8 *h*
*T*2	$S_{0.07}$	¾	¾	¾	0.10(3)	2*a*

Table A.49 Selected bond lengths (Å) in the crystal structure of manaevite-(Ce) from Kovdor

Bond	Bond length (Å)	Bond	Bond length (Å)	Bond	Bond length (Å)
*Z*1–O1 4×	1.720(4)	*X*2–O6	3.114(5)	*Y*1–O6 4×	2.108(4)
		*X*2–O2	2.447(4)	*Y*1–O10	2.086(17)
*Z*2–O7	1.711(6)	*X*2–O5	2.436(4)	<*Y*1–O>	2.104
*Z*2–O2	1.683(4)	*X*2–O4	2.486(4)		
*Z*2–O3	1.681(4)	*X*2–O1	2.469(4)	*Y*2–O8 2×	1.976(3)
*Z*2–O4	1.693(4)	<*X*2–O>	2.514	*Y*2–O11 2×	1.960(4)
<*Z*2–O>	1.692			*Y*2–O4 2×	1.994(3)
		*X*3–O7	2.550(5)	<*Y*2–O>	1.974
*Z*3–O6	1.627(4)	*X*3–O3	2.468(4)		
*Z*3–O8	1.635(4)	*X*3–O11	2.461(4)	*Y*3–O1	1.981(4)
*Z*3–O5	1.639(4)	*X*3–O6	2.911(5)	*Y*3–O11	1.996(4)
*Z*3–O9	1.682(2)	*X*3–O6	2.516(4)	*Y*3–O2	1.951(4)
<*Z*3–O>	1.646	*X*3–O7	2.641(5)	*Y*3–O3	2.041(4)
		*X*3–O10	2.662(2)	*Y*3–O5	2.068(4)
*X*1–O1 4×	2.356(4)	*X*3–O7	2.442(5)	*Y*3–O4	2.082(4)
*X*1–O2 4×	2.550(4)	*X*3–O8	2.592(4)	<*Y*3–O>	2.020
<*X*1–O>	2.453	<*X*3–O>	2.583		
				*T*1–O11 2×	1.669(16)
*X*2–O8	2.368(3)	*X*4–O6 4×	2.373(5)	*T*1–O7 2×	1.98(2)
*X*2–O5	2.401(4)	*X*4–O9 4×	2.613(5)	<*T*1–O>	1.825
*X*2–O3	2.394(3)	<*X*4–O>	2.493		

Table A.50 Powder diffraction data for Milanriederite from the Kombat mine

I_{obs}	d_{obs}	I_{calc}*	d_{calc}**	*h k l*
1	11.11	0.5	11.072	110
3	9.47	1	9.454	101
3	7.84	3	7.829	200
10	5.934	11	5.930	002
2	5.545	2	5.536	220
4	5.233	3	5.227	112
1	4.963	1	4.951	310
5	4.733	5	4.727	202
2	4.564	1, 1	4.569, 4.525	311, 212
9	4.056	5, 9	4.078, 4.047	321, 222
5	3.922	5	3.914	400
4	3.836	3, 1	3.833, 3.801	103, 312
1	3.698	1	3.691	330

(continued)

Table A.50 (continued)

I_{obs}	d_{obs}	I_{calc}*	d_{calc}**	*h k l*
11	3.508	5, 8	3.503, 3.501	322, 420
2	3.367	1	3.358	421
6	3.272	6	3.267	402
5	3.205	5	3.198	412
11	3.080	3, 13	3.089, 3.071	313, 510
11	3.028	8, 6	3.028, 3.015	501, 422
50	2.970	4, 54	2.972, 2.965	511, 004
9	2.929	8	2.923	323
5	2.869	6	2.864	114
100	2.774	1, 100, 29	2.773, 2.769, 2.767	204, 432, 440
2	2.690	3	2.685	530
87	2.617	1, 2, 32, 81	2.621, 2.619, 2.614, 2.611	423, 531, 224, 522
3	2.548	3	2.544	314
30	2.481	47	2.476	620
10	2.459	10	2.455	503
4	2.395	2, 4	2.395, 2.389	541, 602
11	2.368	14, 2	2.363, 2.361	404, 612
10	2.344	2, 3, 8	2.345, 2.342, 2.337	105, 523, 414
4	2.315	4	2.311	334
4	2.296	5	2.290	631
3	2.251	3	2.247	215
8	2.223	6, 7	2.221, 2.214	533, 550
2	2.182	4	2.177	711
5	2.162	2, 1, 3	2.159, 2.157, 2.153	305, 613, 434
19	2.143	15, 10	2.139, 2.136	315, 641
5	2.102	8	2.098	623
7	2.082	4, 3, 4	2.080, 2.076, 2.074	543, 524, 712
6	2.061	11	2.056	730
4	2.033	4, 6	2.039, 2.026	642, 731
9	2.015	1, 0.5, 13	2.022, 2.012, 2.010	722, 415, 633
4	1.982	1, 7, 0.5	1.990, 1.977, 1.977	534, 651, 006
3	1.946	2, 3	1.944, 1.943	614, 732
5	1.920	3, 4	1.917, 1.916	741, 206
9	1.904	2, 4, 1, 7	1.902, 1.900, 1.899, 1.899	216, 624, 652, 820
4	1.882	4, 5	1.891, 1.877	505, 515
2	1.840	1, 2	1.846, 1.836	812, 316
2	1.813	2, 2	1.811, 1.808	831, 822

(continued)

Table A.50 (continued)

I_{obs}	d_{obs}	I_{calc}*	d_{calc}**	*h k l*
9	1.779	2, 15	1.778, 1.774	535, 714
1	1.755	1, 0.5	1.752, 1.751	644, 840
3	1.734	1, 5	1.732, 1.729	841, 910
2	1.717	2	1.713	625
9	1.692	16	1.690	734
17	1.676	1, 4, 28	1.681, 1.679, 1.671	921, 842, 436
38	1.638	3, 28, 10, 42	1.635, 1.635, 1.633, 1.633	931, 526, 804, 922
4	1.585	1, 0.5, 10	1.592, 1.590, 1.582	536, 932, 770
12	1.570	1, 9, 14	1.576, 1.568, 1.567	606, 616, 664
4	1.552	2, 3	1.554, 1.551	735, 754
6	1.512	2, 2, 8, 2	1.514, 1.508, 1.507, 1.507	10.0.2, 636, 844, 10.1.2
5	1.486	1, 1, 6	1.488, 1.486, 1.482	10.3.1, 10.2.2, 008
2	1.463	1, 2	1.462, 1.462	871, 646
2	1.456	1, 0.5, 2	1.456, 1.454, 1.450	863, 10.4.0, 10.1.3
1	1.433	3	1.431	10.2.3
6	1.399	1, 15	1.397, 1.396	915, 774
3	1.386	0.5, 0.5, 2, 2	1.386, 1.385, 1.385, 1.384	408, 746, 864, 800
3	1.381	0.5, 4	1.381, 1.380	765, 873
5	1.356	0.5, 2, 16	1.359, 1.355, 1.353	963, 935, 954
3	1.338	2, 1, 4	1.343, 1.336, 1.335	10.6.0, 647, 518

Table A.51 Crystallographic data and refinement parameters for crystal structure of milanriederite from Kombat mine

Crystallographic parameters	
Symmetry	Tetragonal
Space group	*P4/nnc*
Unit cell parameters *a*, *c* (Å)	15.6578(4), 11.8597(5)
Unit cell volume (Å^3)	2907.6(2)
Z	2
Calculated density (g/cm^3)	3.523
Absorption coefficient (mm^{-1})	3.659
Crystal size (mm)	0.17 × 0.15 × 0.12
Data collection parameters	
2θ range (°)	6.76–55.00
h, *k*, *l* range	−9 → 20, −17 → 19, −15 → 13
Number reflections	6807
Unique reflections (R_{int})	1684 (0.023)

(continued)

Table A.51 (continued)

Crystallographic parameters	
Symmetry	Tetragonal
Unique reflections c $F > 4\sigma F$	1603
Refinement parameters	
R_1 [$F > 4\sigma(F)$], wR_2 [$F > 4\sigma (F)$]	0.0268, 0.0644
R_1, wR_2 all data	0.0291, 0.0652
S	1.247
ρ_{max}, ρ_{min}, e Å^{-3}	0.76, –0.50
Refinment method	Full matrix least-squares on F^2

Table A.52 Atom fractional coordinates, displacement parameters (Å^2) and site occupancy for crystal structure of milanriederite fom Kombat mine

Site	Occupancy	*x*/*a*	*y*/*b*	*z*/*c*	U_{ani}	Multiplicity
*X*1	Ca	¾	¼	¼	0.0082(4)	4*c*
*X*2	Ca	0.68883(3)	0.45368(4)	0.37840(5)	0.0066(2)	16*k*
*X*3	$Ca_{0.78}REE_{0.22}$	0.67999(3)	0.60096(3)	0.60991(4)	0.0113(2)	16*k*
*X*4	$Ca_{0.40}Na_{0.10}$	¾	¾	0.1488(2)	0.0088(4)	4*e*
*Z*1	Si	¾	¼	½	0.0057(3)	4*d*
*Z*2	Si	0.54011(5)	0.31972(5)	0.37228(6)	0.0056(2)	16*k*
*Z*3	Si	0.58469(5)	0.65077(5)	0.36374(6)	0.0067(2)	16*k*
*Y*1	$Fe^{3+}{}_{0.47}Mg_{0.03}$	¾	¾	0.4571(2)	0.0113(6)	4*e*
*Y*2	Al	½	½	½	0.0053(2)	8*f*
*Y*3	$Mg_{0.56}Al_{0.25}$ $Fe^{3+}{}_{0.19}$	0.61226(4)	0.37875(5)	0.12579(6)	0.0041(3)	16*k*
O1	O	0.7776(1)	0.3272(1)	0.4147(2)	0.0084(4)	16*k*
O2	O	0.6172(1)	0.3385(1)	0.28190(1)	0.0094(4)	16*k*
O3	O	0.7238(1)	0.4551(1)	0.5727(1)	0.0084(4)	16*k*
O4	O	0.5594(1)	0.3951(1)	0.4686(1)	0.0085(4)	16*k*
O5	O	0.8277(1)	0.4887(1)	0.3202(1)	0.0108(4)	16*k*
O6	O	0.6226(1)	0.7249(1)	0.4436(18)	0.0138(5)	16*k*
O7	O	0.4436(1)	0.3264(1)	0.3205(2)	0.0123(4)	16*k*
O8	O	0.5920(1)	0.5604(1)	0.4308(1)	0.0085(4)	16*k*
O9	O	0.6467(1)	0.6467(1)	¼	0.0104(6)	8*h*
O10	(OH, O)	¾	¾	0.6356(4)	0.023(1)	4*e*
O11	(OH, O)	0.5608(1)	0.4986(1)	0.6351(1)	0.0090(4)	16*k*

Table A.53 Selected bond length (Å) in the crystal structure of milanriederite fom Kombat mine

Bond	Bond length (Å)	Bond	Bond length (Å)	Bond	Bond length (Å)
*Z*1–O1 4×	1.635(2)	*X*2–O3	2.369(2)	*X*4–O9 4×	2.582(3)
		*X*2–O6	3.068(2)	<*X*4–O>	2.446
*Z*2–O7	1.634(2)	*X*2–O2	2.412(2)		
*Z*2–O2	1.641(2)	*X*2–O5	2.433(2)	*Y*1–O6 4×	2.039(2)
*Z*2–O3	1.638(2)	*X*2–O4	2.468(2)	*Y*1–O10	2.117(6)
*Z*2–O4	1.671(2)	*X*2–O1	2.458(2)	<*Y*1–O>	2.055
<*Z*2–O>	1.646	<*X*2–O>	2.487		
				*Y*2–O8 2×	1.909(2)
*Z*3–O6	1.612(2)	*X*3–O7	2.391(2)	*Y*2–O11 2×	1.864(2)
*Z*3–O8	1.627(2)	*X*3–O3	2.425(2)	*Y*2–O4 2×	1.924(2)
*Z*3–O5	1.623(2)	*X*3–O11	2.477(2)	<*Y*2–O>	1.899
*Z*3–O9	1.663(1)	*X*3–O6	2.494(2)		
<*Z*3–O>	1.631	*X*3–O6	2.910(2)	*Y*3–O1	1.963(2)
		*X*3–O7	2.596(2)	*Y*3–O11	1.981(2)
*X*1–O1 4×	2.337(2)	*X*3–O10	2.5962(8)	*Y*3–O2	1.957(2)
*X*1–O2 4×	2.527(2)	*X*3–O7	2.534(2)	*Y*3–O3	2.022(2)
<*X*1–O>	2.432	*X*3–O8	2.610(2)	*Y*3–O5	2.063(2)
		<*X*3–O>	2.593	*Y*3–O4	2.103(2)
*X*2–O8	2.341(2)			<*Y*3–O>	2.015
*X*2–O5	2.347(2)	*X*4–O6 4×	2.310(2)		

Table A.54 Powder X-ray diffraction data for Ti-rich modraite

I_{meas}	d_{meas}, Å	I_{calc}	d_{calc}, Å	*hkl*	I_{meas}	d_{meas}, Å	I_{calc}	d_{calc}, Å	*hkl*
7	5.91	12	5.91	002	3	2.154	2	2.153	305
3	5.50	2	5.50	220	4	2.145	3	2.144	434
3	4.706	3	4.710	202	12	2.133	12	2.132	315
4	4.507	0.8	4.508	212	10	2.124	9	2.124	514
2	4.054	2	4.056	321	3	2.103	0.5	2.104	721
14	4.031	14	4.030	222	8	2.088	4	2.088	623
8	3.892	3	3.891	400	6	2.069	4	2.069	543
1	3.670	0.5	3.669	330	4	2.064	3	2.063	712
15	3.481	10	3.481	420	13	2.044	6	2.043	730
9	3.252	11	3.251	402	5	2.028	2	2.027	642
4	3.079	4	3.078	313	8	2.014	4	2.014	731
11	3.054	8	3.052	510	8	2.000	6	1.999	633
13	3.010	7	3.010	501	7	1.9655	4	1.9650	651

(continued)

Table A.54 (continued)

I_{meas}	d_{meas}, Å	I_{calc}	d_{calc}, Å	*hkl*	I_{meas}	d_{meas}, Å	I_{calc}	d_{calc}, Å	*hkl*
42	2.959	45	2.958	004	8	1.9309	6	1.9310	732
4	2.913	5	2.911	323	4	1.9226	3	1.9220	713
4	2.856	3	2.856	114	3	1.9121	4	1.9110	206
100	2.754	100	2.755	432	4	1.9062	3	1.9050	741
4	2.668	2	2.669	530	3	1.8981	2	1.8970	216
26	2.606	38	2.605	224	7	1.8888	6	1.8880	652
58	2.598	32	2.597	522	3	1.8839	2	1.8840	505
6	2.535	3	2.535	314	3	1.8709	2	1.8700	515
3	2.500	2	2.501	611	12	1.7979	7	1.7980	822
34	2.462	33	2.461	620	8	1.7657	4	1.7660	714
9	2.445	9	2.443	503	3	1.7187	6	1.7180	910
9	2.375	5	2.375	602	9	1.6811	2	1.6810	734
10	2.355	7	2.355	404	16	1.6659	9	1.6660	436
13	2.328	10	2.328	414	12	1.6294	21	1.6290	526
3	2.303	4	2.302	334	26	1.6238	12	1.6230	922

Table A.55 Crystallographic data and refinement parameters for Ti-rich vesuvianite

Crystal data	
Formula	Ca_{19}(Al, Fe, Ti, Mg)[SiO_4]$_{10}$ [Si_2O_7]$_4$(OH, O, F)
Formula weight (g/mol)	2968.3
Absorption, μ (mm^{-1})	3.259
Density, D_x (g/cm^3)	3.460
Crystal system	Tetragonal
Space group	*P*4/*nnc*
a (Å)	15.5326(2)
c (Å)	11.8040(2)
V (Å^3)	2847.87(8)
Z	2
Crystal size (mm)	0.13 × 0.18 × 0.25
Data collection	
Diffractometer	SMART APEX2 CCD
Radiation; wavelength (Å)	Mo*K*α; 0.7107
Data collection method	ω-scans
Temperature (K)	120
F(000)	2938
Theta range for data collection (°)	1.85 to 62.83

(continued)

Table A.55 (continued)

Crystal data	
Index ranges	$-38 < h < 38$
–	$-38 < k < 38$
–	$-29 < l < 29$
Reflections collected	498719
Unique reflections/R_{int}	11247/5.84
Observed reflections ($I_{obs} > 2\sigma(I)$)	10270
Refinement	
Refinement method	Full-matrix least-squares on F; weighting scheme: $w = 1/(\sigma^2F + 0.002116F^2)$
Goodness-of-fit on F	1.02
Final R indices	$R = 3.12$; $wR = 5.65$
$\Delta\rho_{min}/\Delta\rho_{max}$	−0.68/1.04

Table A.56 Selected interatomic distances for Ti-rich modraite

Bond		Bond length, Å	Bond		Bond length, Å
*X*1	O1	2.3407(4) × 4	*Y*1a	O6	2.0670(6) × 4
	O2	2.5425(5) × 4		O10	2.363(2)
Mean		2.4416	*Mean*		2.126
*X*2	O8	2.3230(5)	*Y*1a	*Y*1b	0.3813(16)
	O5	2.3382(5)	*Y*1b	O10	1.982(2)
	O3	2.3714(4)		O6	2.0853(5) × 4
	O2	2.4240(5)	*Mean*		2.064
	O5	2.4258(5)	*Y*2	O8	1.8730(4) × 2
	O4	2.4571(5)		O11	1.8751(4) × 2
	O1	2.4662(5)		O4	1.9386(5) × 2
	O6	2.9406(5)	*Mean*		1.8956
Mean		2.4683	*Y*3	O11	1.8757(5)
*X*3	O7	2.3452(5)		O2	1.8865(5)
	O3	2.4549(5)		O1	1.9463(5)
	O11	2.4572(5)		O3	1.9503(5)
	O7	2.4737(5)		O5	1.9787(5)
	O6	2.4966(5)		O4	2.0831(4)
	O7	2.5594(5)	*Mean*		1.9534
	O10	2.5703(2)	*Z*1	O1	1.6402(4) × 4
	O8	2.6157(4)	*Mean*		1.6402

(continued)

Table A.56 (continued)

Bond		Bond length, Å	Bond		Bond length, Å
	O6	2.9873(6)	*Z*2	O7	1.6106(5)
Mean		2.5511		O2	1.6427(5)
*X*4	O6	2.2944(5) × 4		O3	1.6430(5)
	O9	2.6266(5) × 4		O4	1.6719(5)
Mean		2.4605	*Mean*		1.6421
*X*4	*Y*1a	0.909(2)	*Z*3	O6	1.6076(5)
	*Y*1b	1.2904(9)		O8	1.6218(5)
	*X*4	2.5213(7)		O5	1.6389(5)
				O9	1.6577(3)
			Mean		1.6315

References

Agilent Technologies (2012) CrysAlisPro

Agilent Technologies (2014) CrysAlis CCD and CrysAlis RED. Oxford Diffraction Ltd., Yarnton, Oxfordshire

Aksenov SM, Chukanov NV, Rusakov VS et al (2016) Towards a revisitation of vesuvianite-group nomenclature: the crystal structure of Ti-rich vesuvianite from Alchuri, Shigar Valley, Pakistan. Acta Crystallogr Sect B Struct Sci Cryst Eng Mater 72:744–752. https://doi.org/10.1107/S2052520616010246

Aksenov SM, Bykova EA, Rastsvetaeva RK et al (2018) Microporous crystal structure of labuntsovite-Fe and high-pressure behavior up to 23 GPa. Acta Crystallogr Sect B Struct Sci Cryst Eng Mater 74:1–11. https://doi.org/10.1107/S205252061700498X

Allen FM (1985) Structural and chemical variations in vesuvianite Ph.D. thesis. Harvard University, Cambridge Massachusetts

Allen FM, Burnham CW (1992) A comprehensive structure-model for vesuvianite: symmetry variations and crystal growth. Can Mineral 30:1–18

Amelin YV, Larin AM, Tucker RD (1997) Chronology of multiphase emplacement of the Salmi rapakivi granite-anorthosite complex, Baltic Shield: implications for magmatic evolution. Contrib Miner Petrol 127:353–368. https://doi.org/10.1007/s004100050285

Anenburg M, Mavrogenes JA (2018) Carbonatitic versus hydrothermal origin for fluorapatite REE-Th deposits: experimental study of REE transport and crustal "antiskarn" metasomatism. Am J Sci 318:335–366. https://doi.org/10.2475/03.2018.03

Annenkova GA et al (1972) Minerals. Handbook. Moscow

Antao SM (2021) Crystal chemistry of six grossular garnet samples from different well-known localities. Minerals 11:767. https://doi.org/10.3390/min11070767

Antonov AA (2003) Mineralogy of the rodingites of the Bazhenov hyperbasite massif. St. Petersburg

Arem JE (1973) Idocrase (vesuvianite)—A 250-year puzzle. Mineralo Record 4:164–174

Arem JE (1970) Crystal chemistry and structure of idocrase. Ph.D. thesis. Harward

Arem JE, Burnham CW (1969) Structural variations in idocrase. Am Miner 54:1546–1550

Armbruster T, Gnos E (2000a) "Rod" polytypism in vesuvianite: crystal structure of a low-temperature P4nc vesuvianite with pronounced octahedral cation ordering. Schweiz Mineral Petrogr Mitt 80:109–116

Armbruster T, Gnos E (2000b) P4/n and P4nc long-range ordering in low-temperature vesuvianites. Am Miner 85:563–569

Armbruster T, Gnos E (2000c) Tetrahedral vacancies and cation ordering in low-temperature Mn-bearing vesuvianites: indication of a hydrogarnet-like substitution. Am Miner 85:570–577

T. L. Panikorovskii and N. V. Chukanov, *Vesuvianite-Supergroup Minerals*, Springer Mineralogy, https://doi.org/10.1007/978-3-032-06970-2

Armbruster T, Gnos E, Dixon R et al (2002) Manganvesuvianite and tweddillite, two new Mn3+-silicate minerals from the Kalahari manganese fields, South Africa. Mineral Mag 66:121–135. https://doi.org/10.1180/0026461026610018

Aubry AY, Dusausoy AL, Protas J (1969) Détermination et étude de la structure cristalline de l'henritermiérite, hydrogrenat de symétrie quadratique. Bull Soc Fr Minéral 92:126–133

Avdontceva MS, Krzhizhanovskaya MG, Krivovichev SV, Yakovenchuk VN (2015) High-temperature order–disorder phase transition in nacaphite, Na2CaPO4F. Phys Chem Miner 42:671–676. https://doi.org/10.1007/s00269-015-0753-x

Balagansky VV, Basalaev AA, Belyaev OA et al (1996) Geological map of the Kola region 1: 500000: KI KSC RAS. Apatity

Balassone G, Talla D, Beran A et al (2011) Vesuvianite from Somma-Vesuvius volcano (southern Italy): chemical, X-ray diffraction and single-crystal polarized FTIR investigations. Periodico di Mineralogia 80:369–384. https://doi.org/10.2451/2011PM0026

Balassone G, Bellatreccia F, Mormone A et al (2012) Sodalite-group minerals from the Somma Vesuvius volcanic complex, Italy: a case study of K-feldspar-rich xenoliths. Mineral Mag 76:191–212. https://doi.org/10.1180/minmag.2012.076.1.191

Balassone G, Kahlenberg V, Altomare A et al (2014) Nephelines from the Somma-Vesuvius volcanic complex (Southern Italy): crystal-chemical, structural and genetic investigations. Mineral Petrol 108:71–90. https://doi.org/10.1007/s00710-013-0290-6

Balassone G, Petti C, Mondillo N et al (2019) Copper Minerals at Vesuvius Volcano (Southern Italy): a mineralogical review. Minerals 9:730. https://doi.org/10.3390/min9120730

Balassone G, Panikorovskii TL, Pellino A et al (2022) The complex mechanism of Ti4+ incorporation into litidionite from the Somma-Vesuvius volcano, Italy. Mineral Mag 86:222–233. https://doi.org/10.1180/mgm.2022.1

Balassone G, Panikorovskii TL, Pellino A et al (2024) Enricofrancoite, KNaCaSi4 O10, a new Ca–K–Na silicate from Somma-Vesuvius volcano, southern Italy. MinMag 88:277–287. https://doi.org/10.1180/mgm.2024.9

Barr TL, Seal S, Wozniak K, Klinowski J (1997) ESCA studies of the coordination state of aluminium in oxide environments. J Chem Soc Faraday Trans 93:181–186. https://doi.org/10.1039/A604061F

Barth FW (1963) Contributions to the mineralogy of Norway: 22. Vesuvianite from Kristiansand, other occurrences in Norway, the general formula of vesuvianite. Nor Geol Tidsskr 43:457–472

Batieva ID, Bel'kov IV (1984) Saharayok alkaline massif, rocks and minerals: GI KSC RAS. Apatity 133

Batieva ID, Belkov IV, Bogdanova AN, Fedotova EP (1987) Rare-earth vesuvianite from alkaline gabbroids of the Saharyok massif (Kola Peninsula). Mineral Associations and Minerals of the Magmatic Complexes of the Kola Peninsula 10–14

Baur WH (1974) The geometry of polyhedral distortions. Predictive relationships for the phosphate group. Acta Crystallogr Sect B: Struct Crystallogr Cryst Chem 30:1195–1215. https://doi.org/10.1107/S0567740874004560

Belinskii VB (1985) Geochemical features of tin-bearing skarn deposits in Eastern Yakutia. Ph.D. thesis

Bellatreccia F, Cámara F, Ottolini L et al (2005a) Wiluite from Ariccia, Latium, Italy: occurrence and crystal structure. Can Mineral 43:1457–1468. https://doi.org/10.2113/gscanmin.43.5.1457

Bellatreccia F, Della Ventura G, Ottolini L et al (2005b) The quantitative analysis of OH in vesuvianite: a polarized FTIR and SIMS study. Phys Chem Miner 32:65–76. https://doi.org/10.1007/s00269-004-0437-4

Berzelius JJ (1821) Die Anwendung des Löthrohrs in der Chemie und Mineralogie. Nürnberg

Blatov VA, Shevchenko AP, Serezhkin VN (2000) TOPOS 3.2: a new version of the program package for multipurpose crystal-chemical analysis. J Appl Crystallogr 33:1193–1193. https://doi.org/10.1107/S0021889800007202

Bocharnikova TD, Kholodnov VV, Shagalov ES (2011) Composition and sources of fluids in skarns of mineral mines of the Kusinsko-Kopansky intrusive complex (Southern Urals). Lithosphere 5:124–130

Borovikova EY (2005) Crystal chemistry of minerals of the Vesuvianite group of different origin. The study by the methods of vibrational (infrared and Raman scattering) and Mössbauer spectroscopy. Ph.D. thesis

Borovikova EY, Kurazhkovskaya VS (2006) The effect of fluorine on the formation of ordered and disordered modifications of vesuvianite: a study by IR spectroscopy. Zapiski RMO 135:89–95

Bosi F, Biagioni C, Oberti R (2019a) On the chemical identification and classification of minerals. Minerals 9:591. https://doi.org/10.3390/min9100591

Bosi F, Hatert F, Hålenius U et al (2019b) On the application of the IMA–CNMNC dominant-valency rule to complex mineral compositions. Mineral Mag 83:627–632. https://doi.org/10.1180/mgm.2019.55

Brauns R (1883) Über die Ursache der anomalen Doppelbrechung einiger regulär krystallisirender Salze. Neues Jahrbuch Für Mineralogie, Geologie und Paläontologie 1:102–111

Brauns R (1891) No Die Optischen Anomalien der Krystalle. Hirzel, S., Leipzig

Britvin SN, Antonov AA, Krivovichev SV et al (2003) Fluorvesuvianite, Ca19(Al, Mg, Fe2+)13[SiO4]10[Si2O7]4O(F,OH)9, a new mineral species from Pitkäranta, Karelia, Russia: description and crystal structure. Can Mineral 41:1371–1380. https://doi.org/10.2113/gscanmin.41.6.1371

Bruker AXS GmbH (2009) Topas. General profile and structure analysis software for powder diffraction data

Bruker-AXS (2014) APEX2 Version 2014.11-0. Madison, Wisconsin, USA

Burns RG (1994) Mineral Mössbauer spectroscopy: correlations between chemical shift and quadrupole splitting parameters. Hyperfine Interact 91:739–745. https://doi.org/10.1007/BF02064600

Butek J, Spišiak J, Milovská S (2021) Garnet-Vesuvianite Equilibrium in Rodingites from Dobšiná (Western Carpathians). Minerals 11:189. https://doi.org/10.3390/min11020189

Cairncross B, Beukes N, Gutzmer J (1997) The manganese adventure; The South African Manganese Fields. Associated Ore and Metal Cooperation Limited, Marshalltown, Johannesburg

Calas G, Galoisy L, Cormier L (2021) The color of glass. In: Encyclopedia of glass science, technology, history, and culture. Wiley, pp 677–691

Chatterjee ND (1962) Vesuvianite-epidote paragenesis as a product of greenschist facies of regional metamorphism in the Western Alps. Beiträge Zur Mineralogie und Petrographie 8:432–439. https://doi.org/10.1007/BF01082095

Cheng X, Ling M, Li X et al (2024) Chlorite, quartz and pyrite trace element behavior during Cu mineralization in the Hongshi deposit, NW China: Implications for hydrothermal fluid evolution and ore genesis. Ore Geol Rev 169:106100. https://doi.org/10.1016/j.oregeorev.2024.106100

Chumakov AA, Morozov AI, Ginzburg IV (1948) Vesuvian from Western Keivy (on the Kola Peninsula). Dokl Acad Sci USSR 11:1099–1101

Chukanov NV (2014) Infrared spectra of mineral species. Extended library

Chukanov NV, Aksenov SM, Rastsvetaeva RK et al (2015) Calcinaksite, KNaCa(Si4O10)H2O, a new mineral from the Eifel volcanic area, Germany. Mineral Petrol 109:397–404. https://doi.org/10.1007/s00710-015-0376-4

Chukanov NV, Panikorovsky TL, Chervonnyi AD (2018) On the relationships between crystal-chemical characteristics of vesuvianite-group minerals and their IR spectra. Zapiski RMO 147:112–128

Chukanov NV, Panikorovskii TL, Goncharov AG et al (2019) Milanriederite, (Ca,REE)19Fe3+Al4(Mg,Al,Fe3+)8Si18O68(OH,O)10, a new vesuvianite-group mineral from the Kombat Mine, Namibia. Eur J Mineral 31:637–646. https://doi.org/10.1127/ejm/2019/0031-2856

Chukanov NV, Vigasina MF (2020) Vibrational (Infrared and Raman) spectra of minerals and related compounds. Springer International Publishing, Cham

Chukanov NV, Vigasina MF, Zubkova NV et al (2020) Extra-framework content in sodalite-group minerals: complexity and new aspects of its study using infrared and Raman spectroscopy. Minerals 10:363. https://doi.org/10.3390/min10040363

Chukanov NV, Aksenov SM, Rastsvetaeva RK (2021) Structural chemistry, IR spectroscopy, properties, and genesis of natural and synthetic microporous cancrinite- and sodalite-related materials: a review. Microporous Mesoporous Mater 323:111098. https://doi.org/10.1016/j.micromeso.2021.111098

Chukanov N, Shchipalkina N, Shendrik R et al (2022a) Isomorphism and mutual transformations of S-bearing components in feldspathoids with microporous structures. Minerals 12:1456. https://doi.org/10.3390/min12111456

Chukanov N, Shendrik R, Vigasina M et al (2022b) Crystal chemistry, isomorphism, and thermal conversions of extra-framework components in sodalite-group minerals. Minerals 12:887. https://doi.org/10.3390/min12070887

Chukanov NV, Vigasina MF, Shendrik RY et al (2022c) Nature and isomorphism of extra-framework components in cancrinite- and sodalite-related minerals: new data. Minerals 12:729. https://doi.org/10.3390/min12060729

Chukanov NV, Ermolaeva VN, Varlamov DA, Jonsson E (2023a) The association of henritermierite with Mg-Rich Vesuvianite in Mn Ores: indicator significance and an example of crystal chemical selectivity. In: Marin Y (ed) XIII general meeting of the russian mineralogical society and the fedorov session. Springer International Publishing, Cham, pp 43–49

Chukanov NV, Kazheva ON, Fischer RX, Aksenov SM (2023b) Refinement of the crystal structure of fresnoite, Ba2TiSi2O8, from Löhley (Eifel district, Germany); gladstone-dale compatibility, electronic polarizability and vibrational spectroscopy of minerals and inorganic compounds with pentacoordinated TiIV and a titanyl bond. Acta Crystallogr B Struct Sci Cryst Eng Mater 79:184–194. https://doi.org/10.1107/S2052520622012045

Ciriotti ME, Fascio L, Pasero M (2009) Italian type minerals. Pisa

Clarke F (1895) The constitution of the silicates. Bull United States Geol Surv 125:1–100

Clarke F, Steiger G (1905) On "Californite." Bull United States Geol Surv 262:72–73

Coda A, Giusta DA, Isetti G, Mazzi F (1970) On the structure of vesuvianite. Atti Dell Accademica Delle Scienze di Torino 105:1–22

Crook WW, Oswald SG (1979) New data on cerian vesuvianite from san Benito county, california. Am Miner 64:367–368

Czaja M, Lisiecki R, Chrobak A et al (2018) The absorption- and luminescence spectra of Mn3+ in beryl and vesuvianite. Phys Chem Miner 45:475–488. https://doi.org/10.1007/s00269-017-0934-x

de Caritat P, Hutcheon I, Walshe JL (1993) Chlorite geothermometry: a review. Clays Clay Miner 41:219–239. https://doi.org/10.1346/CCMN.1993.0410210

Deer WA, Howie RA, Zussman J (1962) Ortho- and Ring Silicates. Wiley

Deer WA, Howie RA, Zussman J (1982) Rock-forming minerals, vol 1. A-Orthosilicates. Longmans, Green and co. Ltd., London

Desiraju GR, Steiner T (1999) In the weak hydrogen bond in structural chemistry and biology. Oxford University Press Inc., New York

d'Espinose de la Caillerie J-B (1994) A reassessment of the 29Si MAS-NMR spectra of sepiolite and aluminated sepiolite. Clay Minerals 29:313–318. https://doi.org/10.1180/claymin.1994.029.3.02

Dey A, Choudhury SR, Mukherjee S et al (2019) Origin of vesuvianite-garnet veins in calc-silicate rocks from part of the Chotanagpur Granite Gneiss Complex, East Indian Shield: the quantitative P-T-XCO2 topology in parts of the system CaO-MgO-Al2O3-SiO2-H2O-CO2 (+Fe2O3, F). Am Miner 104:744–760. https://doi.org/10.2138/am-2019-6811

Dolivo-Dobrovolsky DD (2016) MINAL, free software

Dolomanov OV, Bourhis LJ, Gildea RJ et al (2009) OLEX2: a complete structure solution, refinement and analysis program. J Appl Crystallogr 42:339–341. https://doi.org/10.1107/S0021889808042726

Dyrek K, Platonov AN, Sojka Z, Żabiński W (1992) Optical absorption and EPR study of Cu2+ ions in vesuvianite ("cyprine") from Sauland, Telemark, Norway. Eur J Mineral 4:1285–1290. https://doi.org/10.1127/ejm/4/6/1285

Eldjarn K, Kvamsdal LO, Selbekk R, Muller-Sigmünd H (2005) Norske vesuvianer og vesuvianforekomster. Norsk Bergverksmuseum Skrifter 30:30–40

Elmi C, Brigatti MF, Pasquali L et al (2010) Crystal chemistry, surface morphology and X-ray photoelectron spectroscopy of Fe-rich osumilite from Mt. Arci, Sardinia (Italy). Phys Chem Miner 37:561–569. https://doi.org/10.1007/s00269-010-0357-4

Elmi C, Brigatti MF, Pasquali L et al (2011) High-temperature vesuvianite: Crystal chemistry and surface considerations. Phys Chem Miner 38:459–468. https://doi.org/10.1007/s00269-011-0419-2

Erokhin YV (2009) Geochemistry of Vesuvianite of the Bazhenov ophiolite complex. Mater Fedorov Sess 210–212

Erokhin YV (2017) Bazhenovskoe deposit (Central Urals, Russia): mineralogy of rodingites. Mineralo Almanac 22:16–24

Erokhin YV, Ivanov KS, Smirnov VN et al (2023) Gabbroids of the Bazhenov gabbro-hyperbasite massif (Middle Urals): material composition and age. LITHOSPHERE (Russia) 23:367–385. https://doi.org/10.24930/1681-9004-2023-23-3-367-385

Farrugia LJ (2012) WinGX and ORTEP for windows: an update. J Appl Crystallogr 45:849–854. https://doi.org/10.1107/S0021889812029111

Fechtelkord M, Stief F, Buhl J-C (2001) Sodium cation dynamics in nitrate cancrinite: A low and high temperature 23Na and 1H MAS NMR study and high temperature Rietveld structure refinement. Am Miner 86:165–175. https://doi.org/10.2138/am-2001-0118

Fedorov ES (1890) Geological research in the Northern Urals in 1884–1886

Ferraris G, Makovicky E, Merlino S (2014) Crystallography of modular materials. Oxford University Press, Oxford

Filatov SK, Bubova RS (2008) High-temperature crystal chemistry of borates and borosilicates

Fitzgerald S, Rheingold AL, Leavens PB (1986a) Crystal structure of a non-P4/nnc vesuvianite from Asbestos, Quebec. Am Miner 71:1483–1488

Fitzgerald S, Rheingold AL, Leavens PB (1986b) Crystal structure of a Cu-bearing vesuvianite. Am Miner 71:1011–1014

Fitzgerald S, Leavens PB, Rheingold AL, Nelen JA (1987) Crystal structure of a REE-bearing vesuvianite from San Benito County, California. Am Miner 72:625–628

Földvári M (2011) Handbook of thermogravimetric system of minerals and its use in geological practice. Geological Institute of Hungary, Budapest

Frey M, Desmons J, Neubauer F (1999) The new metamorphic map of the Alps. Schweiz Mineral Petrogr Mitt 79:1–215. https://doi.org/10.5169/seals-60194

Gagné OC, Hawthorne FC (2017) Empirical Lewis acid strengths for 135 cations bonded to oxygen. Acta Crystallogr B Struct Sci Cryst Eng Mater 73:956–961. https://doi.org/10.1107/S2052520617010988

Galuskina IO, Galuskin EV, Sitarz M (1998) Atoll hydrogarnets and mechanism of the formation of achtarandite pseudomorphs. Neues Jahrb Mineral Monatsh 2:49–62

Galuskin EV (2000) A new mineral species isostructural With Vesuvianite, from the Sakha Republic, Russian Federation: discussion. Can Mineral 38:763–764

Galuskin EV, Galuskina IO (2000) Wiluite a new 1998 mineral from Sakha-Yakutia, a mineral with more than 200 years of history. Zapiski RMO 129:130–132

Galuskin EV, Galuskina IO (2003) Achtarandite—Sponge hibschite pseudomorph after wadalite-like phase: internal morphology and mechanism of formation. Neues Jahrbuch Für Mineralogie - Abhandlungen 178:63–74. https://doi.org/10.1127/0077-7757/2002/0178-0063

Galuskin EV, Armbruster T, Malsy A et al (2003a) Morphology, composition and structure of low-temperature P4/nnc high-fluorine vesuvianite whiskers from Polar Yakutia, Russia. Can Mineral 41:843–856. https://doi.org/10.2113/gscanmin.41.4.843

Galuskin EV, Galuskina IO, Sitarz M, Stadnicka K (2003b) Si-deficient, OH-substituted, boron-bearing vesuvianite from the Wiluy River, Yakutia, Russia. Can Mineral 41:833–842. https://doi.org/10.2113/gscanmin.41.4.833

Galuskin EV (2005) Mineraly Grupy wezuwianu ze skal achtarandytowych (rzeka Wiluj, Jakucja). Widawnictwo Uniwersytetu Slaskiego, Katowice

Galuskin EV, Galuskina IO, Dzierżanowski P (2005) Chlorine in vesuvianites. Mineral Pol 36:51–61

Galuskin EV, Galuskina IO, Stadnicka K et al (2007a) The crystal structure of Si-deficient, OH-substituted, boron-bearing vesuvianite from the Wiluy River, Sakha-Yakutia, Russia. Can Mineral 45:239–248. https://doi.org/10.2113/gscanmin.45.2.239

Galuskin EV, Janeczek J, Kozanecki M et al (2007b) Single-crystal Raman investigation of vesuvianite in the OH region. Vib Spectrosc 44:36–41. https://doi.org/10.1016/j.vibspec.2006.06.022

Gaskelberg LA, Luk'yanova NV, Gaskelberg VG, et al (1978) Joint report on the additional study of the Cave structure and composition of the new geological map at a scale of 1: 200,000. Apatity. Unpublished report. FSI TFGI North-West Federal District 1:1–7

Gaudefroy C, Orliac M, Permingeat F, Permingeat F (1969) L'henritermiérite, une nouvelle espèece minérale. Bull Soc Fr Minéral 92:185–190

Geiger CA, Stahl A, Rossman GR (1999) Raspberry-red grossular from Sierra de Cruces Range, Coahuila. Mexico. Ejm 11:1109–1114. https://doi.org/10.1127/ejm/11/6/1109

Geiger CA, Armbruster T, Khomenko V, Quartieri S (2000) Cordierite I: the coordination of Fe2+. Am Miner 85:1255–1264. https://doi.org/10.2138/am-2000-8-918

Gilg HA, Lima A, Somma R et al (2001) Isotope geochemistry and fluid inclusion study of skarns from Vesuvius. Mineral Petrol 73:145–176. https://doi.org/10.1007/s007100170015

Giuseppetti G, Mazzi F (1983) The crystal structure of a vesuvianite with P4/n symmetry. Tschermaks Mineralogische und Petrographische Mitteilungen 31:277–288

Gnos E, Armbruster T (2006) Relationship among metamorphic grade, vesuvianite "rod polytypism" and vesuvianite composition. Am Miner 91:862–870. https://doi.org/10.2138/am.2006.1973

Goldschmidt VM (1918) Atlas der Krystallformen. Carl Winters Universitätsbuchhandlung, Heidelberg

Grew ES, Krivovichev SV, Hazen RM, Hystad G (2016) Evolution of structural complexity in boron minerals. Can Mineral 54:125–143. https://doi.org/10.3749/canmin.1500072

Grew ES, Locock AJ, Mills SJ et al (2013) IMA report: nomenclature of the garnet supergroup. Am Miner 98:785–810. https://doi.org/10.2138/am.2013.4201

Grigoryev DP (1991) The science of minerals in St. Petersburg in the year of the founding of the mineralogical society. Zapiski RMO 119:100–107

Groat LA, Hawthorne FC, Ercit TS (1992a) The chemistry of Vesuvianite. Can Mineral 30:19–48

Groat LA, Hawthorne FC, Ercit TS (1992b) The role of fluorine in vesuvianite: a crystal-structure study. Can Mineral 30:1065–1075

Groat LA, Hawthorne FC, Ercit TS, Putnis A (1993) The symmetry of vesuvianite. Can Mineral 31:617–635

Groat LA, Hawthorne FC, Ercit TS (1994a) The incorporation of boron into the vesuvianite structure. Can Mineral 32:505–523

Groat LA, Hawthorne FC, Ercit TS (1994b) Excess Y-group cations in the crystal structure of vesuvianite. Can Mineral 32:497–504

Groat LA, Bismayer U, Güttler B (1995a) A ferroelastic phase transition in vesuvianite. Phase Transitions 55:217–227. https://doi.org/10.1080/01411599508200435

Groat LA, Hawthorne FC, Rossman GR, Ercit TS (1995b) The Infrared-spectroscopy of Vesuvianite in the OH Region. Can Mineral 33:609–626

Groat LA, Hawthorne FC, Lager GA et al (1996) X-ray and neutron crystal-structure refinements of a boron-bearing vesuvianite. Can Mineral 34:1059–1070

Groat LA (1988) The crystal chemistry of vesuvianite, Ph.D. thesis. Manitoba

Groat LA, Hawthorne FC, Ercit TS, Grice JD (1998) Wiluite, Ca19(Al,Mg,Fe,Ti)13(B,Al,)5Si18O68(O,OH)10, a New mineral species isostructural with Vesuvianite, from the Sakha Republic, Russian Federation : discussion. Can Mineral 36:1301–1304

Groat LA, Hawthorne FC, Ercit TS, Grice JD (2000) Wiluite, Ca19(Al,Mg,Fe,Ti)13(B,Al,□)5Si18O68(O,OH)10, a new mineral species isostructural with vesuvianite, from the Sakha Republic, Russian Federation. (Discussion). Can Mineral 38:765–766

Groat LA, Evans RJ (2012) Crystal chemistry of Bi- and Mn-bearing vesuvianite from Långban, Sweden. Am Miner 97:1627–1634. https://doi.org/10.2138/am.2012.4091

Groat LA, Evans RJ, Cempírek J et al (2013) Fe-rich and As-bearing vesuvianite and wiluite from Kozlov, Czech Republic. Am Miner 98:1330–1337. https://doi.org/10.2138/am.2013.4358

Gurzhiy VV, Tyumentseva OS, Krivovichev SV et al (2016) Mixed Uranyl Sulfate–Selenates: evolution of structural topology and complexity versus chemical composition. Cryst Growth des 16:4482–4492. https://doi.org/10.1021/acs.cgd.6b00611

Hack AC, Mavrogenes JA (2006) A cold-sealing capsule design for synthesis of fluid inclusions and other hydrothermal experiments in a piston-cylinder apparatus. Am Miner 91:203–210. https://doi.org/10.2138/am.2006.1898

Halama R, Savov IP, Garbe-Schönberg D et al (2013) Vesuvianite in high-pressure-metamorphosed oceanic lithosphere (Raspas Complex, Ecuador) and its role for transport of water and trace elements in subduction zones. Eur J Mineral 25:193–219. https://doi.org/10.1127/0935-1221/2013/0025-2281

Hålenius U (2000) Detection and measurement of Cr3 + at the octahedral Y3-site. Periodico di Mineralogia 69:35–48

Hålenius U, Bosi F, Gatedal K (2013) Crystal structure and chemistry of skarn-associated bismuthian vesuvianite. Am Miner 98:566–573. https://doi.org/10.2138/am.2013.4310

Hålenius U, Panikorovskii TL, Shilovskikh VV (2016) Electronic d-d transitions and intervalence charge transfer processes in 3d-cations in cyprine and chemically related vesuvianite group minerals. In: Book of abstracs EMC 2016. Rimini, p 406

Han S, Chen J, Liao C et al (2025) Hydrothermal Vesuvianite: a U-Pb Dating Mineral. Lithosphere 2025:lithosphere_2024_193. https://doi.org/10.2113/2025/lithosphere_2024_193

Haüy RJ (1801) Traité de minéralogie. Paris

Hawthorne FC (2002) The use of end-member charge-arrangements in defining new mineral species and heterovalent substitutions in complex minerals. Can Mineral 40:699–710. https://doi.org/10.2113/gscanmin.40.2.699

Hawthorne FC, Gagné OC (2024) New ion radii for oxides and oxysalts, fluorides, chlorides and nitrides. Acta Crystallogr B Struct Sci Cryst Eng Mater 80:326–339. https://doi.org/10.1107/S2052520624005080

Henmi C, Kusachi I, Henmi K (1994) Vesuvianite from Kushiro, Hiroshima Prefecture, Japan. In: IMA Abstracts. Pisa, p 172

Himmelberg GR, Miller TP (1980) Uranium- and thorium-rich vesuvianite from the Seward Peninsula, Alaska. Am Miner 65:1020–1025

Hochella MF, Liou JG, Keskinen MJ, Kim HS (1982) Synthesis and stability relations of magnesium idocrase. Econ Geol 77:798–808. https://doi.org/10.2113/gsecongeo.77.4.798

Hoisch TD (1985) The solid solution chemistry of vesuvianite. Contrib Miner Petrol 89:205–214

Horváth L, Gault RA (1990) The mineralogy of Mont Saint-Hilaire Quebec. Mineralogical Record 284–359

Horváth L, Pfenninger-Horváth E, Spertini F (2013) The Jeffrey mine, Asbestos, Québec, Canada. Mineral Rec 44:375–417

Huve L, Delmotte L, Martin P et al (1992) 19F MAS-NMR study of structural fluorine in some natural and synthetic 2:1 layer silicates. Clays Clay Miner 40:186–191. https://doi.org/10.1346/CCMN.1992.0400208

Innes J, Chaplin RC (1986) Ore bodies of the Kombat mine, South West Africa/Namibia. In: Anheusser CR, Maske S (eds) Mineral deposits of southern Africa, Geological Society of South Africa. pp 1789–1805

Ito J, Arem JE (1970) Idocrase: synthesis, phase relations and crystal chemistry. Am Miner 55:880–912

Ivanyuk GY, Yakovenchuk VN, Pakhomovsky YA (2002) Kovdor. Laplandia minerals, Apatity

Ivanyuk GY, Goryainov PM, Pakhomovsky YA, et al (2009) Self-organization of ore complexes. Synergistic principles of forecasting and exploration of mineral deposi

Ivanyuk G, Kalashnikov A, Pakhomovsky Y et al (2017) Subsolidus evolution of the magnetite-Spinel-UlvöSpinel solid solutions in the Kovdor Phoscorite-Carbonatite Complex, NW Russia. Minerals 7:215. https://doi.org/10.3390/min7110215

Ivanyuk G, Konopleva N, Yakovenchuk V et al (2018a) Three-D mineralogical mapping of the Kovdor Phoscorite-Carbonatite Complex, NW Russia: III. Pyrochlore supergroup minerals. Minerals 8:277. https://doi.org/10.3390/min8070277

Ivanyuk GY, Pakhomovsky YA, Panikorovskii TL et al (2018b) Three-D Mineralogical Mapping of the Kovdor Phoscorite-Carbonatite Complex, NW Russia: II. Sulfides. Minerals 8:292. https://doi.org/10.3390/min8070292

Ivashchenko VI (2021) Rare-Metal (In, Bi, Te, Se, Be) Mineralization of Skarn Ores in the Pitkäranta Mining District, Ladoga Karelia, Russia. Minerals 11:124. https://doi.org/10.3390/min11020124

Jannasch P (1884) Zur Kenntniss des Zusammensetzung des Vesuvians. Neues Jahrbuch Für Mineralogie, Geologie, und Paläontologie 1:266–270

Jannasch P, Weingarten P (1896) Uber die chemische Zusammensetzunugn und Konstitution des Vesuvians und des Wiluits. Z Anorg Allg Chem 11:40–48

Justiniano CAB, Rodríguez MG, Panikorovskii TL et al (2025) Vesuvianite as a key tool for the reconstruction of skarn formation conditions: an example from the Sauce Chico Complex, Argentina. Amer Mineralo. https://doi.org/10.2138/am-2024-9588

Kappeler MA (1723) Prodromus crystallographiae de crystallis improprie sic dictis commentarium. Heinrich Rennward Wyssing, Lucerna

Kapral R, Showalter K (eds) (1995) Chemical waves and patterns. Springer, Netherlands, Dordrecht

Karwowski L, Kraczka J, Pieczka A, Żabiński W (1996) Vesuvianite from the Garby Izerskie mt., Lower Silesia, Poland. Mineralogia Polonica 27:23–31

Kashinskii AA (1909) Vesuvianite from Karmankul'skii cordon. Zapiski RMO 2:77–79

Kaverin SV, Krasnova NV, Tarasenko YuN (1988) Mineralogy of apomellitic rocks of the Kovdor massif. In: Mineralogy and geochemistry. Issues of the genesis of endogenous deposits. Leningrad, pp 46–62

Kelemen SR, George GN, Gorbaty ML (1990) Direct determination and quantification of sulphur forms in heavy petroleum and coals: 1. The X-ray photoelectron spectroscopy (XPS) approach. Fuel 69:939–944. https://doi.org/10.1016/0016-2361(90)90001-7

Keusen HR (1972) Mineralogie und Pétrographie des metamorphen Ultramafitit-Komplexes vom Geisspfad (Penninische Alpen). Schweiz Mineral Petrogr Mitt 52:307–320

Kholodnov VV, Fershtater GB, Ronkin YuL et al (2010) Sm-Nd and Rb-Sr ages of gabbroids, granitoids, and titanomagnetite ores from layered intrusions of the Kusa-Kopan Complex (South Urals). Dokl Earth Sci 432:732–736. https://doi.org/10.1134/S1028334X10060048

Kim KJ, Koh TY, Kim CS, Lee YB (2014) Effects of tetrahedral Fe2+ on the structural, magnetic and electronic properties of solution-based titanomagnetite thin films. J Korean Phys Soc 64:93–97. https://doi.org/10.3938/jkps.64.93

Klaproth MH (1797) Untersuchung des Vesuvians, Vesuvian aus Sibirien. In: Beiträge zur chemischen Kenntniss der Mineralkörper. Rottmann, Berlin, pp 33–38

Kobayashi S, Kaneda H (2010) Rodingite with Ti- and Cr-rich vesuvianite from the Sartuohai chromium deposit, Xinjiang, China. J Mineral Petrol Sci 105:112–122. https://doi.org/10.2465/jmps.081224

Koksharov NI (1853) Material for geology of Russia. Saint-Petersburg

Kononova VA (1960) On a metamict variety of vesuvianite from an alkaline pegmatite in southwest Tuva. Dokl Acad Sci USSR 130:129–132

Kornetova VA, Osolodkina GA, Borisovskiy CE (1986) Minerals of voids in rodingites. New Data USSR Minerals 33:78–86

Kraczka J, Żabiński W (2003) Mössbauer study of iron in some vesuvianites. Mineral Pol 34:37–44

Kravchinsky VA, Konstantinov KM, Courtillot V et al (2002) Paleomagnetism of East Siberian traps and kimberlites: two new poles and palaeogeographic reconstructions at about 360 and 250 Ma. Geophys J Int 148:1–33. https://doi.org/10.1046/j.0956-540x.2001.01548.x

Krivovichev SV (2013) Structural complexity of minerals: information storage and processing in the mineral world. Mineral Mag 77:275–326. https://doi.org/10.1180/minmag.2013.077.3.05

Krivovichev SV, Krivovichev VG, Hazen RM et al (2022) Structural and chemical complexity of minerals: an update. MinMag 86:183–204. https://doi.org/10.1180/mgm.2022.23

Krivovichev SV, Panikorovskii TL, Yakovenchuk VN et al (2021) Trigonal variation in the garnet supergroup: the crystal structure of nikmelnikovite, Ca12Fe2+Fe3+3Al3(SiO4)6(OH)20, from Kovdor massif, Kola Peninsula, Russia. Mineral Mag 85:620–626. https://doi.org/10.1180/mgm.2021.55

Krivovichev SV, Zolotarev AA, Panikorovskii TL et al (2013) Crystal chemistry of low-symmetry vesuvianite from Monetnaya Dacha (Middle Ural, Russia). Vestnik SPbSU Ser 7:1–13

Kuleshov VN (2012) A superlarge deposit—Kalahari manganese ore field (Northern Cape, South Africa): geochemistry of isotopes (δ13C and δ18O) and genesis. Lithol Miner Resour 47:217–233. https://doi.org/10.1134/S0024490212030054

Kunz GF (1903) Californite (vesuvianite); a new ornamental stone. Am J Sci 16:397–398

Kurazhkovskaya VS, Borovikova EY, Alferova MS (2005) Infrared spectra, unit cell parameters and optical character of boron-bearing vesuvianites and wiluites. Zapiski RMO 134:45–54

Kurbatov SM (1914) Vesuvianite from the bank. Big Kaznakhty in the Kumachinsky Mountains in the Southern Urals. News Russian Acad Sci 8:901–914

Kurbatov SM (1922) Vesuvianite from Russian deposits II. Vesuvian from the Serafimov mine in the Bridge section of the Monetary dacha in the Urals. News Russian Acad Sci 16:411–424

Kurbatov SM, Solodovnikova LL (1923) Vesuvianite from Russian deposits III. Chrome vesuvian from the vicinity of the Upper-Neyvinskogo plant in the Urals. News Russian Acad Sci 17:115–128

Kurbatov SM (1924) Vesuvianite from Russian deposits. IV. Vesuvians of the Shishim Mountains in the Southern Urals. News Russian Acad Sci 19:43–52

Kurbatov SM (1925a) Vesuvianite from Russian deposits. VI. Vesuvians from asbestos mines on the border of Kamenskaya and Monetnaya dachas in the Urals (description of the deposit). News Russian Acad Sci 19:451–482

Kurbatov SM (1925b) Vesuvianite from Russian deposits. V. Vesuvians from asbestos mines on the border of Kamenskaya and Monetnaya dachas in the Urals (description of the deposit). News Russian Acad Sci 19:43–52

Kurbatov SM (1946) Vesuvianite from USSR localities. Leningrad State Univiversity Publ, Leningrad

Kurbatov SM (1948) To the question of the chemical composition and structure of pomegranate and vesuvianite. Vestnik SPbSU 93:18–34

Lager GA, Xie Q, Ross FK et al (1999) Hydrogen-atom position in P4/nnc vesuvianite. Can Mineral 37:763–768

Larin AM, Amelin YV, Neymark LA (1991) Age and genesis of complex skarn ores in the Pitkäranta Ore Province. Geol Rudn Mestorozh 15–33

Larraz G, Orera A, Sanz J et al (2015) NMR study of Li distribution in Li7–xHxLa3Zr2O12 garnets. J Mater Chem A 3:5683–5691. https://doi.org/10.1039/C4TA04570J

Li H, Ru J, Yin W et al (2009) Removal of phosphate from polluted water by lanthanum doped vesuvianite. J Hazard Mater 168:326–330. https://doi.org/10.1016/j.jhazmat.2009.02.025
Libowitzky E (1999) Correlation of O–H stretching frequencies and O–HO hydrogen bond lengths in minerals. Hydrog Bond Res 1059:103–115. https://doi.org/10.1007/978-3-7091-6419-8_7
Lindberg BJ, Hamrin K, Johansson G et al (1970) Molecular Spectroscopy by Means of ESCA II. Sulfur compounds. Correlation of electron binding energy with structure. Phys Scr 1:286–298. https://doi.org/10.1088/0031-8949/1/5-6/020
Lindström G (1888) Tvenne idokrasanalyser. Geologiska Föreningen i Stockholm Förhandlingar 10:286–289. https://doi.org/10.1080/11035898809445413
Liu S (2016) Device for removing industrial rust
Lussier AJ, Aguiar PM, Michaelis VK et al (2009) The occurrence of tetrahedrally coordinated Al and B in tourmaline: an 11B and 27Al MAS NMR study. Am Miner 94:785–792. https://doi.org/10.2138/am.2009.3000
Lyachovich VV (1954) New data on the mineralogy of the Wiluy Akhtarandite deposit. Works East-Sib Branch USSR Acad Sci 1:85–116
Lyachovich VV (1955) About new varieties of wiluite. Mineral Zbornik of L'vov 9:128–144
Machatschki F (1932) Zur Formel des Vesuvian. Zeitschrift Für Kristallographie - Crystalline Materials 81:148–152. https://doi.org/10.1524/zkri.1932.81.1.148
Macrae CF, Bruno IJ, Chisholm JA et al (2008) Mercury CSD 2.0 new features for the visualization and investigation of crystal structures. J Appl Crystallogr 41:466–470. https://doi.org/10.1107/S0021889807067908
Malczewski D, Dziurowicz M (2015) 222 Rn and 220 Rn emanations as a function of the absorbed α-doses from select metamict minerals. Am Miner 100:1378–1385. https://doi.org/10.2138/am-2015-4917
Manning PG (1968) Optical absorption spectra of octahedrally bonded Fe3+ in vesuvianite. Can J Earth Sci 5:89–92. https://doi.org/10.1139/e68-009
Manning PG, Tricker MJ (1975) Optical absorption and Mössbauer spectral studies of iron and titanium site-populations in vesuvianites. Can Mineral 13:259–265
Mikhailova JA, Kalashnikov AO, Sokharev VA et al (2016) 3D mineralogical mapping of the Kovdor phoscorite–carbonatite complex (Russia). Miner Deposita 51:131–149. https://doi.org/10.1007/s00126-015-0594-z
Mikhailova JA, Ivanyuk GY, Kalashnikov AO et al (2018) Three-D mineralogical mapping of the kovdor phoscorite-carbonatite complex, NW Russia: I. Forsterite. Minerals 8. https://doi.org/10.3390/min8060260
Moiseev MM, Panikorovskii TL, Aksenov SM et al (2020) Insights into crystal chemistry of the vesuvianite-group: manaevite-(Ce), a new mineral with complex mechanisms of its hydration. Phys Chem Miner 47:18. https://doi.org/10.1007/s00269-020-01086-7
Momma K, Izumi F (2011) VESTA 3 for three-dimensional visualization of crystal, volumetric and morphology data. J Appl Crystallogr 44:1272–1276. https://doi.org/10.1107/S0021889811038970
Murzin VV, Palyanova GA, Varlamov DA, Shanina SN (2020) Gold-bearing Rodingites of the Agardag Ultramafic Massif (South Tuva, Russia) and problems of their genesis. Geol Ore Deposits 62:204–224. https://doi.org/10.1134/S107570152002004X
Mushketov IV (1877) Materials for the study of the geognostic structure and ore riches of the Zlatoust mining district in the Southern Urals. Tipography of the Imperial Academy of Sciences, SPb
Myasnikov VS (1940) On titanium vesuvianite from the Perovskite and Akhmatovskaya mines in the Southern Urals. Dokl Akad Nauk SSSR 28:445–448
Nickel EH, Grice JD (1998) The IMA commission on new minerals and mineral names: procedures and guidelines on mineral nomenclature, 1998. Mineral Petrol 64:237–263. https://doi.org/10.1007/BF01226571
Normand C, Williams-Jones AE (2007) Physicochemical conditions and timing of rodingite formation: evidence from rodingite-hosted fluid inclusions in the JM Asbestos mine, Asbestos, Québec. Geochem Trans 8:11. https://doi.org/10.1186/1467-4866-8-11

Nyunt TT, Theye T, Massonne HJ (2009) Na-rich vesuvianite in jadeitite of the Tawmaw jade district, northern Myanmar. Periodico di Mineralogia 78:5–18. https://doi.org/10.2451/2009PM0009

Ohkawa M, Yoshiasa A, Takeno S (1992) Crystal chemistry of vesuvianite: site preferences of square-pyramidal coordinated sites. Am Miner 77:945–953. https://doi.org/10.1144/gsjgs.155.2.0353

Ohkawa M (1994) Crystal chemistry and structure of vesuvianite. Ph.D. thesis. Hiroshima

Ohkawa M, Armbruster T, Galuskin EV (2009) Structural investigation of low-symmetry vesuvianite collected from Tojyo, Hiroshima, Japan: implications for hydrogarnet-like substitution. J Mineral Petrol Sci 104:69–76. https://doi.org/10.2465/jmps.080507

Olejniczak Z, Żabiński W (1996) 27Al NMR study of white vesuvianite from Piz Lunghin, Switzerland. Mineral Pol 27:41–45

Olesch M (1979) Natürliche und synthetische Fe-haltige Vesuviane. Fortschritte in Mineralogie 57:114–115

Oleynikov BV (1979) Geochemistry and ore genesis of platform basites. Novosibirsk

Oleynikov BV, Pavlushin AD (2000) Achtarandite—Mineralogical question with 200 year history. Nauka i Obrazovanie 3:91–94

Orlov YuL, Martyanov NN (1960) Rare earth vesuvian from the Yenisei Ridge. Works Mineralo Museum 11:187–190

Palache C, Bauer LH (1930) On the occurrence of beryllium in the zinc deposits of Franklin, New Jersey. Am Miner 15:30–33

Pallas PS (1793) Nordische Beyträge zur physikalischen und geographischen Erd- und Volkerbeschreibung. Leipzig

Pan YM, Fleet ME (1992) Mineral chemistry and geochemistry of vanadian silicates in the Hemlo gold deposit, Ontario, Canada. Contrib Miner Petrol 109:511–525. https://doi.org/10.1007/BF00306553

Panikorovskii TL (2025a) Iron state in vesuvianite supergroup minerals. Zapiski RMO (in press)

Panikorovskii TL (2025b) Garnet-wiluite intergrowths from Wiluy, Yakutia. Zapiski RMO (in press), Russia

Panikorovskii TL, Britvin SN (2013) On the use of differential scanning calorimetry to study the structural features of organic and inorganic compounds. SPSU 1–13

Panikorovskii TL, Zolotarev AA, Krivovichev SV, Antonov AA (2014) Epitaxial intergrowths of wiluite and grossular: a single-crystal X-ray diffraction study and modular interpretation. In: IMA Abstracts. Johannesburg, p 332

Panikorovskii TL, Krivovichev SV, Galuskin EV et al (2016a) Si-deficient, OH-substituted, boron-bearing vesuvianite from Sakha-Yakutia, Russia: a combined single-crystal, 1H MAS-NMR and IR spectroscopic study. Eur J Mineral 28:931–941. https://doi.org/10.1127/ejm/2016/0028-2570

Panikorovskii TL, Krivovichev SV, Yakovenchuk VN et al (2016b) Crystal chemistry of Na-bearing vesuvianite from fenitized gabbroid of the Western Keivy (Kola peninsula, Russia). Zapiski RMO 145:83–95

Panikorovskii TL, Krivovichev SV, Zolotarev AAJr, Antonov AA (2016c) Crystal chemistry of low-symmetry (P4nc) vesuvianite from the Kharmankul' Cordon (South Urals, Russia). Zapiski RMO 145:94–104

Panikorovskii TL, Shilovskikh VV, Zolotarev AA (2016d) Relaxation structure of cation-ordered P4/n ve- suvian from Stara Kamenica skarn (Sudetes, Poland). Proc FNS 293–296

Panikorovskii TL, Zolotarev AAJr, Krivovichev SV et al (2016e) Crystal chemistry of Cu-bearing vesuvianites ("cyprine") from Kleppan (Norway). Zapiski RMO 145:131–142

Panikorovskii TL, Chukanov NV, Aksenov SM et al (2017a) Alumovesuvianite, Ca19Al(Al,Mg)12Si18O69(OH)9, a new vesuvianite-group member from the Jeffrey mine, asbestos, Estrie region, Québec, Canada. Mineral Petrol 111:833–842. https://doi.org/10.1007/s00710-017-0495-1

Panikorovskii TL, Chukanov NV, Rusakov VS et al (2017b) Vesuvianite from the Somma-Vesuvius complex: new data and revised formula. Minerals 7:248. https://doi.org/10.3390/min7120248

Panikorovskii TL, Mazur AS, Bazai AV et al (2017c) X-ray diffraction and spectroscopic study of wiluite: implications for the vesuvianite-group nomenclature. Phys Chem Miner 44:577–593. https://doi.org/10.1007/s00269-017-0885-2

Panikorovskii TL, Shilovskikh VV, Avdontseva EY et al (2017d) Magnesiovesuvianite, Ca19Mg(Al,Mg)12Si18O69(OH)9, a new vesuvianite-group mineral. J Geosci (Czech Republic) 62:25–36. https://doi.org/10.3190/jgeosci.229

Panikorovskii TL, Shilovskikh VV, Avdontseva EY et al (2017e) Cyprine, Ca19Cu2+(Al,Mg,Mn)12Si18O69(OH)9, a new vesuvianite-group mineral from the Wessels mine, South Africa. Eur J Mineral 29:295–306. https://doi.org/10.1127/ejm/2017/0029-2592

Panikorovskii TL, Yakovenchuk VN, Ivanyuk GY, Krivovichev SV (2017f) Structural evolution of the vesuvian from the skarns of the Kovdor massif (Kola Peninsula). Trudy FNS 149–150

Panikorovskii TL, Mikhailova JA, Pakhomovsky YA et al (2021) Zr-Rich Eudialyte from the Lovozero Peralkaline Massif, Kola Peninsula, Russia. Minerals 11:982. https://doi.org/10.3390/min11090982

Panikorovskii TL, Galuskina IO, Bocharov VN et al (2022) Merohedral mechanism twining growth of natural cation-ordered tetragonal grossular. Crystals 12:1638. https://doi.org/10.3390/cryst12111638

Panikorovskii TL, Yakovenchuk VN, Krivovichev SV (2024) An X-ray diffraction method for estimating the temperature of vesuvianite crystallization. Geol Ore Depos 66:1022–1033. https://doi.org/10.1134/S1075701524700156

Panikorovskii TL, Shilovskikh VV (2025) Compositional relationships between Cr-bearing garnet and vesuvianite from Bazhenovskoe deposit, Midle Urals. Zapiski RMO (in press), Russia

Pavese A, Prencipe M, Tribaudino M, Aagaard SS (1998) X-ray and neutron single-crystal study of P4/n vesuvianite. Can Mineral 36:1029–1037

Peters T (1961) Differentialthermoanalyse von Vesuvian. Schweiz Mineral Petrogr Mitt 41:325–334

Peters T (1965) A water-bearing andradite from the Totalp serpentine (Davos, Switzerland). Am Miner 50:1482–1486

Phillips BL, Allen FM, Kirkpatrick RJ (1987) High-resolution solid-state 27Al NMR spectroscopy of Mg-rich vesuyianite. Am Miner 72:90–94

Platonov AN, Taran MN, Macuk SS (1993) Optical Spectra of Cr3+ in Vesuvianite. Mineral Mag 15:55–59

Platonov AN, Żabiński W, Sachanbiński M (1995) Optical absorption spectra of Mn3+ ions in vesuvianites from Lower Silesia, Poland. Eur J Mineral 7:1345–1352. https://doi.org/10.1127/ejm/7/6/1345

Popov VA, Rassomakhin MA (2023) Mineral assemblages of the Shishim mine in the Southern Urals. МИНЕРАЛОГИЯ (MINERALOGY) 9:23–44. https://doi.org/10.35597/2313-545X-2023-9-1-2

Prendel RO (1887) About wiluite. Proc Imperial St-Petersburg Mineral Soc 2:48–62

Putz H, Brandernburg K (2014) Diamond—Crystal and molecular structure visualization

Rimskaya-Korsakova OM, Krasnova NI (2002) Geology of the deposits of Kovdor Massif. Petersburg University Press, St. Petersburg, St

Rivas-Mercury JM, Pena P, de Aza AH, Turrillas X (2008) Dehydration of Ca3Al2(SiO4)y(OH)4(3−y) (0 < y < 0.176) studied by neutron thermodiffractometry. J Eur Ceram Soc 28:1737–1748. https://doi.org/10.1016/j.jeurceramsoc.2007.12.038

Rogkala A, Petrounias P, Koutsovitis P et al (2022) Rodingites from the Veria-Naousa ophiolite (Greece): mineralogical evolution, metasomatism and petrogenetic processes. Geochemistry 82:125860. https://doi.org/10.1016/j.chemer.2021.125860

Romée de Lisle J-B (1783) Cristallographie ou Description des formes propres à tous les corps du règne minéral dans l'état de combinaison saline, pierreuse ou métallique

Romer RL (1992) Vesuvianite-new tool for the U-Pb dating of skarn ore deposits. Mineral Petrol 46:331–341. https://doi.org/10.1007/BF01173571

Rucklidge JC, Kocman V, Whitlow SH, Gabe EJ (1975) The crystal structures of three Canadian vesuvianites. Can Mineral 13:15–21

Rusakov VS, Kovalchuk RV, Borovikova EY, Kurazhkovskaya VS (2006) State of iron atoms in high vesuvianites according to Mössbauer spectroscopy data. Zapiski RMO 135:91–100

Sapozhnikov AN, Bolotina NB, Chukanov NV et al (2023) Slyudyankaite, Na28Ca4(Si24Al24O96)(SO4)6(S6)1/3(CO2)2H2O, a new sodalite-group mineral from the Malo-Bystrinskoe lazurite deposit, Baikal Lake area, Russia. Am Miner 108:1805–1817. https://doi.org/10.2138/am-2022-8598

Sbrana A, Cioni R, Marianelli P et al (2020) Volcanic evolution of the Somma-Vesuvius complex (Italy). J Maps 16:137–147. https://doi.org/10.1080/17445647.2019.1706653

Seredkin MV (2001) Petrology of iron ore and phlogopite deposits of the Kovdor massif. Ph.D. thesis

Shannon EV (1922) Note on the cyprine from Franklin Furnace, New Jersey. Am Miner 7:140–142

Shannon RD (1976) Revised effective ionic radii and systematic studies of interatomic distances in halides and chalcogenides. Acta Crystallogr A 32:751–767. https://doi.org/10.1107/S0567739476001551

Sheldrick GM (2008) A short history of SHELX. Acta Crystallogr A 64:112–122. https://doi.org/10.1107/S0108767307043930

Shendrik RYu, Chukanov NV, Panikorovskii TL et al (2025) Radical anion $S_2^{\cdot-}$ in minerals. Geol Ore Depos (in press)

Singh PS, Bastow T, Trigg M (2005) Structural studies of geopolymers by 29Si and 27Al MAS-NMR. J Mater Sci 40:3951–3961. https://doi.org/10.1007/s10853-005-1915-x

Sjögren H (1895) Analyser pa tvenne vesuvian-varieter och vesuvianens kemiska constitution I allmänhet. Geol För I Stockholm Forhandlingar 17:267–271

Smart MM, Moore CA, McMillen CD, Kolis JW (2023) Hydrothermal synthesis and crystal structure of Vesuvianite compounds, Ca19Al13Si18O71(OH)7 and Sr19Fe12Ge19O72(OH)6. Crystals 13:1257. https://doi.org/10.3390/cryst13081257

Stalder HA, Aufdenblatten M (1995) Granat (Grossular) aus dem Täschtal VS. Schweizer Strahler 10:267–279

Stativko V, Rat'kovskiy G, Kuznetsov A (2024) Geochemistry of vesuvianite from contact-metasomatic rocks of mineral mines in the Southern Urals. Vestnik of geosciences 0:30–41. https://doi.org/10.19110/geov.2024.11.3

Steudel R, Chivers T (2019) The role of polysulfide dianions and radical anions in the chemical, physical and biological sciences, including sulfur-based batteries. Chem Soc Rev 48:3279–3319. https://doi.org/10.1039/c8cs00826d

Stifeeva MV, Panikorovskii TL, Larin AM et al (2024) Vesuvianite, a New U-Pb geochronometer mineral for dating ore deposits. Dokl Earth Sci 518:1694–1699. https://doi.org/10.1134/S1028334X2460292X

Strunz H, Nickel EH (2001) Mineralogical tables. Schweizerbart, Stuttgart

Tanaka T, Akizuki M, Kudoh Y (2002) Optical properties and crystal structure of triclinic growth sectors in vesuvianite. Mineral Mag 66:261–274. https://doi.org/10.1180/0026461026620027

Taran MN, Koch-Müller M (2011) Optical absorption of electronic Fe–Ti charge-transfer transition in natural andalusite: the thermal stability of the charge-transfer band. Phys Chem Miner 38:215–222. https://doi.org/10.1007/s00269-010-0397-9

Tricker MJ, Manning PG (1979) Can electron exchange processes in vezuvianites be detected by Mössbauer spectroscopy? J De Physique 40:477–479

Tricker MJ, Vaishnava PP, Manning PG (1981) 57Fe Mössbauer spectroscopic studies of electron-hopping processes in vesuvianites. J Inorg Nucl Chem 43:1169–1174. https://doi.org/10.1016/0022-1902(81)80012-7

Trommsdorff V (1968) Mineralreaktionen mit Wollastonit und Vesuvian in einem Kalksilikatfels der alpinen Disthenzone (Claro, Tessin). Schweiz Miner Petrogr Mitt 48:655–666. https://doi.org/10.5169/seals-37788

Uher P, Bačík P, Rybnikova O et al (2024) Modraite, IMA 2023-108a. CNMNC Newsletter 79, Eur J Mineral 36

Valley JW, Peacor DR, Bowman JR et al (1985) Crystal chemistry of a Mg-vesuvianite and implications of phase equilibria in the system CaO–MgO–Al2O3–SiO2–H2O–CO2. J Metamorph Geol 3:137–153. https://doi.org/10.1111/j.1525-1314.1985.tb00311.x
Vannotti LE, Morton JR (1967) Paramagnetic-resonance Spectra of S_2-in Alkali Halides. Phys Rev 161:282–286. https://doi.org/10.1103/physrev.161.282
Veblen DR, Wiechmann MJ (1991) Domain structure of a low-symmetry vesuvianite from Crestmore, California. Am Miner 76:397–404
Vogel IH (1887) Über die chem. Zusamennensetzung des vesuvians. Dissertation. Göttingen
von Lasaulx A (1883) Über den Manganvesuvian vom Johnsberge bei Jordansmühle in Schlesien und Über den Titanomorphit. Zeitschrift Für Kristallographie und Mineralogie 7:71–73
Walker JR (1993) Chlorite Polytype Geothermometry. Clays Clay Miner 41:260–267. https://doi.org/10.1346/CCMN.1993.0410212
Wang C, Fan Q, Gao Z et al (2015) Macroporous cement concrete capable of adsorbing carbon dioxide and its preparation method
Wares RP, Martin RF (1980) Rodingitization of granite and serpentinite in the Jeffrey Mine, Asbestos, Quebec. Can Mineral 18:231–240
Warren BE, Modell DI (1931) The Structure of Vesuvianite Ca10Al4(Mg,Fe)2Si9O34(OH)4. Zeitschrift Für Kristallographie - Cryst Mater 78:422–432. https://doi.org/10.1524/zkri.1931.78.1.422
Wei Q-D, Yang M, Romer RL et al (2022) *In situ* U-Pb geochronology of vesuvianite by LA-SF-ICP-MS. J Anal at Spectrom 37:69–81. https://doi.org/10.1039/D1JA00303H
Weibull M (1896) I. Studien über Vesuvian. Zeitschrift Für Kristallographie - Cryst Mater 25:1–37. https://doi.org/10.1524/zkri.1896.25.1.1
Werner AG (1795) Über vesuvian. Klaproth Beiträge 1:34
Wherry ET, Chapin WH (1908) Determination of boric acid in insoluble silicates. J Am Chem Soc 30:1687–1701. https://doi.org/10.1021/ja01953a005
Willcott MR (2009) MestRe Nova MestRe Nova. Mestrelab Research S.L. Feliciano Barrera 9B, Bajo, 15706 Santiago de Compostela, Spain. http://www.mestrelab.com. See Web site for pricing information. J Am Chem Soc 131:13180–13180. https://doi.org/10.1021/ja906709t
Xiaoyan T, Suyu W, Yang Y et al (2015) Removal of six phthalic acid esters (PAEs) from domestic sewage by constructed wetlands. Chem Eng J 275:198–205. https://doi.org/10.1016/j.cej.2015.04.029
Xu J, Li G, Fan G et al (2017) Hongheite, IMA 2017–027. CNMNC Newsletter No. 39, October 2017, page 1283. Mineral Mag 81:1279–1286
Xu J, Li G, Fan G et al (2019) Hongheite Ca18(□, Ca)2Fe2+Al4(Fe3+, Mg,Al)8 (□, B)4BSi18O69(O,OH)9:a new vesuvianite group mineral from the world- class Gejiu tin polymetallic orefield, Yunnan Province, China. Acta Geol Sin 93:138–146
Xue X, Stebbins Jonathan F (1993) 23Na NMR chemical shifts and local Na coordination environments in silicate crystals, melts and glasses. Phys Chem Minerals 20. https://doi.org/10.1007/BF00215100
Yang C (2016) A kind of food waste disposal method and apparatus
Yesinowski JP, Rossman GR, Eckert H (1988) Characterization of hydrous species in minerals by high-speed1H MAS-NMR. J Am Chem Soc 110:1367–1375. https://doi.org/10.1021/ja00213a007
Yoshiasa A, Matsumoto T (1986) The crystal structure of vesuvianite from Nakatatsu mine: reinvestigation of the cation site-population and of the hydroxyl groups. Mineralo J 13:1–12. https://doi.org/10.2465/minerj.13.1
Żabiński W, Wacławska I, Paluszkiewicz C (1996) Thermal decomposition of vesuvianite. J Therm Anal 46:1437–1447. https://doi.org/10.1007/BF01979256
Zakaznova-Herzog VP, Nesbitt HW, Bancroft GM, Tse JS (2008) Characterization of leached layers on olivine and pyroxenes using high-resolution XPS and density functional calculations. Geochim Cosmochim Acta 72:69–86. https://doi.org/10.1016/j.gca.2007.09.031
Zepharowich VV (1864) Krystallographische Studien über den Idokras. Sitzungsber, Wien

Zhang Y, Song S, Hollings P et al (2022) In-situ U Pb geochronology of vesuvianite in skarn deposits. Chem Geol 612:121136. https://doi.org/10.1016/j.chemgeo.2022.121136

Zhou B, Sherriff BL, Hartman JS, Wu G (2007) 27Al and 23Na NMR spectroscopy and structural modeling of aluminofluoride minerals. Am Miner 92:34–43. https://doi.org/10.2138/am.2007.2269

Zhukov SI, Stupnikova NI, Pavlenko AS (1961) The absolute age of intrusions of the East-Tuva region and the Yenisei Ridge. Geokhimia 7:547–560